Lecture Notes in Computer Science 9316

Commenced Publication in 1973
Founding and Former Series Editors:
Gerhard Goos, Juris Hartmanis, and Jan van Leeuwen

More information about this series at http://www.springer.com/series/7409

Sarantos Kapidakis · Cezary Mazurek
Marcin Werla (Eds.)

Research and Advanced Technology for Digital Libraries

19th International Conference on Theory and Practice
of Digital Libraries, TPDL 2015
Poznań, Poland, September 14–18, 2015
Proceedings

Springer

Editors
Sarantos Kapidakis
Ionian University
Corfu
Greece

Cezary Mazurek
Poznań Supercomputing
   and Networking Center
Poznań
Poland

Marcin Werla
Poznań Supercomputing
   and Networking Center
Poznań
Poland

ISSN 0302-9743   ISSN 1611-3349 (electronic)
Lecture Notes in Computer Science
ISBN 978-3-319-24591-1   ISBN 978-3-319-24592-8 (eBook)
DOI 10.1007/978-3-319-24592-8

Library of Congress Control Number: 2015949408

LNCS Sublibrary: SL3 – Information Systems and Applications, incl. Internet/Web, and HCI

Springer Cham Heidelberg New York Dordrecht London

Printed on acid-free paper

Springer International Publishing AG Switzerland is part of Springer Science+Business Media
(www.springer.com)

Preface

We are proud to present the proceedings of TPDL 2015, the 19th International Conference on Theory and Practice of Digital Libraries, held in Poznań, Poland, during September 14–18, 2015, organized by the Poznań Supercomputing and Networking Center (PSNC).

The International Conference on Theory and Practice of Digital Libraries (TPDL), formerly known as European Conference on Research and Advanced Technology on Digital Libraries (ECDL), constitutes a leading scientific forum on digital libraries that brings together researchers, developers, content providers, and users in the field of digital libraries. The advent of the technologies that enhance the exchange of information with rich semantics is of particular interest in the community. Information providers inter-link their metadata with user-contributed data and offer new services for the development of a web of data and addressing the interoperability and long-term preservation challenges.

TPDL 2015 had the general theme "Connecting Digital Collections" and was focused on four major topics:

- Connecting digital libraries
- Practice of digital libraries
- Digital libraries in science
- Users, communities, personal data

There was also a special call for industry submissions, dedicating the "Systems and Products" conference track for them.

There were 44 full paper and six short paper submissions in the main call. All submissions were independently reviewed in a triple peer review process, initially by four members of the Program Committee. A senior Program Committee member subsequently coordinated a discussion among the four reviewers. The selection stage that followed compared the paper evaluations and finalized the conference program. As a result, 22 submissions were accepted as papers and some of the rest of the submissions were redirected for evaluation as potential posters or demonstrations. These redirected submissions were re-evaluated together with the 11 additional poster and demonstration submissions. Finally, 15 poster/demo submissions were accepted. The dedicated "Systems and Products" call brought an additional six accepted submissions, which are not included in this proceedings volume, but published in a separate booklet and distributed among conference participants.

The most popular topics of submissions were user interfaces and user experience, user studies for and evaluation of digital library systems and applications, applications of digital libraries, infrastructures supporting content processing, social–technical perspectives of digital information, interoperability and information integration, and digital humanities. Regarding the number of accepted papers, the top five countries of authors were: USA, The Netherlands, Greece, Germany, and Brazil. Regarding the number of submitted papers, the top five countries of authors were: Greece, Germany, USA, New Zealand, and The Netherlands.

Beside submitted contributions, three keynote speakers were invited to present their views on crucial aspects of digital libraries. The opening keynote, given by David Giaretta, was focused on issues related to long-term data preservation. The second keynote speaker, Joseph Cancellaro, showed the user perspective in the context of retrieval of digital audio assets, and Costis Dallas in the closing keynote shared his thoughts about the scholarly practice related to access and use of humanities data, in light of his work conducted within the DARIAH-EU community. Another part of the conference was a discussion panel organized by Vittore Casarosa aiming to discuss open access to research data. Abstracts of all keynote speeches and the panel are included in the conference proceedings.

Around the main conference several side activities were organized, together creating a five-day long series of events focused on digital libraries. The overall program began with five tutorials:

- Automatic Methods for Disambiguating Author Names in Bibliographic Data Repositories
- Building Digital Library Collections with Greenstone 3
- Catmandu – A (Meta)Data Toolkit
- Dynamic Data Citation – Enabling Reproducibility in Evolving Environment
- Mappings, Application Profiles and Extensions for Cross-Domain Metadata in the Europeana Context and Beyond

Following the main conference, several workshops were organized:

- 5th International Workshop on Semantic Digital Archives (SDA 2015)
- Cloud-based Services for Digital Libraries
- Extending, Mapping, and Focusing the CRM
- Kick-Off Workshop of the IMPACT-OPF MOOC on Digitization and Digital Preservation
- Networked Knowledge Organization Systems and Services (NKOS)

In this context we are very grateful to the tutorial chairs, Giorgio Maria Di Nunzio and Giannis Tsakonas, for their hard work on attracting and evaluating the proposed tutorials. Also the workshop chairs, Trond Aalberg and Antoine Isaac, gave essential feedback to the conference tutorials. We would also like to thank all members of the conference Program Committee and especially the posters and demos chairs, José Borbinha and Preben Hansen, who did great work in evaluating a significant number of submissions and creating a very interesting program for the conference.

Of course the conference could not happen without proper publicity, which was assured by the publicity chairs, Marcos Goncalves, Raul Palma, Shigeo Sugimoto and Hussein Suleman, and by the conference media partners: Coalition for Networked Information and Digital Meets Culture.

For the Poznań Supercomputing and Networking Center, the operator of the Polish National Research and Education Network PIONIER and the main organizer of the conference, it was a great occasion to actively support knowledge exchange and net-working in the worldwide digital libraries research community. Within the broad range of PSNC research and development activities, those related to the digital libraries domain have been very dynamic since 1999 and have resulted in a number of national and international projects in which PSNC and its Digital Libraries and Knowledge Platforms Department are continuously involved.

We hope that you will enjoy the proceedings and will be inspired to participate in the following editions of the TPDL conference.

September 2015

Sarantos Kapidakis
Cezary Mazurek
Marcin Werla

Organizations

General Chairs

Cezary Mazurek   PSNC, Poland
Marcin Werla   PSNC, Poland

Program Chair

Sarantos Kapidakis   Ionian University, Greece

Organizing Chair

Damian Niemir   PSNC, Poland

Workshops Chairs

Trond Aalberg   Norwegian University of Science and Technology, Norway
Antoine Isaac   VU University Amsterdam, The Netherlands

Posters and Demos Chairs

José Borbinha   IST/INESC-ID, Portugal
Preben Hansen   Stockholm University, Sweden

Tutorials Chairs

Giorgio Maria Di Nunzio   University of Padova, Italy
Giannis Tsakonas   University of Patras, Greece

Publicity Chairs

Marcos Goncalves   Universidade Federal de Minas Gerais, Brazil
Raul Palma   PSNC, Poland
Shigeo Sugimoto   University of Tsukuba, Japan
Hussein Suleman   University of Cape Town, South Africa

Senior Program Committee

Trond Aalberg   Norwegian University of Science and Technology, Norway

Maristella Agosti	University of Padua, Italy
Thomas Baker	DCMI Ltd., USA
Janusz Bień	University of Warsaw, Poland
Jose Borbinha	IST/INESC-ID, Portugal
George Buchanan	City University London, UK
Donatella Castelli	CNR - ISTI, Italy
Stavros Christodoulakis	Technical University of Crete, Greece
Panos Constantopoulos	Athens University of Economics and Business, Greece
Sally Jo Cunningham	Waikato University, New Zealand
Erik Duval	K.U.Leuven, Belgium
Edward Fox	Virginia Polytechnic Institute and State University, USA
Geneva Henry	Rice University, USA
Martin Klein	Los Alamos National Laboratory, USA
Stefanos Kollias	NTUA, Greece
Laszlo Kovacs	MTA SZTAKI
Carl Lagoze	University of Michigan, USA
Ronald Larsen	University of Pittsburgh, USA
Clifford Lynch	CNI, USA
Carlo Meghini	CNR - ISTI, Italy
Erich Neuhold	University of Vienna, Austria
Christos Papatheodorou	Ionian University, Greece
Andreas Rauber	Vienna University of Technology, Austria
Thomas Risse	L3S Research Center, Germany
Giannis Tsakonas	University of Patras, Greece

Program Committee

Robert Allen	Yonsei University, Korea
David Bainbridge	University of Waikato, New Zealand
Christoph Becker	University of Toronto, Canada
Maria Bielikova	Slovak University of Technology in Bratislava, Slovakia
Tobias Blanke	University of Glasgow, UK
Pável Calado	IST/INESC-ID, Portugal
José H. Canós	Universitat Politècnica de València, Spain
Vittore Casarosa	CNR - ISTI, Italy
Lillian Cassel	Villanova University, USA
Fabio Crestani	University of Lugano, Italy
Theodore Dalamagas	IMIS-"Athena" R.C., Greece
Lois Delcambre	Portland State University, USA
Giorgio Maria Di Nunzio	University of Padua, Italy
Boris Dobrov	Research Computing Center of Moscow State University, Russia
J. Stephen Downie	The University of Illinois at Urbana-Champaign, USA
Fabien Duchateau	Université Claude Bernard Lyon 1 - LIRIS, France

Floriana Esposito	Università Aldo Moro Bari, Italy
Pierluigi Feliciati	Università degli studi di Macerata, Italy
Nicola Ferro	University of Padua, Italy
Schubert Foo	Nanyang Technological University, Singapore
Nuno Freire	The European Library, The Netherlands
Ingo Frommholz	University of Bedfordshire, UK
Norbert Fuhr	University of Duisburg-Essen, Germany
Richard Furuta	Texas A&M University, USA
Emmanouel Garoufallou	Alexander TEI of Thessaloniki, Greece
Manolis Gergatsoulis	Ionian University, Greece
C. Lee Giles	Pennsylvania State University, USA
Marcos Goncalves	Federal University of Minas Gerais, Brazil
Jane Greenberg	Drexel University, USA
Preben Hansen	Stockholm University, Sweden
Bernhard Haslhofer	AIT, Austria
Annika Hinze	University of Waikato, New Zealand
Nikos Houssos	National Documentation Centre/National Hellenic Research Foundation, Greece
Jane Hunter	University of Queensland, Australia
Antoine Isaac	Europeana and VU University Amsterdam, The Netherlands
Adam Jatowt	Kyoto University, Japan
Jaap Kamps	University of Amsterdam, The Netherlands
Michael Khoo	Drexel University, USA
Claus-Peter Klas	GESIS, Germany
Alexandros Koulouris	TEI of Athens, Greece
Michał Kozak	Poznań Supercomputing and Networking Center, Poland
Alberto Laender	Federal University of Minas Gerais, Brazil
Ray Larson	University of California, Berkeley, USA
Fernando Loizides	Cyprus University of Technology, Cyprus
Zinaida Manžuch	Vilnius University, Poland
Bruno Martins	IST - Instituto Superior Técnico, Portugal
Dana Mckay	Swinburne University of Technology, Australia
Andras Micsik	MTA SZTAKI, Hungary
Agnieszka Mykowiecka	IPI PAN, Poland
Wolfgang Nejdl	L3S and University of Hannover, Germany
Michael Nelson	Old Dominion University, USA
David Nichols	University of Waikato, New Zealand
Ragnar Nordlie	Oslo and Akershus University College, Nowray
Kjetil Nørvåg	Norwegian University of Science and Technology, Norway
Pasquale Pagano	CNR - ISTI, Italy
Raul Palma	Poznan Supercomputing and Networking Center, Poland
Ioannis Papadakis	Ionian University, Greece

Maggy Pezeril	University of Montpellier, France
Dimitris Plexousakis	FORTH, Greece
Edie Rasmussen	University of British Columbia, Canada
Laurent Romary	Inria and HUB-ISDL, France
Mike Rosner	UM, Malta
Seamus Ross	University of Toronto, Canada
Raivo Ruusalepp	National Library of Estonia
Heiko Schuldt	University of Basel, Switzerland
Timos Sellis	RMIT University, Australia
Michalis Sfakakis	Ionian University, Greece
Frank Shipman	Texas A&M University, USA
Nicolas Spyratos	University of Paris South, France
Shigeo Sugimoto	University of Tsukuba, Japan
Hussein Suleman	University of Cape Town, South Africa
Tamara Sumner	University of Colorado at Boulder, USA
Atsuhiro Takasu	National Institute of Informatics, Japan
Manfred Thaller	Universität zu Köln, Germany
Chrisa Tsinaraki	European Union - Joint Research Center (EU - JRC), Italy
Yannis Tzitzikas	University of Crete and FORTH-ICS, Greece
Pertti Vakkari	University of Tampere, Finland
Felisa Verdejo	Universidad Nacional de Educacion a Distancia, Spain
Jan Weglarz	Poznan University of Technology, Poland
Iris Xie	University of Wisconsin-Milwaukee, USA
Maja Žumer	University of Ljubljana, Slovenia

Additional Reviewers

Assante, Massimiliano	Kotzinos, Dimitri	Stefanidis, Kostas
Bikakis, Nikos	Manghi, Paolo	Students, Threephd
Candela, Leonardo	Manguinhas, Hugo	Students, Twophd
Chandrasekar, Prashant	Mckay, Dana	Williams, Kyle
Coro, Gianpaolo	Moro, Robert	Wu, Jian
Fafalios, Pavlos	Papadakis, Manos	Zagganas, Kostis
Kalogeros, Eleftherios	Papadakos, Panagiotis	
Kanellos, Ilias	Salem, Joseph	

Outlines of Keynote Presentations

Data - Unbound by Time or Discipline – Challenges and New Skills Needed

David Giaretta

Giaretta Associates Ltd., Yetminster, Dorset, UK
david@giaretta.org

Abstract. We live in an exciting information age, where the deluge of data enables the 4th paradigm to be used by the greatest number of scientists who have ever lived, able to connect to hundreds of thousands of sources of information which are encoded digitally and used in an ever changing technological network.

To take advantage of these opportunities presents challenges. The most obvious involves simply coping with the volumes of data with which one has some familiarity, from familiar sources.

However in order to combine data from multitudes of unfamiliar sources, covering a variety of disciplines, created over timescales which are long compared to technological and even many conceptual and terminological cycles there are new challenges both for the researchers and the infrastructure needed to support them.

This presentation will focus on these challenges raised by the need to ensure we can deal with the unfamiliar and outline the resources, both human and technical, which will be needed to address them.

1 Opportunities

The term "4th paradigm" was coined by Jim Gray and colleagues to express the idea that in addition to the empirical, theoretical and computational paradigms we now have data exploration enabled by the vast amount of data that is being produced. This has been explored in the literature as a source for scientific progress. However there are far broader opportunities which those who fund the research are interested in.

The Riding the Wave report provided a vision for 2030 which addressed the question, as part of the EU Digital Agenda, "How Europe can gain from the rising tide of scientific data".

The starting point was the observation that "A fundamental characteristic of our age is the raising tide of data – global, diverse, valuable and complex. In the realm of science, this is both an opportunity and a challenge."

The vision was of "a scientific e-Infrastructure that supports seamless access, use, re-use and trust of data. In a sense, the physical and technical infrastructure becomes invisible and the data themselves become the infrastructure – a valuable asset, on which science, technology, the economy and society can advance."

2   Challenges

An underlying challenge was sustaining the availability and usability of the digitally encoded information across disciplines and over time. An associated, fundamental, question was "who pays and why". While data is newly created and of obvious use there will be resources available, but as the Blue Ribbon Task Force pointed out, the value of much data is potential – it may be useful in the future, but this is not certain.

Resources are needed to address the many V's[1] which are normally discussed in terms of big data – but which are also relevant to small data, since as noted[2] the real revolution, which is the mass democratisation of the means of access, storage and processing of data – small as well as big.

In this presentation I divide these Vs into two groups. The first consists of Volume, Velocity, Variety and Volatility which are ones more related to data management – i.e. issues which arise even if the data is being used by the researchers who created it and over just a few years. The other group consists of Veracity, Validity and Value, which this presentation will focus on for the following reasons.

Veracity, including Understandability and Authenticity, is vital for using data from unfamiliar sources and with which the researcher is unfamiliar – otherwise how can a researcher use the data and trust that it is what it is claimed to be? The challenge will be exacerbated by the data management "Vs" noted previously, in particular scaling with Variety.

Validity (including correctness, data quality and legality) is vital interest to researchers if they wish to undertake scientifically useful work.

Value (or potential value) must be identified in order to justify keeping the data in the long term – and even in the short term (related to Volatility) – because keeping data requires resources. The minimum, relatively easily identified, costs are those for storage which tends to scale with Volume and are very front-loaded. Other costs, which are less obvious and more uncertain are those associated with maintaining Veracity and Validity.

3   Solutions

The bulk of the presentation will look at practical solutions to the challenges presented by the second group of V's. These solutions involve underlying consistent concepts, technology and widely agreed procedures, all supported by skilled and well trained humans, across the whole lifecycle of data from conception through to and including curation.

They will help put in place the data infrastructure which can be used across disciplines and across time for the benefit of science, technology, the economy and society.

---

[1] http://insidebigdata.com/2013/09/12/beyond-volume-variety-velocity-issue-big-data-veracity/

[2] http://www.theguardian.com/news/datablog/2013/apr/25/forget-big-data-small-data-revolution

Digital Audio Asset Archival and Retieval:
A Users Perspective

Joseph Cancellaro

Interactive Arts and Media Depatment
Columbia College Chicago
jcancellaro@colum.edu

Abstract. Both academically and professionally, the problem of coding and formatting audio for archiving and retrieval is constantly present. Many factors weigh in on how to build naming conventions, search criteria, and meta tags for audio assets, particularly in large scale non-linear virtual environment productions. The tendency to build new libraries for each project is most intensive but allows a user to constantly be aware of the assets available and who created them. This requires inventing or reusing a new data retrieval and archiving engine or platform for each project. Needless to say, this approach is impractical and inefficient. The music industry has solved some of these issues through the convenience of having stylized "brands" of music to label. The user can decide whether he/she likes the music and move on to the next piece of similar character. Examples of engines like this include Pandora, Rdio, Spotify and of course Apple and Google. These music streaming automated music recommendation services function moderately well until they don't. In sound effects, this solution is more complex. When there are thousands of samples of differing footsteps, squeaks, knocks, thumps, etc. to manage per category a more refined Content Management System (CMS) is required. Again, the impractical issue of building from scratch is introduced. One method being looked at is Fourier transform based pattern recognition analysis and algorithms. This method uses discrete Fourier transform based pattern classifiers, defined and correlated by a designer, to map, or compare against a predefined data vector specifying a particular sound pattern. This research has already been experimented with but usually not in the form of a searchable library tool. An added layer to this method includes the use of artificial intelligent programming as part of a machine learning tool to help expedite searching and filing of data assets. This is currently being experimented with in the Interactive Arts and Media Department where students and faculty are working on solutions to some of these problems. The landscape of audio, which includes music, sound and silence is massively comprehensive in size and complexity. Methods and strategies for solving for a universal identifying and archival tool are constantly on the minds of composers, sound designers and all who deal with audio assets in linear and non-linear environments. My discussion will raise some of the issues surrounding classifying audio and storage as well as problems encountered by sound designers and composers in the field.

The Era of the Post-repository: Scholarly Practice, Information and Systems in the Digital Continuum

Costis Dallas

Abstract. Research in the arts and humanities is often associated with the world of the solitary scholar, surrounded by dusty books, manuscripts, or artefacts. As early as 1959, C.P. Snow lamented the "gulf of mutual incomprehension" separating humanities scholars from scientists. Yet, the wide-ranging changes in scholarly practice associated with digital technology, the crisis of disciplinarity, the rise of new methodological and theoretical frameworks, the increased risks to the longevity of cultural resources, and the emergence of new fields of contestation around their interpretation and value, casts a different light on Snow's notion of the "two cultures", introducing new challenges and opportunities.

For information researchers and computer scientists engaged with the conceptualization, design and development of digital infrastructures, tools and services in the domain of the arts, humanities and cultural heritage, understanding the nature and direction of these changes is of paramount importance. The advanced field of digital humanities is only part of the story. In fact, humanists working with big data, crafting their own schemas and encoding formalisms, engaging in ontological modeling, and scripting their own analytical and representational tools are but a small minority among an increasing number of scholars producing influential, highly cited research merely facilitated by digital technology – what may be called *digitally-enabled* humanists. As indicated by a survey conducted by the Digital Methods and Practices Observatory Working Group of DARIAH-EU, digitally-enabled humanists use frequently applications such as word processors and spreadsheets besides repositories to organize and curate research resources, controlled vocabularies and classification systems that are more often homegrown than standard, and a variety of readily available online services and social media for information discovery, collaboration and dissemination. And, in tandem with changes in scholarly practice, the rise of computational intelligence and social and participatory media, as well as the increasing availability of humanities and heritage resources in the networked and mobile digital environment at a time of globalization, bring about new important stakeholders in their representation and interpretation, such as descendant and source communities, amateurs engaging in citizen science, and culture and heritage publics.

Established wisdom on digital infrastructures for the arts and humanities is shaped by a notion of centralized custodial control, replicating the traditional structures of the physical archive, library and museum: in other words, on the notion that research resources can be curated and preserved in the future in large-scale, centralized digital repositories. This becomes problematic as financial means grow increasingly scarce, and as the cultural record broadens to include a proliferation of born digital resources, grey literature, outcomes of independent and commercial research, fruits of self-publication, remix and social media interaction, and manifestations of community and personal memory. In fact, research on emerging digital research practices in a discipline such as archaeology shows how the availability of multimodal, interactive, real time recording and documentation technologies, and the plurality of research actors, interpretations and uses of archaeological knowledge give rise to multiple kinds of densely interconnected digital resources (e.g., GIS, LiDAR, formatted data, 3D models, annotations, interpretive narratives, video documentation, blogs, and social media interactions) and intertwine the ostensibly distinct processes of data recording and interpretation. The shift towards a digital infrastructure for humanities and heritage resources, in tandem with these changes, brings about a rising "curation crisis" which calls for a radical reconsideration of priorities in the specification and design of digital infrastructures.

Central to this reconsideration is the concept of the records continuum, originally advanced by Australian archival scholars to indicate the limits of a lifecycle approach in dealing with the capabilities and challenges of digital information. Criticizing the custodial notion of archives as data mortuaries, continuum thinking calls for a unified approach to recordkeeping capable of attending to records from the point of creation to their "pluralizing" interpretation and use by diverse communities. It resonates with a call for a radical re-examination of the theory and practice of digital curation, based on the recognition that curation of research resources facilitated by ubiquitous pervasive digital technology takes place increasingly "in the wild", involves multiple stakeholders "exercising the archive" beyond data custodians, concerns not merely information resources *qua* digital objects but also their evolving epistemic content and context, and thus requires a rethink of the requirements, affordances and priorities of digital infrastructures.

The promise of going beyond traditional repositories to deploy a digital infrastructure which explicitly focuses on the provision of curation capabilities is demonstrated by the Metadata and Object Repository (MORe), a system deployed by the Digital Curation Unit, IMIS-Athena Research Centre to support the dynamic evolution and continuous semantic enrichment of heterogeneous metadata and registry descriptions of arts and humanities resources and collections. MORe has been used extensively for Europeana metadata aggregation in the CARARE and LoCloud projects, supporting semi-automated and manual digital curation activities, and leveraging workflows of external services such as historic names gazeteers and SKOS vocabularies. It supports the curation of resources "in the wild" such as Wikimedia assets, and connects with client systems, including Omeka-based LoCloud Collections, and a Metadata Entry Tool that could support "sheer curation" on a digital tablet at the point of creation.

Systems such as MORe herald a new approach to digital infrastructures, beyond the architecture and functionalities of traditional repositories such as Fedora or DSpace. Yet, a key challenge remains how to address the fact that arts and humanities scholars, amateur researchers, memory institutions, collectors, and online users and curators of digital information assets will continue to employ a bricolage of digital tools and methods available "at hand", some of which may be imprisoned within technical or commercially-controlled silos. An overarching vision for future infrastructures might thus call for a radically expanded version of custodial repositories, combining open cloud storage of dynamic, potentially intelligent and self-documenting information objects with curation-enabled, distributed information systems and orchestrated, user-configurable services accessible to multiple interfaces of end-user tools and applications in the continuum.

Open Access to Research Data: Is it a Solution or a Problem?

Vittore Casarosa

CNR - ISTI, Italy
Panel held in connection with TPDL 2015

Panel Introduction

In the past few years, we have witnessed a (slow) paradigm shift about the way in which research results are being published and disseminated. More and more we have seen the push for publishing research results as Open Access (OA) "digital publications" and more recently the push (especially from the European Commission) towards "Open Science". This means not only the OA publication of research results, but also the OA publication of the "input to research", i.e. the raw material underlying the research process, generically identified as Research Data.

The main argument in favour of Open Access is that most of research is being done with public funds, with research results and research data being produced in the public interest, and therefore they should remain publicly available. Availability should be restricted only by legitimate reasons, such as privacy protection or intellectual copyright. Of course, Open Access does not prevent commercial exploitation and protection of the research results and the research data, with patents and copyrights.

Following the recommendations of the European Commission, to ensure open access, publication should be done either by self archiving the material in an online repository, or by open access publication in peer-reviewed open access journals, which very often charge "Article Processing Charges" to the authors, to offset the cost of making the content of the journal freely available. The first alternative is commonly indicated as "Green OA" and the second one as "Gold OA". The diagram below, borrowed from the European Commission, summarizes these concepts.

The push towards open access to research data has only increased the number of issues generally encountered with open access to scientific publications. More than ten years ago, a report from OECD [1] identified and categorized the main issues related to Open Access.

- Technological issues: Broad access to research data, and their optimum exploitation, requires appropriately designed technological infrastructure, broad international agreement on interoperability, and effective data quality controls;
- Institutional and managerial issues: While the core open access principle applies to all science communities, the diversity of the scientific enterprise suggests that a variety of institutional models and tailored data management approaches are most effective in meeting the needs of researchers;

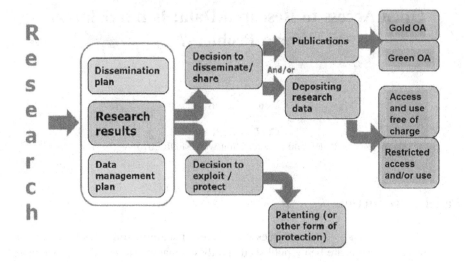

- Financial and budgetary issues: Scientific data infrastructure requires continued, and dedicated, budgetary planning and appropriate financial support. The use of research data cannot be maximized if access, management, and preservation costs are an add-on or after-thought in research projects;
- Legal and policy issues: National laws and international agreements directly affect data access and sharing practices, despite the fact that they are often adopted without due consideration of the impact on the sharing of publicly funded research data;
- Cultural and behavioural issues: Appropriate reward structures are a necessary component for promoting data access and sharing practices. These apply to those who produce and those who manage research data.

Panel Objectives

Given the breadth and depth of all the issues, it should be clear that the main objective of the panel is not to solve the issues of Open Access. From one point of view, Open Access to Research Data is the "solution" to achieve a "better and more efficient science (Science 2.0)". From another point of view, Open Access to Research Data brings with it so many issues and problems that it might become an impediment to the dissemination of research results. The panel will try to stimulate a discussion and an exchange of ideas among the panellists, which is expected to trigger a wider discussion with the audience, touching (some of) the benefits and issues mentioned before. The panellists will bring to the table their experience in many of the issues mentioned before, such as infrastructures and institutional repositories, data curation in libraries and archives, long term preservation of data, education and training for data producers and data curators, and so on.

As it is often the case in this type of events, most probably at the end of the panel there will be even more questions than answers, but hopefully it will have contributed to gain a more global view and a better understanding of the issues related to the actual implementation of Open Access to Research Data.

Panel coordinator
Vittore Casarosa (CNR-ISTI, Italy)
Panel participants
David Giaretta (Alliance for Permanent Access)
Steve Griffin (University of Pittsburgh, Pittsburgh, USA)
Herbert Maschner (University of Southern Florida, Tampa, USA)
Cezary Mazurek (Poznan Supercomputing and Networking Center, Poznan, Poland)
Andy Rauber (Technical University of Vienna, Vienna, Austria)
Anna Maria Tammaro (University of Parma, Parma, Italy)

Reference

[1] Arzberger, P., et al.: Promoting access to public research data for scientific, economic, and social development. Data Sci. J. **3** (2004)

Contents

User Studies for and Evaluation of Digital Library Systems and Applications

Applications of Digital Libraries

Digital Humanities

Social-Technical Perspectives of Digital Information

Poster and Demo Papers

Interoperability and Information Integration

Web Archive Profiling Through CDX Summarization

Sawood Alam[1]([✉]), Michael L. Nelson[1], Herbert Van de Sompel[2],
Lyudmila L. Balakireva[2], Harihar Shankar[2], and David S.H. Rosenthal[3]

[1] Computer Science Department, Old Dominion University, Norfolk, VA, USA
{salam,mln}@cs.odu.edu
[2] Los Alamos National Laboratory, Los Alamos, NM, USA
{herbertv,ludab,harihar}@lanl.gov
[3] Stanford University Libraries, Stanford, CA, USA
dshr@stanford.edu

Abstract. With the proliferation of public web archives, it is becoming more important to better profile their contents, both to understand their immense holdings as well as support routing of requests in the Memento aggregator. To save time, the Memento aggregator should only poll the archives that are likely to have a copy of the requested URI. Using the CDX files produced after crawling, we can generate profiles of the archives that summarize their holdings and can be used to inform routing of the Memento aggregator's URI requests. Previous work in profiling ranged from using full URIs (no false positives, but with large profiles) to using only top-level domains (TLDs) (smaller profiles, but with many false positives). This work explores strategies in between these two extremes. In our experiments, we gained up to 22 % routing precision with less than 5 % relative cost as compared to the complete knowledge profile without any false negatives. With respect to the *TLD-only* profile, the registered domain profile doubled the routing precision, while complete hostname and one path segment gave a five fold increase in routing precision.

Keywords: Web archives · Profiling · CDX Files · Memento

1 Introduction

The number of public web archives supporting the Memento protocol [17] natively or through proxies continues to grow. The Memento Aggregator [12], the Time Travel Service[1], and other services, both research and production, need to know which archives to poll when a request for an archived version of a file is received. In previous work, we showed that simple rules are insufficient to accurately model a web archive's holdings [3,4]. For example, simply routing requests for *.uk URIs to the UK National Archives is insufficient: many other archives hold *.uk URIs, and the UK National Archives holds much more than just *.uk URIs. This is true for the many other national web archives as well.

---

[1] http://timetravel.mementoweb.org/.

© Springer International Publishing Switzerland 2015
S. Kapidakis et al. (Eds.): TPDL 2015, LNCS 9316, pp. 3–14, 2015.
DOI: 10.1007/978-3-319-24592-8_1

In this paper we examine strategies for producing *profiles* of web archives. The idea is that profiles are a light-weight description of an archive's holdings to support applications such as coordinated crawling between archives, visualization of the archive's holdings, or routing of requests to the Memento Aggregator. It is the latter application that is the focus of this paper.

An archive profile has an inherent trade-off in its size vs. its ability to accurately describe the holdings of the archive. If a profile records each individual original URI (URI-R in Memento terminology) the size of the profile can grow quite large and difficult to share, query, and update. On the other hand, an aggregator making routing decisions will have perfect knowledge about whether or not an archive holds archived copies of the page, or mementos (URI-Ms in Memento terminology). On the other hand, if a profile contains just the summaries of top-level domains (TLDs) of an archive the profile size will be small but can result in many unnecessary queries being sent to the archive. For example, the presence of a single memento of bbc.co.uk will result in the profile advertising .uk holdings even though this may not be reflective of the archive's collection policy.

In this paper we examine various policies for generating profiles, from the extremes of using the entire URI-R to just the TLD. Using the CDX files[2] of the UK Web Archive (covering 10 years and 0.5 TB) and the ODU copy of the Archive-It (covering 14 years and 1.8 TB), we examine the trade-offs in profile size and routing precision for three million URIs requests.

2   Related Work

Query routing is common practice in various fields including meta-searching and search aggregation. Memento query routing was explored in the two efforts described below, but they explored extreme cases of profiling. We believe that an intermediate approach that gives flexibility with regards to balancing accuracy and effort can result in better and more effective routing.

Sanderson et al. created exhaustive profiles [13] of various IIPC member archives by collecting their CDX files and extracting URI-Rs from them (we denote it as *URIR Profile* in this paper). This approach gave them complete knowledge of the holdings in each participating archive, hence they can route queries precisely to archives that have any mementos (URI-M) for the given URI-R. It is a resource and time intensive task to generate such profiles and some archives may be unwilling or unable to provide their CDX files. Such profiles are so big in size (typically, a few billion URI-R keys) that they require special infrastructure to support fast lookup. Acquiring fresh CDX files from various archives and updating these profiles regularly is not easy.

Many web archives tend to limit their crawling and holdings to some specific TLDs, for example, the British Library Web Archive prefers sites with .uk TLD. AlSum et al. created profiles based on TLD [3,4] in which they recorded *URI-R*

---

[2] CDX files are created as an index of the WARC [10] files generated from the Heritrix web crawler; see [8] for a description of the CDX file format.

Count and *URI-M Count* under each TLD for twelve public web archives. Their results show that they were able to retrieve the complete TimeMap [17] in 84 % of the cases using only the top 3 archives and in 91 % of the cases when using the top 6 archives. This simple approach can reduce the number of queries generated by a Memento aggregator significantly with some loss in coverage.

3   Methodology

In this study we used CDX files to generate profiles, but profile generation is not limited to only CDX processing, it can also be done by sampling URI sets and querying the live archives or by using full-text searching feature provided by some archives. To deal with periodic updates of profiles, smaller profiles are generated with new data and these small profiles are merged into the base profile. Without an option to merge smaller profiles to build a large profile gradually, updates will require a complete reprocessing of the entire dataset, including the dataset previously processed. In these two cases the statistical measures such as URI-R count cannot have absolute values, so we use the sum of URI-M counts (as "frequency") from all the profiles under each *URI-Key* and keep track of the number of profiles they came from (as "spread") as an indication of the holdings as shown in Fig. 1.

URI-Key is a term we introduced to describe the keys generated from a URI based on various policies. So far we have created policies that can be classified in two categories, *HmPn* and *DLim*. Policies of the generic form *HmPn* mean that the keys will have a maximum of "m" segments from the hostname and a maximum of "n" segments from the path. A *URI-Key* policy with only one hostname segment and no path segments (*H1P0*) is called *TLD-only* policy (as discussed in Sect. 2). *H3P0* policy covers most of the registered domains (that have one or two segments in their suffix [11], such as .com or .co.uk). If the number of segments are not limited, they are denoted with an "x", for example, *HxP1* policy covers any number of hostname segments with maximum of one path segment and *HxPx* means any number of hostname and path segments. Note that the *HxPx* policy is not the same as the *URIR* policy (as discussed in Sect. 2) because it strips off the query parameters from the URI, while the *URIR* policy stores complete URIs. Policies of the generic form *DLim* are based on the registered domain name, the number of segments in sub-domain, path, and query sections of a URI and the initial letter of the path. A generic template for this category of *URI-Keys* can be given as

```
1  @context https://oduwsdl.github.io/contexts/archiveprofile.jsonld
2  @id http://www.webarchive.org.uk/ukwa/
3  @about {"name": "UKWA 1996 Collection", "type": "urikey#H3P1", "...": "..."}
4  com,dilos)/region {"frequency": 14, "spread": 2}
5  edu,orst)/groups {"frequency": 3, "spread": 1}
6  uk,ac,rpms)/ {"frequency": 124, "spread": 1}
7  uk,co,bbc)/images {"frequency": 152, "spread": 3}
```

Fig. 1. Sample profile in CDXJ format

```
1   URI: https://www.news.BBC.co.uk/images/Logo.png?width=200&height=80&rotate=90#top
2   Canonical URL: news.bbc.co.uk/images/Logo.png?height=80&rotate=90&width=200
3   SURT URL: uk,co,bbc,news)/images/Logo.png?height=80&rotate=90&width=200
4   Registered Domain: uk,co,bbc)/
5   Segment Counts: {subdomain:1, path: 2, query: 3}
6   Path Initial: i
7   URI-Keys:
8              H1P0: uk)/
9              H3P0: uk,co,bbc)/
10             HxP1: uk,co,bbc,news)/images
11             DDom: uk,co,bbc)/
12             DPth: uk,co,bbc)/1/2
13             DIni: uk,co,bbc)/1/2/3/i
```

Fig. 2. Illustration of URI-Key

"`registered_domain)/[#subdomain[/#path[/#query[/path_initial]]]]`".
The *DDom* policy includes only the registered domain name in Sort-friendly
URI Reordering Transform (SURT) [14] format, while *DSub*, *DPth*, *DQry*, and
DIni policies also include sections of the template up to #subdomain, #path,
#query, and path_initial respectively.

Figure 2 illustrates the process of generating *URI-Keys* from a URI. URIs are
first canonicalized then go through SURT. For *HmPn* policies, query section and
fragment identifier of the URI are removed (if present), then depending on the
values of "m" and "n" any excess portions from the SURT URL are chopped off.
The hostname segments are given precedence over the path segments in a way
that no path segment is added until all the hostname segments are included,
hence uk,co)/images is an invalid *URI-Key*, but uk,co,bbc,news)/images
would be valid if the hostname is news.bbc.co.uk or www.news.bbc.co.uk. For
DLim policies, the registered domain name is extracted with the help of the
Public Suffix list (which is updated periodically). Then depending on the individual policies, segments from zero or more sections (such as sub-domain and
path) of the URI are counted, and if necessary, the initial letter of the first path
segment is extracted (replaced with a "-" if not alphanumeric). These values are
then placed inside the above template to form the key.

4   Implementation

We have implemented a *URI-Key* Generator, more than one CDX Profiler, and
a script to merge profiles and published the code on GitHub[3]. We have also
made the code available for analyzing and benchmarking the profiles. We plan
to collect all the profiles generated from various places in a public repository,
but our script currently generates local files and publishes them in the form of
a public Gist[4] if configured to do so.

Our initial implementation used JSON [5] and JSON-LD [15] for profile serialization. It was good for small profiles, but for large profiles any data format that
has a single root node (such as JSON, XML, or YAML) introduces many scale
challenges. To make frequent lookup in these single root node profiles efficient,

---
[3] https://github.com/oduwsdl/archive_profiler.
[4] https://gist.github.com/.

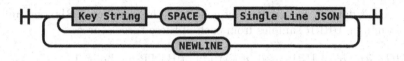

Fig. 3. Railroad diagram: CDXJ file format

they need to be completely loaded into memory, which becomes a bottleneck. A single malformed character can make these profiles unusable. Hence a more linear key-value(s) based data formats such as CDX or ARFF [18] is more suitable in such cases as they allow an arbitrary split of data and enable easier profile merging. We have come up with a similar sort and index friendly file format "CDXJ" that is a fusion of CDX and JSON formats as illustrated in Fig. 3 and utilized in Fig. 1. The new CDXJ format also reduces the number of keys in a profile as it allows partial key lookup as opposed to the JSON format where we had to store all the intermediate smaller keys for higher level statistics.

Our initial implementation built the complete data structure in memory before serializing it to a file. We encountered a limitation in Python's dictionary implementation that degrades performance significantly when the number of keys in the dictionary is large. Hence, to make it scalable, we experimented with different profile generation optimization techniques such as preprocessing CDX files with standard Unix utilities (like grep, sort, uniq, sed, and awk) or using key-value databases (file based or in memory) for intermediate processing. The latter approaches involve more steps and setup, but scale well.

5   Evaluation

We generated 23 different profiles (17 *HmPn* policies, five *DLim* policies, and one *URIR* policy) for each archive dataset to measure their resource requirement and routing efficiency. To perform the analysis we prepared two types of datasets, archive profiles and query URI-Rs. For profiles, we used two archives:

- *Archive-It Collections* – We acquired the complete holdings of Archive-It [9] before 2013 and indexed the collections (in CDX format) to create a replica of the service. The archive has 2,952 collections with more than 5.3 billion URI-Ms and about 1.9 billion unique URI-Rs. Our Archive-It replica has more than 1.9 million ARC/WARC files that take about 230 TB disk space in compressed format. We created *URI-Key* profiles with various policies for the entire archive from the CDX files.
- *UK Web Archive* – We acquired a publicly available CDX index dataset from UKWA [16]. The dataset has separate CDX files for each year. We created individual *URI-Key* profiles from each of the early 10 years of CDX files (from year 1996 to 2005) with different profiling policies. We also created a combined profile by incrementally accumulating data for each successive year to analyze the growth. These 10 years of CDX files have about 1.7 billion URI-Ms and about 0.7 billion unique URI-Rs.

A second dataset was created by collecting three million URIs; one million random unique URI-R samples from each of these three sources:

- *DMOZ Archive* – URIs used in a study of HTTP methods [1].
- *IA Wayback Access Log* – URIs extracted from the access log used in a study of links to the Internet Archive (IA) content [2].
- *Memento Aggregator Access Log* – URIs extracted from the access log used in a previous archive profiling study [4].

5.1   Profile Growth Analysis

In commonly used CDX files each entry corresponds to a URI-M[5]. The length of each line in a CDX file depends on the length of the URI-R in it. Our experiment shows that the average number of bytes per line (α) in our dataset is about 275, which means every one gigabyte of CDX file holds about 3.9 million URI-Ms. Figure 4(a) can be used to estimate α in a CDX file. Equation 1 can be used to quickly estimate number of URI-Ms (C_m) in a large collection if total size of CDX files in bytes (S_c) is known.

$$C_m = \frac{S_c}{\alpha} \tag{1}$$

Figure 4(b) shows the relationship between URI-M Count (C_m) and URI-R Count (C_r) in two ways; (1) for each year of UKWA CDX files individually as if they were separate collections and (2) accumulated ten consecutive years of data one year at a time while each time it recalculates total number of unique URI-Rs visited. The ratio of URI-M Count to URI-R Count (γ) as shown in Eq. 2 is indicative of the average number of revisits per URI-R for any given time period. The value of γ varies from one archive to the other because some archives perform shallow archiving while others revisit old URI-Rs regularly. In our dataset the value of γ is 2.46 for UKWA and 2.87 for Archive-It. The accumulated trend better accounts for how archives actually grow over time. It follows Heaps' Law [6] as shown in Eq. 3 where URI-Rs are analogous to unique words in a corpus. K and β are free parameters that are affected by the value of γ, but the actual values are determined empirically. For the UKWA dataset $K = 2.686$ and $\beta = 0.911$.

$$\gamma = \frac{C_m}{C_r} \tag{2}$$

$$C_r = KC_m^\beta \tag{3}$$

URI-Keys are used as lookup keys in the *URI-Key* profile. The number of *URI-Keys* is a function of URI-Rs, not URI-Ms, hence increasing URI-Ms without introducing new URI-Rs does not affect the number of *URI-Keys*, instead, it

---

[5] In our dataset Archive-It has 0.71 % non-HTTP entries in their CDX files while UKWA has no non-HTTP entries.

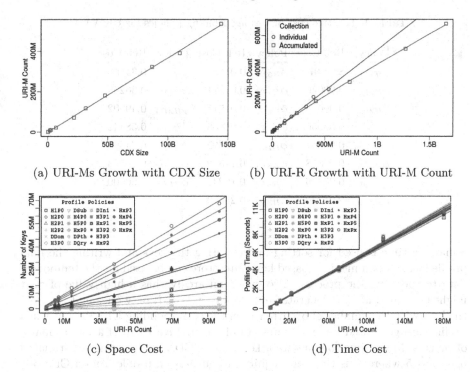

(a) URI-Ms Growth with CDX Size   (b) URI-R Growth with URI-M Count

(c) Space Cost   (d) Time Cost

Fig. 4. Growth costs analysis for different profiling policies

only changes the value. Figure 4(c) shows the number of unique *URI-Keys* (C_h) generated for different numbers of unique URI-Rs (C_r) on different profiling policies. Every profiling policy follows a straight line with a slope value (ϕ_{policy}) and zero *Y-axis* intersect because for zero URI-Rs there will be zero *URI-Keys*. For a given profile policy, we define the ratio of *URI-Key Count* to *URI-R Count* as *Relative Cost* (ϕ_{policy}) as shown in Eq. 4. The *Relative Cost* (ϕ_{policy}) varies from one archive to the other based on their crawling policy. Archives that crawl only a few URIs from each domain will have relatively higher *Relative Cost* than those who crawl most of the URIs from each domain they visit. Table 1 lists ϕ_{policy} values for UKWA dataset.

$$\phi_{policy} = \frac{C_h}{C_r} \qquad (4)$$

Figure 4(d) illustrates the time required to generate profiles with different policies and different data sizes. Previously we used memory based, but now we use a file based profiling, which scales better and allows for distributed processing. Our experiment shows that the profiling time is mostly independent of the policy used, but we found that *DLim* policies take slightly more time than *HmPn* policies because they require more effort to extract the registered domain based on the public suffix list. We found that about 95 % of the profiling time is spent on generating keys from URIs and storing them in a temporary file while the

Table 1. Relative cost of various profiling policies for UKWA.

Policy	Rel. Cost	Policy	Rel. Cost	Policy	Rel. Cost
ϕ_{H1P0}	8.5e-07	ϕ_{H5P0}	0.01368	ϕ_{H3P3}	0.34668
ϕ_{H2P0}	0.00026	ϕ_{HxP0}	0.01371	ϕ_{HxP2}	0.36298
ϕ_{H2P1}	0.00038	ϕ_{DPth}	0.01577	ϕ_{HxP3}	0.49902
ϕ_{H2P2}	0.00056	ϕ_{DQry}	0.01838	ϕ_{HxP4}	0.58442
ϕ_{DDom}	0.00857	ϕ_{DIni}	0.06892	ϕ_{HxP5}	0.64365
ϕ_{H3P0}	0.00858	ϕ_{H3P1}	0.11812	ϕ_{HxPx}	0.70583
ϕ_{DSub}	0.00876	ϕ_{HxP1}	0.16247	ϕ_{URIR}	1.00000
ϕ_{H4P0}	0.01340	ϕ_{H3P2}	0.25379		

remaining time is used for sorting and counting the keys and writing the final profile file. Hence a memory based key-value store can be used for the temporary data to speed up the process. Also, when an archive has a high value of γ, it might be a good idea to generate keys from each URI-R only once and multiply the keys with the number of occurrences of the corresponding URI-Rs, but when profiles are generated on small sub-sets of the archive there is a lower chance of revisits. Mean time to generate a key for one CDX entry τ can be estimated using Eq. 5 where T is the time required to generate a profile from a CDX file with C_m URI-Ms. The value of τ depends on the processing power, memory, and I/O speed of the machine. On our test machine it was between 5.7e-5 to 6.2e-5 s per URI-M (wall clock time). As a result, we were able to generate a profile from a 45GB CDX file with 181 million URI-Ms and 96 million unique URI-Rs in it in three hours.

$$\tau = \frac{T}{C_m} \tag{5}$$

Figure 5 illustrates the correlation among the number of *URI-Keys* Fig. 5(a) and profile size on disk Fig. 5(b) for various collection sizes with various profiling policies. The policies are sorted in the increasing order of their resource requirement. If generated profiles are compressed (using gzip [7] with the default compression level), for bigger profiles they use about 15 times less storage than uncompressed profiles, but result in a similar growth trend. These figures are helpful in identifying the right profiling policy depending on the available resources such as storage, memory, computing power. Here are some common observations in these figures:

- For host segments less than three, path segments do not make a significant difference as they are not included unless all the host segments of the URI are already included.
- Keeping either the hostname or path segments constant while increasing the other shows growth in the value, but the growth rate decreases as the segment count increases.
- *DDom* profile shows quite the same results as *H3P0* profile.

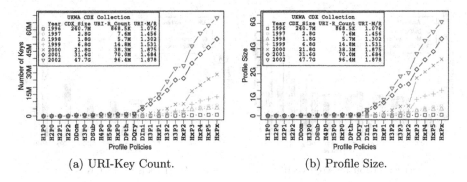

(a) URI-Key Count.   (b) Profile Size.

Fig. 5. Resource requirement for various profiling policies and collection sizes

- The last data-point (HxPx) shows a different trend, it is not just one path segment ahead of its predecessor (HxP5), but any path segments more than 5 are included in it.
- Growth due to path segments is significantly faster than the growth due to hostname segments.
- If a single path segment is to be included, it is better to include all host segments as it will not cause any significant resource overhead, but will provide better details.

5.2 Routing Efficiency

To analyze the routing efficiency of profiles, we picked eight policies from the 23 policies we have used to generate profiles for both the archives in our dataset. These policies are *TLD-only (H1P0)*, *H3P0*, *HxP1*, and all the five variations of the *DLim* policies. We then examined the presence of resources in the archives for our three query URI sample sets, each containing one million unique URI-Rs. Based on the profiling policy, a query URI-R is transformed into a *URI-Key* then it is looked up in the *URI-Key* profile keys to predict its presence.

To establish a baseline, we only check to see if the lookup key is present in the profile and do not use the statistical values (such as their frequency) present in the profile. This brings false positives in the result, but no false negatives, hence we do not miss any URI-Ms from the archive for the query URI-R. Table 2 is a good indicator of why archive profiles are useful: since both archives have < 5 % of any of the query sets, there would be many unnecessary queries if an aggregator simply broadcasted all queries to all archives.

$$\text{Routing Precision}_{policy} = \frac{|\text{URI-R Present in Archive}|}{|\text{URI-R Predicted by Profile}_{policy} \text{ in Archive}|} \quad (6)$$

As illustrated in Fig. 6(a) and (b), from both archives it can be seen that the *Routing Precision* (as defined in Eq. 6) grows when a profile with higher *Relative*

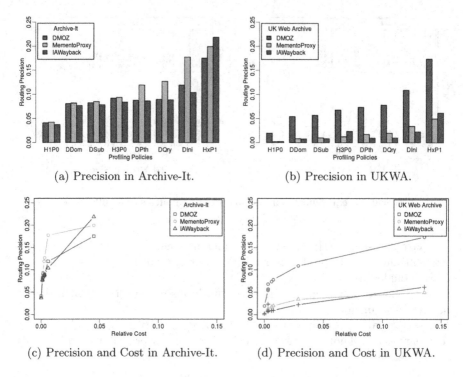

(a) Precision in Archive-It.   (b) Precision in UKWA.

(c) Precision and Cost in Archive-It.   (d) Precision and Cost in UKWA.

Fig. 6. Routing precision of different profiling policies in different archives.

Cost is chosen. Figure 6 shows that *DDom* profile doubles the precision as compared to the *TLD-only* profile and the *HxP1* profile brings five fold increment in the routing precision. These figures also show that inclusion of sub-domain count does not affect the results much as there are not many domains that are utilizing more than one sub-domain. Figure 6(c) and (d) show that there is significant gain in *Routing Precision* with little increment in *Relative Cost*. The difference in the trends of Fig. 6(c) and (d) can be understood by means of the following factors:

- Table 2 shows that less than 0.3 % of the sample URIs from MementoProxy and IAWayback logs are archived in UKWA. This affects the growth in *Routing Precision* (on *Y-axis*) from one profile to the next.
- UKWA uses a shallow crawling policy which results in higher *Relative Cost*. This increases the distance (on *X-axis*) from one profile to the next.

A *URI-Key* profile is more robust than a *URIR* profile because it can predict the presence of resources in an archive that were added in the archive after the profile was generated. For example, if a new image `Large-Logo.png` is added under `bbc.co.uk/images`, it is very likely that the archives that crawl BBC will capture the new image. Without any update the *URI-Key* profile illustrated in

Table 2. Presence of the sample query URI-Rs in each archive.

Archive	DMOZ	MementoProxy	IAWayback
Archive-It	4.097 %	4.182 %	3.716 %
UK Web Archive	1.912 %	0.179 %	0.231 %

Fig. 1 will be able to predict the presence of the new resource, but a *URIR* profile will need an update to include the new URI-R before it can predict it.

6   Future Work and Conclusions

In this paper we have examined the space and precision trade-offs in different policies for producing profiles of web archives. We defined the term "URI-Key" to refer to the keys generated from a URI based on various policies that are used to track the distribution of holdings of an archive at different hostname and path depths or different segment counts. We found that the growth of the profile with respect to the growth of the archive follows Heaps Law, but the values of free parameters are archive dependent. We implemented a *URI-Key* generator and profiler scripts and published the code. We used CDX files from ODU's Archive-It replica and the UK Web Archive for generating profiles, and evaluated the profiles using a query set of three million URIs, created from one million URIs from each of DMOZ, IA Wayback access logs, and Memento Aggregator access logs. With precision defined as correctly predicting that the requested URI is present in the archive, we gained up to 22 % routing precision without any false negatives with less than 5 % relative cost as compared to the complete knowledge profile (*URIR* profile) that has both routing precision and relative cost 100 %. The registered domain profile doubles the routing precision with respect to the *TLD-only* profile, while a profile with complete hostname and one path segment gives five fold routing precision. We found that less than 5 % of the queried URIs are present in each of the individual archives. As a result, looking each URI up in every archive is wasteful. Hence, even a small improvement in routing precision can save a lot of resources and time in Memento aggregation.

Going forward, we plan to study the trade-off between the routing precision and recall by utilizing the statistical values stored against each key in the profile. We plan to develop a more sophisticated non-absolute statistical property to predict distribution of the holdings for various lookup keys that remains useful after merging multiple profiles. We plan to combine results of more than one type of profiles (such as *Time* and *URI-Key*) to improve the routing precision and recall. We also plan to create profiles of live archives from non-CDX sources, such as URI and keyword sampling to generate *URI-Key*, Language, Time, and Media Type profiles. We would also like to generate classification based profiles such as news, social media, game, and art. Finally, we would like to implement a production ready profile based Memento query routing system to be used in aggregators.

Acknowledgements. This work is supported in part by the International Internet Preservation Consortium (IIPC). Andy Jackson (BL) helped us with the UKWA datasets. Kris Carpenter (IA) and Joseph E. Ruettgers (ODU) helped us with the Archive-It data sets. Ilya Kreymer contributed to the discussion about CDXJ profile serialization format.

References

1. Alam, S., Cartledge, C.L., Nelson, M.L.: Support for Various HTTP Methods on the Web. Technical report. arXiv:1405.2330 (2014)
2. AlNoamany, Y., AlSum, A., Weigle, M.C., Nelson, M.L.: Who and what links to the Internet Archive. Int. J. Digit. Libr. **14**(3–4), 101–115 (2014)
3. Alsum, A., Weigle, M.C., Nelson, M.L., Van de Sompel, H.: Profiling web archive coverage for top-level domain and content language. In: Aalberg, T., Papatheodorou, C., Dobreva, M., Tsakonas, G., Farrugia, C.J. (eds.) TPDL 2013. LNCS, vol. 8092, pp. 60–71. Springer, Heidelberg (2013)
4. AlSum, A., Weigle, M.C., Nelson, M.L., Van de Sompel, H.: Profiling web archive coverage for top-level domain and content language. Int. J. Digit. Libr. **14**(3–4), 149–166 (2014)
5. Crockford, D.: The application/json media type for JavaScript Object Notation (JSON). RFC 4627 (2006)
6. Egghe, L.: Untangling Herdan's law and Heaps' law: mathematical and informetric arguments. J. Am. Soc. Inform. Sci. Technol. **58**(5), 702–709 (2007)
7. Gailly, J., Adler, M.: GZIP File Format (2013). http://www.gzip.org/
8. Internet Archive: CDX File Format. http://archive.org/web/researcher/cdx_file_format.php (2003)
9. Internet Archive: Archive-It - Web Archiving Services for Libraries and Archives (2006). https://www.archive-it.org/
10. ISO 28500: WARC (Web ARChive) file format (2009). http://www.digitalpreservation.gov/formats/fdd/fdd000236.shtml
11. Mozilla Foundation: Public Suffix List (2015). https://publicsuffix.org/
12. Sanderson, R.: Global web archive integration with memento. In: Proceedings of the 12th ACM/IEEE-CS Joint Conference on Digital Libraries, pp. 379–380. ACM (2012)
13. Sanderson, R., Van de Sompel, H., Nelson, M.L.: IIPC Memento Aggregator Experiment (2012). http://www.netpreserve.org/sites/default/files/resources/Sanderson.pdf
14. Sigursson, K., Stack, M., Ranitovic, I.: Heritrix User Manual: Sort-friendly URI Reordering Transform (2006). http://crawler.archive.org/articles/user_manual/glossary.html#surt
15. Sporny, M., Kellogg, G., Lanthaler, M.: A JSON-based serialization for linked data. W3C Recommendation (2014)
16. UK Web Archive: Crawled URL Index JISC UK Web Domain Dataset (1996–2013) (2014). doi:10.5259/ukwa.ds.2/cdx/1
17. Van de Sompel, H., Nelson, M.L., Sanderson, R.: HTTP Framework for Time-Based Access to Resource States - Memento. RFC 7089, December 2013
18. Weka: Attribute-Relation File Format (ARFF) (2009). http://weka.wikispaces.com/ARFF

Quantifying Orphaned Annotations in Hypothes.is

Mohamed Aturban[✉], Michael L. Nelson, and Michele C. Weigle

Department of Computer Science, Old Dominion University,
Norfolk, VA 23529, USA
{maturban,mln,mweigle}@cs.odu.edu
http://www.cs.odu.edu/

Abstract. Web annotation has been receiving increased attention
recently with the organization of the Open Annotation Collaboration
and new tools for open annotation, such as Hypothes.is. In this paper,
we investigate the prevalence of *orphaned annotations*, where a live Web
page no longer contains the text that had previously been annotated
in the Hypothes.is annotation system (containing 6281 highlighted text
annotations). We found that about 27 % of highlighted text annotations
can no longer be attached to their live Web pages. Unfortunately, only
about 3.5 % of these orphaned annotations can be reattached using the
holdings of current public web archives. For those annotations that are
still attached, 61 % are in danger of becoming orphans if the live Web
page changes. This points to the need for archiving the target of annotations at the time the annotation is created.

Keywords: Web annotation · Web archiving · HTTP

1 Introduction

Annotating web resources helps users share, discuss, and review information
and exchange thoughts. Haslhofer et al. [9] define annotation as associating extra
pieces of information with existing web resources. Annotation types include commenting on a web resource, highlighting text, replying to others' annotations,
specifying a segment of interest rather than referring to the whole resource,
tagging, etc.

In early 2013, Hypothes.is[1], an open annotation tool, was released and is
publicly accessible for users to annotate, discuss, and share information. It provides different ways to annotate a web resource: highlighting text, adding notes,
and commenting on and tagging a web page. In addition to that, it also allows
users to share an individual annotation URI with each other as an independent web resource. The annotation is provided in JSON format and includes
the annotation author, creation date, target URI, annotation text, permissions,
tags, comments, etc.

---

[1] http://hypothes.is.

© Springer International Publishing Switzerland 2015
S. Kapidakis et al. (Eds.): TPDL 2015, LNCS 9316, pp. 15–27, 2015.
DOI: 10.1007/978-3-319-24592-8_2

One of the well-known issues of the Web is that Web pages are not fixed resources. A year after publication, about 11 % of content shared on social media will be gone [12,13], and 20 % of scholarly articles have some form of reference rot [10]. Lost or modified web pages may result in *orphaned annotations*, which can no longer be attached to their target web pages.

Figure 1 shows a web page http://caseyboyle.net/3860/readings/against. html which has 144 annotations from Hypothes.is. The text with darker highlights indicates more users have selected this part of the page to annotate. The issue here is that all of these annotations are in danger of being orphaned because no copies of the target URI are available in the archives. Figure 2 shows the annotation "Who does that someone have to be?" on the highlighted text "Get someone integrate it into bitcoin/litecoin/*coin" at the target URI http:// zerocoin.org/, created in January 2014. In January 2015, this annotation can no longer be attached to the target web page because the highlighted text no longer appears on the page, as shown in Fig. 3. Although the live Web version of http://zerocoin.org/ has changed and the annotation is orphaned, the original version that was annotated has been archived and is available at Archive.today (http://archive.today/20131201211910/, http://zerocoin.org/). The annotation could be re-attached to this archived resource.

In this paper, we present a detailed analysis of the extent of orphaned highlighted text annotations in the Hypothes.is annotation system as of January, 2015. We also look at the potential for web archives to be used to reattach these

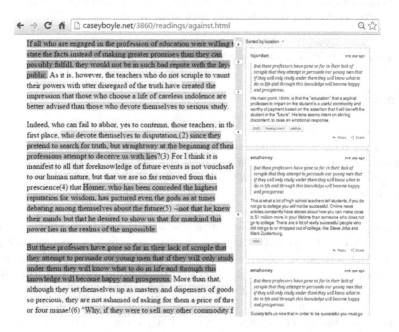

Fig. 1. Using the Hypothes.is browser extension to view the 144 annotations of http://caseyboyle.net/3860/readings/against.html

Fig. 2. http://zerocoin.org/ in January 2014

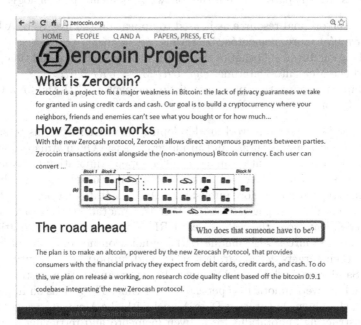

Fig. 3. http://zerocoin.org/ in January 2015

orphaned annotations to archived versions of their original targets. We find that
27 % of the highlighted text annotations at Hypothes.is are orphans, and only
a few can be reattached using web archives. Further, we show that 61 % of the
currently attached annotations could potentially become orphans if their live
Web resources change, because there are no archived versions of the annotated
resources available. Our analysis points to the potential for reducing orphaned
annotations by archiving web resources at the time of annotation.

2   Related Work

Annotation has long been recognized as an important and fundamental aspect
of hypertext systems [11] and an integral part of digital libraries [1], but broad
adoption of general annotation for the Web has been slow. Annotations have been
studied for digital library performance [6,17] and methods have been explored
for aligning annotations in modified documents [4], but typically such studies
are limited to annotation systems specific for a particular digital library. While
orphaned annotations of general web pages have been studied in the context of
Walden's Paths [5,7], our study of Hypothes.is is a more recent evaluation of
annotation and page synchronization in a widely deployed system.

Memento [18] is an HTTP protocol extension that aggregates information
about the resources available in multiple Web archives. We can use Memento to
obtain a list of archived versions of resources, or mementos, available in several
web archives. In this paper, we use the following Memento terminology:

- URI-R - the original resource as it used to appear on the live Web. A URI-R
  may have 0 or more mementos (URI-Ms)
- URI-M - an archived snapshot of the URI-R at a specific date and time, which
  is called the Memento-Datetime, *e.g.*, $URI\text{-}M_i = URI\text{-}R@t_i$
- TimeMap - a resource that provides a list of mementos (URI-Ms) for a URI-R,
  ordered by their Memento-Datetimes

There has been previous work in developing annotation systems to support
collaborative work among users and in integrating the Open Annotation Data
Model [15] with the Memento framework. The Open Annotation Collaboration
(OAC) [9] has been introduced to make annotations reusable through different
systems like Hypothes.is. Before publishing OAC, annotations would not be useful if the annotated web pages were lost because annotations were not assigned
URIs independent from the web pages' URIs. By considering annotations as
first-class web resources with unique URIs, annotation not only would become
reusable if their targets disappear, but also would support interactivity between
systems. Sanderson and Van de Sompel [16] built annotation systems which
support making web annotations persistent over time. They focus on integrating
features in the Open Annotation Data Model with the Memento framework to
help reconstructing annotations for a given memento and retrieving mementos
for a given annotation. They did not focus on the case of orphaned annotations and assumed that the archived resources were available in web archives.

Ainsworth et al. have estimated how much of the web is archived [2]. The result indicated that 35–90 % of publicly accessible URIs have at least one archived copy, although they did not consider annotations in their work, the result might estimate the number of orphaned annotations by factors like how frequently web pages are archived and the archiving process coverage. In other work [8,14] researchers built annotation systems that can deliver a better user experience for specialized users and scholars. The interfaces allow users to annotate multimedia web resources as well as medieval manuscripts in a collaborative way. In this paper, we focus on orphaned annotations and investigate how web archives could be used to reattach these annotations to the original text.

3   Methodology

We performed our analysis on the publicly accessible annotations available at Hypothes.is. The interface allows users to create different types of annotations: (1) making a note by highlighting text and then adding comments and tags about the selected text, (2) creating highlights only, (3) adding comments and tags without highlighting text, and (4) replying to any existing annotations. In January 2015, we downloaded the JSON of all 7744 publicly available annotations from Hypothes.is. Figure 4 shows the JSON of the annotation from Fig. 2 with relevant fields shown in bold. The "updated" field gives the annotation creation date, "source" provides the annotation target URI, "type":"TextQuoteSelector" indicates that it is a highlighted text annotation, "exact" contains the highlighted text, and "text" contains the annotation text itself. We focus only on annotations with highlighted text ("type":"TextQuoteSelector"), leaving 6281 annotations for analysis. To determine how many of those annotations are orphaned, for each annotation we performed the following steps:

- Determine the current HTTP status of the annotation target URIs ("source").
- Compare selected highlighted text ("exact") to the text of the current version of the URI.
- Discover available mementos for the target URI.
- Search for highlighted text within the discovered mementos.

In Table 1, we show the top 10 hosts with annotations at Hypothes.is. Many of these hosts, including the top three, are academic servers and appear to use the system for annotation of scholarly work. Apart from this listing, we did not attempt to make judgements about the content of the annotations or annotation target text in our analysis.

3.1   Determining the HTTP Status

In the first step, the current HTTP status of annotation target URIs can be obtained by issuing HTTP HEAD requests for all URIs. In addition, we extended this to detect "soft" 401, 403, and 404 URIs, which return a 200 OK status but

```json
{
    "updated": "2014-01-13T19:49:33.052047+00:00",
    "target": [
        {
            "source": "http://zerocoin.org/",
            "selector": [
                {
                    "type": "RangeSelector",
                    "startContainer": "/div[1]/div[3]/div[1]/
                                        ul[2]/li[3]/em[1]",
                    "endContainer": "/div[1]/div[3]/div[1]/
                                     ul[2]/li[3]/em[1]",
                    "startOffset": 0,
                    "endOffset": 52
                },
                {
                    "type": "TextQuoteSelector",
                    "prefix": "cation of the Zerocoin protocol",
                    "exact": "Get someone integrate it
                              into bitcoin/litecoin/*coin",
                    "suffix": "Created by Ian Miers @imichaelm"
                },
                {
                    "start": 5522,
                    "end": 5574,
                    "type": "TextPositionSelector"
                }
            ]
        }
    ],
    "created": "2014-01-13T19:49:33.052030+00:00",
    "text": "Who does that someone have to be?",
    "tags": [
        "bitcoin"
    ],
    "uri": "http://zerocoin.org/",
    "user": "acct:rdhyee@hypothes.is",
    "consumer": "00000000-0000-0000-0000-000000000000",
    "id": "SvI30rBYR52gpaCh_IiJgQ",
    "permissions": {
        "admin": [
            "acct:rdhyee@hypothes.is"
        ],
        "read": [
            "group:__world__",
            "acct:rdhyee@hypothes.is"
        ],
        "update": [
            "acct:rdhyee@hypothes.is"
        ],
        "delete": [
            "acct:rdhyee@hypothes.is"
        ]
    }
}
```

Fig. 4. An annotation described in JSON format, available at https://hypothes.is/api/annotations/SvI30rBYR52gpaCh_IiJgQ

actually indicate that the page is not found or is located behind authentication [3]. One technique we used to detect "soft" 4xx is to modify the original URI by adding some random characters. It is likely that the new URI does not exist. After that, we download the content of the original URI and the new one. If the content of both web pages is the same, we consider that the HTTP status of the original URI is "soft" 4xx.

Table 1. The top hosts with annotated pages

Number of annotations	Host
1077	`caseyboyle.net`
886	`rhetoric.eserver.org`
246	`umwblogs.org`
131	`hypothes.is`
111	`dohistory.org`
101	`www9.georgetown.edu`
65	`github.com`
63	`courses.ischool.berkeley.edu`
55	`www.nytimes.com`
46	`www.emule.com`

The returned responses will determine the next action which should be made for every URI. The resulting responses can be categorized into 3 different groups. The first group contains URIs with hostnames `localhost` or URIs which are actually URNs. The second group has URIs with one of the following status codes: "soft" and actual 400, 401, 403, 404, 429 or Connection-Timeout. URIs with 200 status code belong to the third group.

The first group, localhost and URN URIs, were excluded completely from our analysis because these are pages that are not publicly accessible on the live Web. URIs in the second group, soft/actual 4xx and timed-out URIs, have been checked for mementos in the web archives. For URIs with response code 200, we have compared their associated highlighted annotation text with both the current version of the web page and the available mementos in the archives. Even though some annotations are still attached to their live web pages, we are still interested to see if they have mementos to know how likely those annotations are to become orphans if their current web pages change or become unavailable.

3.2 Are Annotations Attached to the Live Web?

The second step is to compare the annotated text ("`exact`") of each annotation target URI that has a 200 HTTP status code with the current version of its web page and see if they match; this can be done by downloading the web page and extracting only the text which will be compared to the highlighted annotation text. We use `curl` to access and download web pages. Then, we extract only the text after cleaning it by removing all HTML tags, extra white-space characters, and others. If the highlighted annotation text is not found in the web page, it is considered *not attached*. For example, as shown in Fig. 3, the annotation text is no longer attached to the web page as the highlighted text, shown in Fig. 2, has been removed from the live web page.

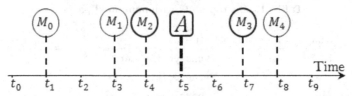

(a) Existing Mementos Before and After the Annotation Creation Date

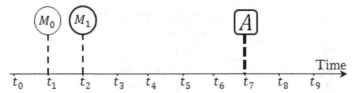

(b) Existing Mementos Only Before the Annotation Creation Date

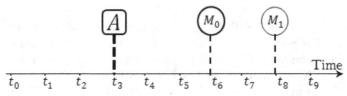

(c) Existing Mementos Only After the Annotation Creation Date

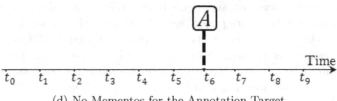

(d) No Mementos for the Annotation Target

Fig. 5. Annotation and memento creation dates

3.3   Discovering Mementos for All Valid URIs

The third step is in discovering mementos for all valid annotation target URIs. For this purpose, we used a Memento Aggregator [18], which provides a TimeMap of the available mementos for a URI-R. It would be a time-consuming task to check all available mementos for a URI-R to see whether they can be used to recover web pages. For example, URIs like http://www.nytimes.com/ or http://www.cnn.com/ have thousands of existing mementos in different archives. The strategy that we use here is effective in terms of execution time. For each URI, we only retrieve the nearest mementos to the annotation's creation date ("updated"). More precisely, we are capturing the closest memento(s) to the date *before* the annotation was created and the closest memento(s) to the date *after* the annotation was created.

Table 2. HTTP status code for all annotation target URIs

Number of annotations	Status code
5432	200
321	Time out
155	404
85	localhost
73	403
60	URN
47	401
46	410
37	Soft 401/403
17	400
8	Soft 404 and others

In Fig. 5(a), the annotation A was created at the time t_5. The closest memento to the date before t_5 was M_2 (captured at t_4) while the closest memento to the date after t_5 was M_3 (captured at t_7). So, for this annotation we picked the two closest mementos which are M_2 and M_3. Figure 5(b) is an example where mementos are only available before the annotation creation date while, in Fig. 5(c), mementos are only available after the annotation creation date. It is also possible that an annotation target has no mementos at all as Fig. 5(d) shows. If there are multiple closest mementos from different archives that share the same creation date (*memento-datetime*), then we consider all of these mementos for two different reasons. First, it is possible that at the time a memento is requested from an archive, there would be a technical problem or server-related issue which may affect returning the requested mementos. Second, we would like to know how different archives could contribute to provide mementos and recover annotation target text.

3.4   Are Annotations Attached to the Selected Mementos?

The final step is to see whether annotated URIs can be recovered by their mementos. The same technique introduced in Sect. 3.2 is used to test mementos. If the annotation target text ("exact") matches the text in the discovered memento, then we consider that this annotation is attached to the memento. Otherwise, we consider that the annotation cannot be attached.

4   Results

We collected 7744 annotations from Hypothes.is. Table 2 shows the results of checking the HTTP status code for the target URIs in each annotation. We find

Table 3. Annotation targets with existing mementos before and after the annotation creation date.

Number of annotations	Attached to live web page	Attached to memento (L)	Attached to memento (R)
902	Yes	Yes	Yes
9	Yes	Yes	No
28	Yes	No	Yes
9	Yes	No	No
31	No	Yes	Yes
4	No	Yes	No
12	No	No	Yes
71	No	No	No

Table 4. Annotation targets with existing mementos only before the annotation creation date

Number of annotations	Attached to live web page	Attached to memento (L)
599	Yes	Yes
11	Yes	No
14	No	Yes
68	No	No

Table 5. Annotation targets with existing mementos only after the annotation creation date

Number of annotations	Attached to live web page	Attached to mementos (R)
0	Yes	Yes
24	Yes	No
0	No	Yes
25	No	No

that 13.5 % of the annotations have URI-Rs that are no longer available on the live Web. In our further analysis, we will focus only on the 6281 annotations that include highlighted text. After checking each annotation, we found that 4566, or 72.7 %, of the highlighted text annotations are still attached to their live web pages. This means that the remaining 27.3 % of the annotations are orphans.

Next for each annotation, we checked the archives for the presence of mementos of the target URI near the annotation creation date. In Table 3 we consider annotations that have mementos both before ("L") and after ("R") the annotation date. "No" under the L and R columns means that annotation cannot be attached to the nearest memento while "Yes" means that the annotation attaches to the nearest memento.

Table 6. Annotation targets with no existing mementos

Number of annotations	Attached to live web
2737	Yes
1289	No

Table 7. Annotation targets recovered by different archives

Archive	Attached to live web	Not attached to live web
web.archive.org	1546 (86.5 %)	23 (20.3 %)
archive.today	249 (13.9 %)	59 (52.2 %)
wayback.archive-it.org	101 (5.65 %)	14 (12.3 %)
wayback.vefsafn.is	74 (4.14 %)	18 (15.9 %)
webarchive.nationalarchives.gov.uk	0 (0 %)	2 (1.76 %)
Total	1786 (110.2 %)	113 (102.5 %)

Table 4 shows the number of annotations that have mementos only on the L side (before annotations were created) of the annotation date, and Table 5 shows the number of annotations that have mementos only on the R side (after the annotation creation date) of the annotation date. Finally, Table 6 illustrates the number of annotations whose targets have no mementos. From these tables, we see that 1715 (27 %) of the annotations can no longer be attached to their live web pages. Unfortunately, the current holdings of web archives only allow 61 % of these to be re-attached. As shown in Table 6, the majority of annotations have no mementos available at all. Those that can no longer be attached to their live web version are lost, but those that are still attached can be recovered if these pages are archived before the annotated text changes.

Table 7 shows the number of annotations that can be recovered using various archives, split by whether or not they are still attached to the live web. As expected `web.archive.org` can be used to recover the most annotations, but for those annotations not attached to the live web, we find that `archive.today`, an on-demand service, can recover more orphaned annotations.

5   Conclusions

In this paper, we analyzed the attachment of highlighted text annotations in Hypothes.is. We studied the prevalence of orphaned annotations, and found that 27 % of the highlighted text annotations are orphans. We used Memento to look for archived versions of the annotated pages and found that orphaned annotations can be reattached to archived versions, if those archived versions exist. We also found that for the majority of the annotations, no memento exists in the archives. This points to the need for archiving pages at the time of annotation.

References

1. Agosti, M., Ferro, N., Frommholz, I., Thiel, U.: Annotations in digital libraries and collaboratories – facets, models and usage. In: Heery, R., Lyon, L. (eds.) ECDL 2004. LNCS, vol. 3232, pp. 244–255. Springer, Heidelberg (2004)
2. Ainsworth, S.G., Alsum, A., SalahEldeen, H., Weigle, M.C., Nelson, M.L.: How much of the web is archived? In: Proceedings of the 11th ACM/IEEE Joint Conference on Digital Libraries (JCDL), pp. 133–136. ACM (2011)
3. Bar-Yossef, Z., Broder, A.Z., Kumar, R., Tomkins, A.: Sic transit gloria telae: towards an understanding of the web's decay. In: WWW 2004: Proceedings of the 13th International Conference on World Wide Web, pp. 328–337 (2004)
4. Brush, A., Bargeron, D., Gupta, A., Cadiz, J.J.: Robust annotation positioning in digital documents. In: Proceedings of the SIGCHI Conference on Human Factors in Computing Systems, pp. 285–292. ACM (2001)
5. Francisco-Revilla, L., Shipman, F., Furuta, R., Karadkar, U., Arora, A.: Managing change on the web. In: Proceedings of the ACM/IEEE Joint Conference on Digital Libraries (JCDL), pp. 67–76. ACM (2001)
6. Frommholz, I., Fuhr, N.: Probabilistic, object-oriented logics for annotation-based retrieval in digital libraries. In: Proceedings of the 6th ACM/IEEE Joint Conference on Digital Libraries (JCDL), pp. 55–64. ACM (2006)
7. Furuta, R., Shipman III, F.M., Marshall, C.C., Brenner, D., Hsieh, H.w.: Hypertext paths and the world-wide web: experiences with walden's paths. In: Proceedings of the 8th ACM Conference on Hypertext, pp. 167–176. ACM (1997)
8. Haslhofer, B., Sanderson, R., Simon, R., Van de Sompel, H.: Open annotations on multimedia web resources. Multimedia Tools Appl. **70**(2), 847–867 (2014)
9. Haslhofer, B., Simon, R., Sanderson, R., Van de Sompel, H.: The open annotation collaboration (OAC) model. In: Proceedings of the IEEE Workshop on Multimedia on the Web (MMWeb), pp. 5–9. IEEE (2011)
10. Klein, M., Van de Sompel, H., Sanderson, R., Shankar, H., Balakireva, L., Zhou, K., Tobin, R.: Scholarly context not found: one in five articles suffers from reference rot. PloS One **9**(12), e115253 (2014)
11. Marshall, C.C.: Toward an ecology of hypertext annotation. In: Proceedings of the 9th ACM Conference on Hypertext and Hypermedia: Links, Objects, Time and Space-Structure in Hypermedia Systems, pp. 40–49. ACM (1998)
12. SalahEldeen, H.M., Nelson, M.L.: Losing my revolution: how many resources shared on social media have been lost? In: Zaphiris, P., Buchanan, G., Rasmussen, E., Loizides, F. (eds.) TPDL 2012. LNCS, vol. 7489, pp. 125–137. Springer, Heidelberg (2012)
13. Salaheldeen, H.M., Nelson, M.L.: Resurrecting my revolution. In: Aalberg, T., Papatheodorou, C., Dobreva, M., Tsakonas, G., Farrugia, C.J. (eds.) TPDL 2013. LNCS, vol. 8092, pp. 333–345. Springer, Heidelberg (2013)
14. Sanderson, R., Albritton, B., Schwemmer, R., Van de Sompel, H.: SharedCanvas: a collaborative model for medieval manuscript layout dissemination. In: Proceedings of the 11th Annual International ACM/IEEE Joint Conference on Digital Libraries (JCDL), pp. 175–184. ACM (2011)
15. Sanderson, R., Ciccarese, P., Van de Sompel, H.: Designing the W3C open annotation data model. In: Proceedings of the 5th Annual ACM Web Science Conference, pp. 366–375. ACM (2013)
16. Sanderson, R., Van de Sompel, H.: Making web annotations persistent over time. In: Proceedings of the 10th ACM/IEEE Joint Conference on Digital Libraries (JCDL), pp. 1–10. ACM (2010)

17. Soo, V.W., Lee, C.Y., Li, C.C., Chen, S.L., Chen, C.c.: Automated semantic annotation and retrieval based on sharable ontology and case-based learning techniques. In: Proceedings of the 2003 ACM/IEEE Joint Conference on Digital Libraries (JCDL), pp. 61–72. IEEE (2003)

18. Van de Sompel, H., Nelson, M.L., Sanderson, R., Balakireva, L.L., Ainsworth, S., Shankar, H.: Memento: Time Travel for the Web. Technical report (2009). arXiv:0911.1112

Query Expansion for Survey Question Retrieval in the Social Sciences

Nadine Dulisch[1]([✉]), Andreas Oskar Kempf[2], and Philipp Schaer[1]

[1] GESIS – Leibniz Institute for the Social Sciences,
50669 Cologne, Germany
{nadine.dulisch,philipp.schaer}@gesis.org
[2] ZBW – German National Library for Economics,
20354 Hamburg, Germany
a.kempf@zbw.eu

Abstract. In recent years, the importance of research data and the need to archive and to share it in the scientific community have increased enormously. This introduces a whole new set of challenges for digital libraries. In the social sciences typical research data sets consist of surveys and questionnaires. In this paper we focus on the use case of social science survey question reuse and on mechanisms to support users in the query formulation for data sets. We describe and evaluate thesaurus- and co-occurrence-based approaches for query expansion to improve retrieval quality in digital libraries and research data archives. The challenge here is to translate the information need and the underlying sociological phenomena into proper queries. As we can show retrieval quality can be improved by adding related terms to the queries. In a direct comparison automatically expanded queries using extracted co-occurring terms can provide better results than queries manually reformulated by a domain expert and better results than a keyword-based BM25 baseline.

Keywords: Scientific data management · Survey question retrieval · Survey question reuse · Query expansion · Co-occurrence analysis · Thesauri · Evaluation

1 Introduction

Digital libraries in academia increasingly include research data sets [4]. In order to facilitate data reuse, a retrieval infrastructure for research data needs to be built up [10]. Taking the example of quantitative social science research, research data for the purpose of reuse mostly consist of survey data, i.e. data collected to capture attitudes and behaviors as well as factual information of a population or population groups. The core method of data collection in survey methodology is the questionnaire. It provides the basis on which respondents' answers are converted into data that can be analyzed statistically. With respect to the reuse

---

Authors are listed in alphabetical order.

© Springer International Publishing Switzerland 2015
S. Kapidakis et al. (Eds.): TPDL 2015, LNCS 9316, pp. 28–39, 2015.
DOI: 10.1007/978-3-319-24592-8_3

scenario of survey data, it is important to keep in mind that other researchers are less interested in the entire survey. Rather they are looking for data on a single studied phenomenon or social construct and how it is translated into individual questions or items as part of the questionnaire.

Regarding the development of a retrieval infrastructure for survey question reuse it is important to keep in mind that in the majority of cases the measured social construct can hardly be derived directly from the question text itself (which is explained in more detail in Sect. 2). They need to be broken down into measurable properties, this way getting operationalized in the form of survey questions or items as part of a questionnaire – the so-called measuring instrument. It is this operationalization process which is the main reason for the great interest in concrete survey questions and which at the same time constitutes a major challenge for the development of retrieval services for the reuse of survey questions. Not only does the exact wording of a question determine whether a survey question really is suitable to allow for conclusions on a phenomenon and therefore is a valid measuring instrument. It also determines whether results of one survey could be compared with results of another survey which pretends to investigate the same phenomenon. Researchers then could resort to questions or whole item batteries that have already been developed by other researchers and used in various studies. Tested and established measuring instruments for the social sciences can be found in special scientific databases like question banks. Provided that a full text documentation of survey data questionnaires exists researchers could find exactly those measuring instruments they need to operationalize their own research interest.

Based on an analysis of the search log files of ZACAT[1], the GESIS online study catalogue system, potential re-users of survey question usually search with keywords related to the phenomenon they are interested in. This stands in contrast to the actual information contained in ZACAT where usually the actual question text itself is stored whereas the underlying social construct could only be found in the study description. For this reason a mere string-based search query is not enough to find well-established survey questions suitable to measure phenomena directly linked to one's own research interest. In this paper we want to find out whether a query expansion system is a suitable tool to support the retrieval of survey questions. This retrieval support is motivated by the idea to improve the reuse of survey questions across different studies and questionnaires.

2 Social Science Survey Data

Social science surveys essentially consist of numerical data. The data files are usually composed of tables which entail numbers or codes, which represent the values of respondents' answers to a survey question. It is for this reason why subject content of a survey can hardly be concluded by the dataset itself, but rather from other pieces of information linked to the study the data had been

---

[1] http://zacat.gesis.org/webview/.

collected for. It could only be derived from the study as a whole [6]. In general, three different levels of survey data structure can be distinguished. On the study level information about the general content of a study is provided. It includes, for example, information about the research fields of the data producers as well as the codebook. The narrower level is the variable level which gives detailed information on the phenomena under study. It contains the different variables which reflect the various dimensions of the definition of the studied phenomenon laying the ground for the entire formulation of the questionnaire. The third level is the question level. It contains the concrete questions or items the respondents have to deal with, which could be formulated rather differently, ranging from question format and statements (see Fig. 1) to tasks which need to be solved. In this article we focus on the questionnaire, which is at the core of each survey.

A constitutive part of questionnaire design is the operationalization process which stands for the translation of a research construct (e.g. antisemitism) into measurable units. The researcher identifies the different dimensions of a construct and defines it referring to relevant research literature, earlier studies or even his or her own qualitative pre-studies in the field. He or she then derives measurable aspects out of this definition which can be included in the different survey questions (e.g. agreement/disagreement with the statement: "Jewish people have too much influence in the world"). So the underlying research construct is encoded, which is why it usually cannot be extracted or derived out of the concrete question text. Only less complex social constructs or manifest variables as for example demographic variables like sex and age or the level of education usually appear in a questionnaire in their literal form.

The reuse scenario we focus on is based on the operationalization process as part of every questionnaire design. The user group we have in mind are social scientists who are planning to create their own surveys and who want to know how to put their research interest into concrete survey questions or items as part

---

F046

INT.: Please display card 46!

Every now and then, one hears different opinions about Jewish people. Some of these are listed on the card. Using the card, would you please tell me to what extent you agree or disagree with these statements?

		Completely disagree						Completely agree	N.A.
		1	2	3	4	5	6	7	
A	Jewish people have too much influence in the world.	o	o	o	o	o	o	o	o
B	I'm ashamed that Germans have committed so many crimes against Jewish people.	o	o	o	o	o	o	o	o
C	Many Jewish people try to take personal advantage today of what happened during the Nazi era and make Germans pay for it.	o	o	o	o	o	o	o	o
D	As a result of their behaviour, Jewish people are not entirely without blame for being persecuted.	o	o	o	o	o	o	o	o

Fig. 1. Excerpt from the German General Social Survey (ALLBUS) 2012

of a questionnaire. To the best of our knowledge there is no literature on the concrete search and information behavior of social scientists during the questionnaire design process. Information services like ZIS[2] illustrate the importance of documentation of social science measurement instruments. The problem at hand is a typical information seeking problem where subject information resources (survey questions) have to be retrieved from a database by a search engine. Similar to other retrieval problems, we have to deal with vagueness and ambiguity of human language (see the following section on related work). This problem gets even more pronounced given the fact that question texts, as well as search queries in survey question databases, tend to be rather short. Typical survey questions in our data set (see Sect. 4) are less than 100 characters long and typical search queries contain less than two words.

3 Related Work

A typical problem that arises during every search-based retrieval task (in contrast to browsing or filter-based tasks) is the so-called language or vocabulary problem [7]: During the formulization of an information need, a searcher can (in theory) use the unlimited possibilities of human language to express him- or herself [1]. This is especially true when expressing information needs in the scientific domain using domain-specific expressions that are very unique and context-sensitive. Every scientific community and discipline has developed its own special vocabulary that is not commonly used by other researchers from other domains. With regard to survey question retrieval, this problem is even more pronounced as it is likely that the underlying topic of a survey question is not directly represented in the question text. In this special setting of short textual documents and a very domain-specific content, this long-known problem becomes even more pronounced.

Hong et al. apply four methods for microblog retrieval [9]: query reformulation, automatic query expansion, affinity propagation as well as a combination of these techniques. To reformulate the query hashtags are extracted from tweets and used as additional information for the query. Furthermore, every two consecutive words of the query are grouped and added to the query. A relevance feedback model is used for automatic query expansion. The respective top ten terms of the top ten documents are selected. The affinity propagation approach is implemented by using a cluster algorithm to group tweets. The idea behind this is that the probability of tweets being relevant is higher for those, which are similar to relevant tweets. It is shown that automatic query expansion is a very effective method, while affinity propagation is less successful.

Microblogging services like Twitter also face the vocabulary problem for short texts. A tweet consists of up to 140 characters, while the question texts used in this work have an average length of 83.57 characters. The latest research in the field of microblog retrieval is therefore relevant for the problem at hand. For instance, pseudo-relevance feedback [14] and document expansion [5] are

---

[2] http://www.gesis.org/en/services/data-collection/zisehes/.

common approaches to address the vocabulary problem [3]. Jabeur et al. analyze two approaches for microblog retrieval [11]. The first approach uses a retrieval model based on Bayesian networks. The influence of a microblogger as well as the temporal distribution of search terms are included in the calculation of the relevance of a tweet. Here, only the usage of topic-specific features improved the results. In the second approach, query expansion (pseudo-relevance feedback) and document expansion methods are implemented. Tweets obtained by these approaches are merged. Additionally, those tweets are extended by contained URLs. Final scores are calculated by applying Rocchio-expansion as well as using the vector space model. Document expansion combined with vector space model improves retrieval results. Automatic query expansion does not increase recall, but significantly increases precision.

To surpass the language problem in digital libraries tools like thesauri and classifications try to control the vagueness of human language by defining a strict rule set and controlled vocabularies. These tools can help in the query formulation phase by actively supporting users in expressing their information need. A wide range of possible query expansion and search term recommender techniques are known in the information retrieval community [15]. These techniques can be categorized (1) global techniques that rely on the analysis of a whole collection, and (2) local techniques that emphasize the analysis of the top-ranked documents [16]. While local techniques generally outperform global techniques, global techniques cannot be applied. In web search engines, the use of query suggestion systems is common, however, the situation in digital library systems is different [13]. This holds true especially in the very special setting of survey question retrieval. In digital library systems the use of knowledge organization systems is common practice. Typically, entire collections are indexed with controlled terms from a domain-specific thesaurus or a classification system. A query expansion system might try to suggest terms that are closely related to both the users initial query term, as well as the knowledge organization system. While theoretically any kind of metadata may be recommended, the most promising approaches [8] use terms from a knowledge organization system.

Commercial tools like Colectica or QuestionPro are software packages that allow questionnaire designers to reuse questions. These tools include so-called question banks that store previously used questions. These question banks only allow basic string-based queries and are therefore not suitable for supporting users in the best possible way. Indexing of research data with controlled vocabularies has begun only recently. An overhauled version of Nesstar, the software system for research data publishing and online analysis owned by the Norwegian Social Science Data Services (NSD), provided an indexing function on the variable level. Initially intended for indexing of research data of social science data archive members of the European Social Science Data Archives Consortium (CESSDA), it has not been implemented so far. Nevertheless, the CESSDA consortium is planning to establish a database for social science survey questions using the European Language Social Science Thesaurus (ELSST). Organizations like the DDI Alliance highly advocate the reusability and the exchange of survey metadata and proposes to use the DDI metadata standard.

4 Test Collection Construction and Experimental Setup

In order to conduct a careful and considerable evaluation on the problem of survey question retrieval, we chose to implement a TREC-style evaluation setup with (1) a document corpus, (2) a set of topics, and (3) a set of relevance assessments corresponding to the set of topics.

The document corpus was extracted from the ALLBUS (German General Social Survey) and SOEP (German Socio-Economic Panel) questionnaires. It contained 16,764 question- and sub-question texts (see Fig. 1 for an example). The textual information was short, compared to normal TREC-style documents like newspaper articles or web pages. The question texts contained in the corpora had an average length of only 83.57 characters. We chose the ALLBUS and SOEP data sets as both are relevant and domain-specific questionnaires that are used in thousands of social science publications worldwide.

We extracted the topic set from log files (between 2012-01-01 and 2013-06-02) of the social science data catalogue ZACAT. The log files of ZACAT were chosen because these queries represent real-world usage patterns of scientists who are looking for questionnaires and survey data sets. Typical TREC topic sets consist of at least 50 topics so we extracted 60 queries from the log files. Since the corpus of question texts consists of German texts solely and the ZACAT queries were predominantly formulated in English, the queries were translated into German (official translation from ALLBUS/SOEP were used if applicable). The frequencies of the log entries show a typical power-law-like skewed distribution. Therefore to select the set of topics the log entries were arranged in descending order by their frequency. Queries were grouped according to their frequency (high = more than 10 log entries; medium = 10 log entries which was the mean value over all log entries; low = 1 log entry). To gather a good mix of different kinds of query topics we selected random log entries from the different groups: 27 common entries (high frequencies), 17 out of the medium frequent and 16 from special (low frequent) entries. Most of the log entries (mostly the high frequency ones) were short keyword queries (1.97 terms on average).

The ground truth was composed out of a set of 12,190 relevance assessments based on a three level rating system (not relevant, (partially) relevant, and very relevant). The relevance assessments were conducted manually by one single person familiar with the domain and with a set of rules and guidelines for the assessments. An example for such a guideline for the topic "democracy" was:

- very relevant: direct questions concerning democracy (e.g. "How content are you with the present state of democracy in Germany?");
- relevant: political questions concerning democracy in a broader sense (e.g. party system, election, etc.);
- not relevant: everything else.

All retrieval experiments were conducted using Lucene with BM25 ranking and language dependent settings (stemmers, stop words, etc.). In all experiments we report on recall at n (R@5 and R@10) and on nDCG at n (nDCG@5 and nDCG@10) calculated using TREC_EVAL. In our setup we understand recall as fraction of the research questions that are relevant to the query and are successfully retrieved within the first n results in the result list. Normalized discounted cumulative gain (nDCG) is a measure of retrieval quality for ranked lists that, in contrast to precision, makes use of graded relevance assessments [12]. NDCG is computed as follows:

$$
\text{nDCG} = Z_i \sum_{j=1}^{R} \frac{2^{r(j)} - 1}{\log(1 + j)}
\tag{1}
$$

Z_i is a constant to normalize the result to the value of 1. $r(j)$ is an integer representing the relevance level of the result returned at rank j where R is the last possible ranking position. For our experiments the relevance levels are 0 = irrelevant; 1 = relevant and 2 = very relevant. NDCG@n is a variation of this calculation where only the top-n results are considered. We use nDCG@5 and nDCG@10. The same two levels are chosen for the calculation of recall at n. Here we simply count the relevant documents (relDocs) among the first 5 or 10 results and divide them by the actual level:

$$
R@n = \frac{|\text{relDocs}|}{n}
\tag{2}
$$

Since all our retrieval results are ranked we focus on the top five and ten results for our evaluation. This is done to simulate the needs of a real world user who is used to inspect only the first few results in a ranked list. All results are tested for statistical significance using a paired t-test.

5   Query Expansion for Survey Question Retrieval

For each approach, queries are expanded as follows: A query consists of one or more clauses which in turn consist of terms (or phrases) and Boolean operators (for query syntax also see the Apache Lucene documentation). For each query term, the corresponding expansion terms are retrieved and connected with the query term by OR-operation. An example query for two query terms qt_1 and qt_2 with their corresponding alternative query terms $qt_{1,alt}$ would look like:

$$(qt_1 \text{ OR } qt_{1,alt} \text{ OR} \ldots) \text{ AND/OR } (qt_2 \text{ OR } qt_{2,alt} \text{ OR} \ldots)$$

For evaluation, we expanded the unprocessed queries from the query logs in order to ensure a realistic setting.

5.1   Thesaurus-Based Expansion

Thesauri are tools to surpass the vocabulary problem: natural language allows expressing things in various ways. One word can have the same (synonyms) or different meaning (polysemes). Thesauri contain various associations and relationships between terms in the form of synonyms, associations, etc. By extending the query with these related terms, documents can be found which do not contain the terms of the original query, but are still relevant with respect to the information need of the user. Since thesauri are not developed automatically but by creating relations manually between terms, this is an intellectual approach. The first approach involves two different thesauri: (1) the *Open Thesaurus* (OT), a general natural language and community-based thesaurus, (2) and the *Thesaurus for the Social Sciences* (TSS), a domain-specific thesaurus developed by a small editorial group of domain experts [17].

The Open Thesaurus contains German synonyms, hypernyms, hyponyms and associations. It does not consist of a domain-specific language and is licensed under the LGPL (GNU Lesser General Public License) License. The TSS contains descriptor-based synonyms, broader-, narrower and related terms as well as preferred terms. For expanding the queries with the Open Thesaurus we limit the terms to synonyms as well as associations. With TSS we limit the terms to synonyms and related/preferred terms. The aim is to reduce the number of suggested expansion terms. As there is no criterion to determine the "degree of relatedness" to the query term, any limited selection would have been too random. Therefore, all determined expansion terms are added to the original query (by logically linking them to the original term with an OR function). These are 19.49 terms on average, which are retrieved from a database. Furthermore, to ensure comparability between both thesauri approaches, similar kinds of relationships have to be used (synonyms, associations/related terms). The goal of using a general thesaurus as well as a domain-specific thesaurus is to compare natural language expansion to domain language expansion.

5.2 Co-occurrence-Based Expansion

As described in the previous section, the thesaurus-based query expansion is an intellectual approach. Statistical approaches to determine terms for expansion have proven to be more applicable [3]. In this paper the statistical method of co-occurrence analysis has been tested. Co-occurrence analysis is a well-established approach, which is for instance used in natural language processing or to support manual coding of qualitative interviews [2]. It serves the purpose of the analysis of term-term relationships. It is assumed that terms, which often occur within the same context, are associated with each other and are, for example, similar in meaning.

Since each similarity measure performs differently depending on the data we used two different metrics: logarithmic Jaccard index, also known as the Jaccard similarity coefficient and cosine similarity. We define the logarithmic Jaccard index as follows:

$$J_{log}(x, y) = \frac{\log(df_{xy})}{\log(df_x + df_y - df_{xy})} \tag{3}$$

The cosine similarity is calculated as follows:

$$c(x, y) = \frac{df_{xy}}{\sqrt{df_x + df_y}} \tag{4}$$

df_x and df_y are the document frequencies of the terms x and y, thus the number of documents these terms occur in. df_{xy} is the number of documents which contain x as well as y. Jaccard index corresponds to intersection/union.

As we want our co-occurrence-based method to be comparable to our previous approach that was based on a thesaurus we again focus on domain-specific vocabularies. For this purpose, we use the social science literature database SOLIS with more than 450,000 literature references from the social sciences. Each reference includes title, abstract and controlled keywords from the TSS. Our system calculates the semantic relatedness between any free text such as titles or abstracts and controlled terms (TSS-terms) for an entire document corpus (stop word removal and stemming is applied). Using co-occurrence measures like Jaccard index or cosine similarity we calculated term suggestions taken from the TSS for every query term. As an example, users who are looking for the string "youth unemployment" in a social science context will get search term suggestions from the thesaurus that are semantically related to the initial query such as "labor market" or "education measure". Another possible suggestion might be "adolescent", which is a controlled term for "youth". The suggestions generated by this approach go beyond simple term completion and can support the search experience. Since ALLBUS and SOEP do not include any annotated entries, we had to use a different corpus to train our term recommender. ALLBUS, SOEP and SOLIS share the same scientific domain and the same domain-specific language which makes this cross-corpora term recommendation plausible.

The term suggestion system generates a ranked list of search term recommendations from which we used the top 20 terms for query expansion. This amount of terms was chosen because the average number of synonyms/related terms of the thesaurus-based query expansion was 19.49.

6   Results

We ran a pretest involving simple keyword-based queries generated directly from the log file entries and a hand-crafted query formulation from a domain expert (Q_{expert}). Although the results are only slightly better, the domain expert's query formulation is chosen as the baseline for our further experiments. We compare four different automatic QE techniques using two thesaurus-based (QE_{ot} and QE_{tss}) and two co-occurrence-based expansions (QE_{jac} and QE_{cos}) to this baseline. These different systems are compared to the baseline using R@5, R@10, nDCG@5, and nDCG@10 (see Table 1).

In general the thesaurus-based approaches were able to increase both precision and recall. QE_{ot} produces more precise results, while also increasing recall compared to the baseline (nDCG@10 + 8.96 %, R@10 + 23.44 %). An expansion with TSS improves retrieval results both in precision and recall as well (nDCG@10 + 14.5 %, R@10 + 13.12 %). As the results show, QE_{ot} produces a better recall than QE_{tss}, while the latter is more precise. The co-occurrence-based approaches produce similar results compared to each other regarding recall, which is also better than the baseline as well as both thesauri (e.g. QE_{cos}:

Table 1. Results of the retrieval test on the survey questions comparing four different query expansion (prefix QE) techniques to the best manual query formulation technique from a pretest (Q_{expert}). Best results are marked in bold font. Statistical significant results are marked with the following confidence level: (*) $\alpha = 0.1$.

	R@5	R@10	NDCG@5	NDCG@10
Q_{expert}	0.0975	0.1502	0.4056	0.3918
QE_{ot}	**0.1308**	0.1854	0.4511	0.4269
QE_{tss}	0.1245	0.1699	**0.4857(*)**	**0.4486**
QE_{jac}	0.1265	**0.1965**	0.3411	0.3471
QE_{cos}	0.1271	0.1938	0.4077	0.3998

R@10 + 29.03 %). The results of QE_{cos} are slightly more precise than the baseline, though less precise than the results of the thesauri. QE_{jac} produces the lowest values for nDCG.

Although the results are not statistically significant, the previously mentioned criteria support the validity of the following results: Regarding the thesaurus-based approaches, both thesauri produce distinctly more precise results as well as a greater amount of relevant documents compared to the baseline. While the nDCG values for both co-occurrence-based approaches are better than the baseline, nDCG values are lower. A positive effect of all QE approaches was a higher amount of relevant retrieved question items. All systems return a higher number of relevant question items, while precision is also increased (except for QE_{jac}). With co-occurrence-based expansion, best results are produced with cosine similarity. Cosine similarity delivers a higher number of relevant survey questions, but with less precision compared to the thesauri methods. The most relevant results are retrieved with logarithmic Jaccard index.

7 Conclusion and Future Work

Our evaluation shows that manual keyword-based search (Q_{expert}) for survey questions suffers from very low recall rates and that our generated approaches can provide better results without losing precision. This demonstrates that for our use case query expansion is an appropriate approach to support survey question retrieval as it provides better recall and precision. This constitutes an important step to facilitate the reuse of survey questions for questionnaire design in the social sciences.

The two different approaches we evaluated in this paper (thesaurus-based and co-occurrence-based expansion) show different results when we compare them to queries manually formulated by a domain expert. Even though our approaches produce better results than the domain expert, the results are not coherent. Generally speaking, the co-occurrence-based approaches were better in increasing recall (R@10) while the thesaurus-based expansions were better in increasing retrieval quality measured by nDCG. This might be related to the different expansion concepts that underlie the current experiment. The concept-relations in a thesaurus are all intellectually curated and hand-crafted by domain experts while the relations we calculated with our co-occurrence methods are statistical values. The co-occurrences show that there is a statistical relatedness between a term and a concept. On average, the statistical methods were able to retrieve more relevant results – although this higher recall comes at the cost of a lower quality. In general, the statistical methods were more liberal in suggesting term-concept relations while the thesauri were stricter. This is an observation that is also true for the two different thesauri used. The Open Thesaurus as a common language thesaurus was better suited to improve both recall and retrieval quality while the Thesaurus for the Social Sciences showed to be too strict on higher recall levels (R@10). Taking the operationalization process as part of every questionnaire design into consideration, this is hardly surprising as survey questions in general do not address a discipline-specific community. A broader and less domain-specific thesaurus seems to be the better tool for the specific problem we faced in this study.

Another aspect that can be observed is the fact that automatic query expansion not only achieves a higher recall but precision is also higher than through intellectual expansion. Consequently, even domain experts would have profited from the implementation of an interactive recommendation system which offers term or query suggestions during the query formulation phase. In domain-specific search scenarios, this has proven to increase retrieval performance and user satisfaction [8]. Therefore, the TREC-style evaluation setting of this paper has to be expanded for an interactive information retrieval setting.

In future work, we would like to do further experiments regarding a combination of intellectually and automatically generated query expansions. First experiments show promising results as we could further improve nDCG values, as well as the number of retrieved documents by combining the different approaches. We would like to implement different topic-related query expansion systems and evaluate the effects of using these specialized recommenders on each topic.

References

1. Blair, D.C.: Information retrieval and the philosophy of language. Annu. Rev. Inform. Sci. Technol. **37**(1), 3–50 (2003). http://dx.doi.org/10.1002/aris.1440370102
2. Brent, E., Slusarz, P.: Feeling the beat - intelligent coding advice from meta-knowledge in qualitative research. Soc. Sci. Comput. Rev. **21**(3), 281–303 (2003). http://ssc.sagepub.com/content/21/3/281
3. Carpineto, C., Romano, G.: A survey of automatic query expansion in information retrieval. ACM Comput. Surv. **44**(1), 1:1–1:50 (2012). http://doi.acm.org/10.1145/2071389.2071390
4. Dallmeier-Tiessen, S., Mele, S.: Integrating data in the scholarly record: community-driven digital libraries in high-energy physics. Zeitschrift für Bibliothekswesen und Bibliographie **61**(4–5), 220–223 (2014). http://zs.thulb.uni-jena.de/receive/jportal_jparticle_00324882

5. Efron, M., Organisciak, P., Fenlon, K.: Improving retrieval of short texts through document expansion. In: Proceedings of the 35th International ACM SIGIR Conference on Research and Development in Information Retrieval, SIGIR 2012, pp. 911–920. ACM, New York (2012). http://doi.acm.org/10.1145/2348283. 2348405
6. Friedrich, T., Kempf, A.: Making research data findable in digital libraries: a layered model for user-oriented indexing of survey data. In: 2014 IEEE/ACM Joint Conference on Digital Libraries (JCDL), pp. 53–56 (2014)
7. Furnas, G.W., Landauer, T.K., Gomez, L.M., Dumais, S.T.: The vocabulary problem in human-system communication. Commun. ACM **30**(11), 964–971 (1987)
8. Hienert, D., Schaer, P., Schaible, J., Mayr, P.: A novel combined term suggestion service for domain-specific digital libraries. In: Gradmann, S., Borri, F., Meghini, C., Schuldt, H. (eds.) TPDL 2011. LNCS, vol. 6966, pp. 192–203. Springer, Heidelberg (2011)
9. Hong, D., Wang, Q., Zhang, D., Si, L.: Query expansion and message-passing algorithms for TREC microblog track. In: Voorhees, E.M., Buckland, L.P. (eds.) Proceedings of The Twentieth Text REtrieval Conference, TREC 2011, Gaithersburg, Maryland, November 15–18, 2011. National Institute of Standards and Technology (NIST) (2011). http://trec.nist.gov/pubs/trec20/papers/Purdue_IR.microblog.update.pdf
10. Hyman, L., Lamb, J., Bulmer, M.: The use of pre-existing survey questions: implications for data quality. In: Proceedings of Q2006, Cardiff, April 2006. http://eprints.port.ac.uk/4300/
11. Jabeur, L.B., Damak, F., Tamine, L., Cabanac, G., Pinel-Sauvagnat, K., Boughanem, M.: IRIT at TREC microblog 2013. In: Voorhees, E.M. (ed.) Proceedings of the Twenty-Second Text REtrieval Conference, TREC 2011, Gaithersburg, Maryland, November 19–22, 2013. NIST Special Publication, vol. 500–302. National Institute of Standards and Technology (NIST) (2013). http://trec.nist.gov/pubs/trec22/trec2013.html
12. Järvelin, K., Kekäläinen, J.: Cumulated gain-based evaluation of IR techniques. ACM Trans. Inf. Syst. **20**(4), 422–446 (2002). http://doi.acm.org/10.1145/582415.582418
13. Lüke, T., Schaer, P., Mayr, P.: A framework for specific term recommendation systems. In: Proceedings of the 36th International ACM SIGIR Conference on Research and Development in Information Retrieval, SIGIR 2013, pp. 1093–1094. ACM, New York (2013). http://doi.acm.org/10.1145/2484028.2484207
14. Miyanishi, T., Seki, K., Uehara, K.: Improving pseudo-relevance feedback via tweet selection. In: Proceedings of the 22nd ACM International Conference on Conference on Information & Knowledge Management, CIKM 2013, pp. 439–448. ACM, New York (2013). http://doi.acm.org/10.1145/2505515.2505701
15. Schaer, P.: Applied informetrics for digital libraries: an overview of foundations, problems and current approaches. Hist. Soc. Res. **38**(3), 267–281 (2013). http://eprints.rclis.org/22630/1/HSR_38.3_Schaer_a.pdf
16. Xu, J., Croft, W.B.: Improving the effectiveness of information retrieval with local context analysis. ACM Trans. Inf. Syst. **18**(1), 79–112 (2000). http://doi.acm.org/10.1145/333135.333138
17. Zapilko, B., Schaible, J., Mayr, P., Mathiak, B.: TheSoz: a SKOS representation of the thesaurus for the social sciences. Semant. Web **4**(3), 257–263 (2013). http://dx.doi.org/10.3233/SW-2012-0081

Multimedia Information Management and Retrieval and Digital Curation

Practice-Oriented Evaluation of Unsupervised Labeling of Audiovisual Content in an Archive Production Environment

Victor de Boer[1,2]($\boxtimes$), Roeland J.F. Ordelman[1,3], and Josefien Schuurman[1]

[1] Netherlands Institute for Sound and Vision, Hilversum, The Netherlands
v.de.boer@vu.nl
[2] The Network Institute, VU University Amsterdam, Amsterdam, The Netherlands
[3] University of Twente, Enschede, The Netherlands

Abstract. In this paper we report on an evaluation of unsupervised labeling of audiovisual content using collateral text data sources to investigate how such an approach can provide acceptable results given requirements with respect to archival quality, authority and service levels to external users. We conclude that with parameter settings that are optimized using a rigorous evaluation of precision and accuracy, the quality of automatic term-suggestion are sufficiently high. Having implemented the procedure in our production work-flow allows us to gradually develop the system further and also assess the effect of the transformation from manual to automatic from an end-user perspective. Additional future work will be on deploying different information sources including annotations based on multimodal video analysis such as speaker recognition and computer vision.

Keywords: Audiovisual access · Information extraction · Thesaurus · Audiovisual archives · Practice-oriented evaluation

1 Introduction

Traditionally, audiovisual content in digital libraries is being labeled manually, typically using controlled and structured vocabularies or domain specific thesauri. From an archive perspective, this is not a sustainable model given (i) the increasing amounts of audiovisual content that digital libraries ingest (quantitative perspective), and (ii) a growing emphasis on improving access opportunities for these data (qualitative perspective). The latter is not only addressed in the context of traditional search, but increasingly in the context of linking within and across collections, libraries, and media. Ultimately, search and linking is shifting from a document-level perspective towards a segment-level perspective in which segments are regarded as individual, 'linkable' media-objects. In this context, the traditional, manual labeling process requires revision to increase both quantity and quality of labels.

© Springer International Publishing Switzerland 2015
S. Kapidakis et al. (Eds.): TPDL 2015, LNCS 9316, pp. 43–55, 2015.
DOI: 10.1007/978-3-319-24592-8_4

In earlier years, we investigated optimization of the labeling process from a "term suggestion" perspective (see e.g., [1]). Here the aim was to improve efficiency and inter-annotator agreement by generating annotation *suggestions* automatically from textual resources related to the documents to be archived. In [2] we defined collateral data[1] to refer to data that is somehow related to the primary content objects, but that is not regarded as metadata, such as subtitles, scripts and program-guide information. Previous work at our archive emphasized the ranking of possibly relevant terms extracted from the collateral text data, leaving the selection of the most relevant terms to the archivist. The proposed term suggestion methods were evaluated in terms of Precision and Recall by taking terms assigned by archivists as 'ground-truth'. The outcome was that a tf.idf approach gave the most optimal performance in combination with an importance weighting of keywords on the basis of a Pagerank-type of analysis of keywords within the structure of the used thesaurus ($F@5 = 0.41$). One important observation of the study was that the inter-annotator agreement was limited, with an average agreement of 44 %.

Although the results were promising, the evidence provided by the study was not conclusive enough to justify adaptations of the archival annotation work-flow and incorporate the suggested methodology. However, as the assumptions that drove the earlier study are still valid and have become even more clear and pressing, we recently took up the topic again. This time however from the perspective of fully *unsupervised* labeling. The main reason for this is that we expect that the efficiency gain of providing suggestions in a supervised labeling approach is too limited in the context of the increasing amounts of data that need labeling. Furthermore, instead of relying on topically condensed text sources such as program guide descriptions used in the previous study, we include a collateral text source more easily available in our production work-flow: subtitles for the hearing impaired. Finally, as inter-annotator agreement is expected to be limited given the earlier study, we wanted to investigate how this agreement relates to an unsupervised labeling scenario that aims to generate labels for improving access to audiovisual collections. This makes our task different from more generic classification or tagging tasks such as done in the MUMIS project [3].

In this paper, we present an evaluation of unsupervised labeling focusing on *the practical usage of the method in an archive production environment*. In Sect. 2 we overview the archival context of the labeling approach. In Sect. 3 we present the automatic term extraction framework that we used for the evaluations described in Sect. 4. Section 5 discusses and concludes the results from the evaluation, followed by some notes on future work.

2   Archival Context

The implementation of innovative processes for automatic content annotation in an archive production work-flow needs to be addressed critically. A key

---

[1] This data is sometimes also referred to as 'context data' but as for example newspaper data can also be regarded as 'context' we prefer the term 'collateral data'.

requirement with respect to this type of innovation is that the archive remains in control of the quality of the automatically generated labels. Not only because of principals of archival reliability and integrity, but also from a service-level point of view. Media professionals use a broadcast archive to search for footage that can be re-used in new productions. The probability that their search process will get disturbed due to incorrect automatic labeling is undesired, despite the fact that the overall number of entry points generated by the automatic tool will increase, potentially having a positive effect on the search process.

Authority, being in control of the quality of the annotation tool, also means having control on parameters of the tool. In the case of automatic term labeling two important variables are: (i) quality, specifically the balance between Precision and Recall –or from a classification perspective: Hits and False Positives versus Misses– that controls the relation between quantity and quality of generated labels, and (ii) the vocabulary that in an archival setting could be closely related to controlled vocabularies or thesauri that are used. In this work, the automatic labeling process is required to output terms that are defined in the Common Thesaurus for Audiovisual Archives[2] (GTAA). The GTAA closely follows the ISO-2788 standard for thesaurus structures and consists of several facets for describing TV programs: subjects, people mentioned, named entities (Corporation names, music bands etc.), locations, genres, producers and presenters. The GTAA contains approximately 160.000 terms and is updated as new concepts emerge on television. For the implementation of unsupervised labeling in the archive's metadata enrichment pipeline, the balance between Precision and Recall, and the matching of candidate terms with the thesaurus have the main focus of attention.

2.1   Data

The general aim of the project for which the evaluation described in this paper was performed, is to label automatically the daily ingest of Radio and Television broadcasts. This data is quite heterogeneous: it contains news broadcasts, documentaries and talk shows but also sports and reality shows. As named-entity extraction tools typically perform better for common entities as opposed to less common ones, we assume that the performance will differ for different genres.

For each program that is ingested also subtitles for the hearing impaired (TT888) –a verbatim account of the (Dutch) speech present in the data– is flowing into the archive. These TT888 files are used as input for the term-extraction pipeline described in Sect. 3. Instead of subtitles also other collateral data such as program guide information or production scripts and auto-cues could be used. As the availability of these data is less stable as is the case for subtitles, we focus on subtitles in the current work-flow.

For evaluation purposes we selected one year of previously ingested programming for which we have manually generated labels, created by professional

---

[2] http://datahub.io/dataset/gemeenschappelijke-thesaurus-audiovisuele-archieven.

archivists. This set will be referred to as 'gold-standard' in our (pilot) experiments. However, as such a gold-standard implies exact matches or terminological consistency, we also asked professional archivist to assess the conceptual consistency (see also [4] about consistency, [1] for the approach that was taken earlier).

As discussed above, we use the internal thesaurus as a reference for extracted terms. The GTAA is available as Linked Open Data [5] and its concepts are identified through URIs. In the production system the extracted terms end-up as URIs identifying GTAA concepts unique IDs, which in turn can also be linked to and from using RDF relations. This allows us to in the near future reuse background information in the Linked Data cloud insofar as it is linked to or from those GTAA concepts. For the evaluation described here, the term-extraction pipeline only used the "subject" and "named-entities" facets of the thesaurus for validation.

3   Automatic Term Extraction

An overview of the term extraction pipeline is presented in Fig. 1. This shows the different steps performed in the algorithm, detailed below. The term-extraction pipeline is set up as a webservice. The webservice takes a single text, such as the subtitles for a television broadcast, as input and outputs a list of relevant thesaurus terms. This set-up allows the service to be re-used for other related tasks such as the extraction of terms from digitized program guides or other collateral data sources.

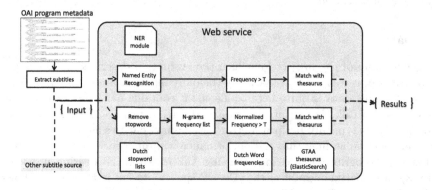

Fig. 1. Overview of the algorithm

The web service is called through a HTTP post request, where the input text is passed in the body as a JSON string. At the same time, parameter settings can be passed in the same HTTP request to override default values for these parameters (see Sect. 3.4 for the parameters).

The output is a JSON object containing a list of thesaurus terms, on the basis of the parameter settings used (if not overridden, the default values are returned). For every term, also a *matching score* is returned (see Sect. 3.3). Within the archive production workflow, the service is called when new programs are ingested. The thesaurus terms provided by the service are then added to the program metadata without manual supervision.

For the experiments described below, the subtitles are derived from an OAI-PMH interface[3] to the archive's database. We retrieves for one or more programs the subtitle information from the OAI response (the program metadata) and remove the temporal metadata and other XML markup from the subtitles so that we end up with a single subtitle text per program. These are then presented one at a time to the service. As the extraction of subject terms and named entities require an individual tuning of parameters, the textual data is processed in two parallel tracks: one for subject terms and one for named entities (see Fig. 1).

3.1 Pre-processing and Filtering

For the subject track, the first pre-processing step is to remove stopwords using a generic list of Dutch stopwords. In the next step, frequencies for 1, 2, and 3-grams are generated. For the uni-grams (single terms) also normalized frequencies are calculated using a generic list of Dutch word frequencies obtained from a large newspaper corpus. In the filtering step, candidate terms (in the form of n-grams) above a certain threshold value of frequency scores are selected. Frequency scores are based upon both the absolute frequency (how often a term occurs in the subtitles) and a relative frequency (normalized by the frequency of the term in the Dutch language, only for 1-grams). The frequency thresholds are parameters of the service. In the next phase, candidate n-gram terms are matched with terms in the thesaurus.

3.2 Named Entity Recognition

In the named-entity track of the algorithm, Named Entities (NEs) are extracted. Pilot studies determined that NEs –more so than non-entity terms– have a high probability of being descriptive of the program, especially if they occur in higher frequencies. For this track, we use a Named Entity Recognizer (NER). The NER is implemented as a separate module in the service and we experimented with different well-performing open-source systems NER systems for this module.

1. **XTAS.** The NER tool from the open-source xTAS text analysis suite.[4]
2. **CLTL.** An open-source NER module developed at the CLTL group[5].

In the current Web service, the NER module to be used is a parameter of the method and can be set to "XTAS" or "CLTL" for the respective services. Both modules are implemented as wrappers around existing services which take

---

[3] http://www.openarchives.org/pmh/.

[4] http://xtas.net/. Specifically, the FROG module was used using default settings.

[5] http://www.cltl.nl/. Here the OpenNER web service was used in combination with the CLTL POS tagger.

as input a text (string) and as output a JSON list of entities and their types. The types used by the web service are *person, location, organization* or *misc.* Internal NE types from the individual modules are mapped to these four types

3.3 Vocabulary Matching

The previous phases yield candidate terms to be matched against the thesaurus of five categories: subjects (from the subject track) and persons, places, organizations, and miscellaneous (from the NE track). The next step in the algorithm identifies the concepts in the thesaurus that match these terms. As there can be many candidate terms at this stage and the GTAA thesaurus is fairly sizable with some 160.000 concepts, we need to employ a method for matching terms to thesaurus concepts that is scalable.

For this, the thesaurus has been indexed in an ElasticSearch instance[6]. ElasticSearch is a search engine that indexes documents for search and retrieval. In our case, thesaurus concepts are indexed as documents, with preferred and alternative labels as document fields. The concept schemes (facets or "axes" in the GTAA) are represented as different ElasticSearch *indices* which allows for fast search for term matches across and within a concept scheme. When searching for concepts matching a candidate term, ElasticSearch will respond with candidate matches and a *score* indicating the quality of the match between candidate term and the document. In our algorithm, we employ a threshold on this score, resulting in an additional parameter. In this final step, the different categories of candidate terms are matched to a specific concept scheme. For example, persons are matched to the "Persoonsnamen" (Person names) concept scheme in the GTAA thesaurus and both the subject terms and MISC are mapped to the "Onderwerpen" (Subject) concept scheme.

3.4 Parameters

The algorithm parameters are shown in Table 1. This table shows the parameter name, the default value and the description. All default values can be overridden in the HTTP POST request. These default values were determined in pilot experiments (Sect. 4.1) and the experiment described in Sect. 4.2 was used to determine optimal values for a number of these parameters for a specific task.

4   Experiments

4.1   Pilot Experiments

We performed a number of pilot experiments to fine-tune the setup of the main experiment. In one of these pilot experiments, we compared the output of an

---

[6] http://www.elastic.co/products/elasticsearch.

Table 1. Parameters and default values for the service

Nr.	Parameter name	Default	Description
P1	tok.min.norm.freq	4×10^{-6}	Threshold on normalized freq for 1-gram
P2	tok.max.gram	3	Maximum N for topic N-grams
P3	tok.min.gram	2	Minimum N for topic N-grams (excl. 1)
P4	tok.min.token.freq	2	threshold on absolute freq for 1-gram
P5	repository	cltl	NER module (xtas or cltl)
P6	ne.min.token.freq	2	Threshold on absolute freq for all NEs
P7	ne.organization.min.score	8	Threshold on ElasticSearch matching score
P8	ne.organization.min.token.freq	2	Threshold on absolute freq for
P9	ne.person.min.score	8	Threshold on matching score for persons
P10	ne.person.min.token.freq	1	Threshold on absolute freq for persons
P11	ne.location.min.score	8	Threshold on matching score for locations
P12	ne.location.min.token.freq	2	Threshold on absolute freq for locations
P13	ne.misc.min.score	8	Threshold on matching score for misc
P14	ne.misc.min.token.freq	2	Threshold on absolute frequency for misc

earlier version of the algorithm to a gold-standard of existing manual annotations (see Sect. 2.1). The results showed that although there was some overlap[7], comparing to this gold standard was not deemed by the experts to be an informative evaluation, since many "false positives" identified by the algorithm were identified to be interesting nonetheless. Therefore in subsequent experiments, we presented the extracted terms to domain experts for evaluation. In this way, only precision of the suggested terms can be determined (no "recall"). This pilot also suggested that the correctness of suggested terms should be determined on a scale rather than correct or incorrect.

In a second pilot experiment, we presented extracted terms for random programs to four in-house experts and asked them to rate this on a five point Likert-scale [6]. The results were used to improve the matching algorithm and to focus more on the named entities rather than the generic terms since the matching here seemed to result in more successful matches. Lastly, in feedback to this pilot the experts indicated that for some programs the term extraction was considerably less useful than for others. This was expected but in order to reduce the amount of noise from programming that from an archival perspective has a lesser degree of annotation priority, we selected programs with a high priority[8]. For the main experiment we sampled from this subset rather than from the entire collection. From this evaluation we derived default parameter values shown in Table 1 which provided 'reasonable' results (not too many obvious errors) including for example the value for P4, P6, P8, P10, P12 and P14 (minimum frequencies for terms to be considered a candidate term).

---

[7] For this non-optimized variant, recall was 21 %.

[8] This prioritization is done by archivists independently of this work. It is in use throughout the archive and mostly determined by potential (re)use by archive clients.

In the main experiment, the goal was twofold: (i) to determine the quality of the algorithm and (ii) to determine optimal values for other system parameters.

4.2 Experimental Setup

For the main experiment, we randomly selected 18 individual broadcasts from five different Dutch television shows designated as being of high-priority by the archivist. These shows are the evening news broadcast (4 videos), two talk shows (3+4 videos), a documentary show (4 videos) and a sports news show (3 videos). For these videos, we presented four evaluators with (a) the video, (b) the existing metadata (which did not include descriptive terms) ad (c) the terms generated by the algorithm using different parameter settings. The evaluators were asked to indicate the relevance of the terms for the video on a five-point Likert scale:

    0: Term is totally irrelevant or incorrect,
    1: Term is not relevant,
    2: Term is somewhat relevant,
    3: Term is relevant,
    4: Term is very relevant

Parameter Settings. Parameters P1-P4 were set to their default values as listed in Table 1. These were established in the pilot experiments and proved reasonable for this specific task. For P5, we used both values, so both NER modules are evaluated. Some terms were found by both modules, and other terms were found by only one of the two. Evaluators did not see the origin of the terms. P6 was fixed to 2, as were the thresholds on the NE specific frequencies (P7, P9, P11, P13). For the Elasticsearch matching scores, we used a bottom threshold of 9.50 and presented all terms with a score higher than that value to the evaluators. We retain the scores so that in the evaluation we can compare the quality for threshold values of 9.50 and higher. The pilot studies showed that with thresholds below 9.50, mostly incorrect terms were added. The scores were also not available to the evaluators to avoid an evaluation bias. For the 18 videos, a total of 289 terms for XTAS and 222 terms for CLTL were presented to the evaluators.

4.3 Results

One of the evaluators (Eval4) finished all 18 videos. Table 2 shows the statistics for the four evaluators including the average score given for all terms. This shows that there is quite some disagreement among the averages. To measure inter-annotator agreement, we calculated the Pearson-coefficient between the pairs of evaluators. The results are shown on the right in Table 2. The agreement between Eval1 and Eval2 is rather low at 0.45, but for the other pairings it is on a more acceptable level. For most of the subsequent evaluations, we use the average score for an extracted term given by the evaluator.

Table 2. Evaluator results (left) and inter-annotator agreement matrix (right)

			Agreement		
Evaluator	Nr. evaluated	Avg. score	Eval2	Eval3	Eval4
Eval1	8	1.31	0.45	0.66	0.69
Eval2	14	2.21	0.63	0.74	
Eval3	6	1.57	0.73		
Eval4	18	1.64			

Named Entity Modules. To determine the difference in quality of the two NER modules, we separated the scores for the two values (CLTL and XTAS) and determined the average score. If all terms are considered (respectively 289 and 222 terms for XTAS and CLTL), the average score for XTAS is 1.79 and that for CLTL is slightly higher at 1.94. We can also plot the average scores of the two modules given a single threshold on the matching scores for the terms (in this case we use a single value for the threshold parameters P7, P9, P11 and P13). This is shown in Fig. 2.

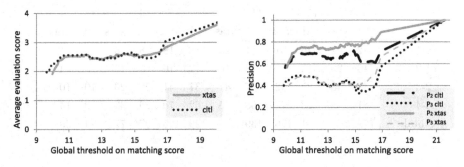

Fig. 2. Average scores (left) and precision graphs (right) for the global threshold values on matching score for the two NER modules

This figure shows that the performance of the two modules is very comparable. It shows that at very low thresholds (< 10), the performance for both modules indeed drops considerably. Investigation of the data shows that below 10, mostly terms with average score 0 are added, which corresponds with findings from the pilot study. Furthermore, the graph shows that increasing the threshold, increases the average evaluation score for both modules. However, there is only a slight gain between 10 and 16. Based on these results, we concluded that the choice of NER module is of no great consequence to the overall quality of the results

Global Precision Values. Other than averages, we also determined precision values by setting cutoff points to the average score. Specifically, we calculate P_N which we define as the *precision, given that a term with a score of N or higher is considered "correct"*. We calculate this for $N = 2$ and $N = 3$, which corresponds to minimum scores of "somewhat relevant" and "relevant" respectively. Figure 2 shows these values for the different global threshold values. Here, we can see that the P_2 values are around 0.7 to 0.8 for most threshold values (not considering very high values where very few terms are added. The more strict version of P_3 hovers around 0.4, which is considerably low. To get an even better insight in the hits and misses of the two versions of the algorithm, for different values of the threshold we list the number of terms evaluated in four bins (0-1, 1-2, 2-3, 3-4) . These are shown in Table 3 for both CLTL and XTAS. This table shows for example that given a threshold on the matching score of 11, the algorithm extracts a total of 155 terms when using the XTAS tool. In that case, 18 extracted terms receive an evaluation between 0-1 and 116 receive an average evaluation between 2 and 4 (41+75).

Table 3. Frequencies of terms in average evaluation bins for six threshold values.

Score bin	Threshold											
	10		10.5		11		12		14		16	
	cltl	xtas	cltl	xtas	cltl	xtas	cltl	xtas	cltl	xtas	cltl	xtas
0-1	42	62	26	31	21	23	18	18	8	5	2	1
1-2	16	20	15	19	13	16	12	16	8	13	3	4
2-3	40	48	37	42	37	41	37	41	22	26	10	10
3-4	81	88	73	78	68	75	62	70	29	33	9	14
Total	179	218	151	170	139	155	129	145	67	77	24	29

Individual Score Parameters. In the previous paragraphs, we have used a global value for the parameters P7, P9, P11 and P13. We now look at optimal values for each of these. For this, we weigh the Precision for each axis (Named Entity class corresponding to one of the four parameters) against an estimated recall. For this estimated Recall we assume that the total number of correct items for that NE class is the total number to be found. This means that the maximum Recall is 1.0 (which is found at threshold values 9.5). This is of course an incorrect assumption but it does give us a gradually increasing Recall when the threshold is lowered. Given that there are not significantly more false negatives than are found, this is a reasonable estimate for the true Recall. After calculating the Recall, we then calculated the F1 measure, which is the weighted average between Precision and Recall. All three values are calculated with the assumption that an average evaluation of 2 or higher is "correct", we therefore get P_2, $R_{est,2}$ and $F1_{est,2}$. The maximum value for $F1_{est,2}$ is an indication for the optimum value of the threshold. These optimal values are presented in Table 4. This shows that the optimal threshold values are approximately 10 for person and 12 for locations and miscellaneous (regardless of the NER module). For organizations, the two modules present different values. This might reflect an artifact in the data

Table 4. "Optimal" values for the threshold parameters for the four NE categories for both NER modules. At these values the $F1_{est,2}$ is maximized.

	Threshold		P_2		$R_{est,2}$		$F1_{est,2}$	
	cltl	xtas	cltl	xtas	cltl	xtas	cltl	xtas
P7 (person)	10.12	10.12	0.58	0.54	0.88	0.83	0.7	0.65
P9 (organization)	10.56	12.05	0.8	0.76	0.89	0.85	0.84	0.8
P11 (location)	12.19	12.19	0.82	0.79	1.00	1.00	0.90	0.88
P13 (misc)	12.15	12.15	0.75	0.83	1.00	1.00	0.86	0.91

4.4 Result Summary

The evaluation results indicate that the agreement between evaluators is not very high but at least acceptable. Using their assessments as ground-truth we saw that precision values of around 0.7 to 0.8 are obtained in a less strict evaluation where terms should minimally "somewhat relevant" (P_2). When we apply a stricter evaluation that requires a term to be "relevant", performance drops to around 0.4. Concerning parameter settings, thresholds in the range of 10 for person and 12 for locations and miscellaneous provides optimal results. With respect to the two NER modules we have seen that the choice of NER module is of no significant consequence to the overall quality of the results.

5 Discussion and Conclusion

In this paper we reported on an evaluation of automatic labeling of audiovisual content in an archive production environment. The aim was to evaluate if an unsupervised labeling approach based on subtitles using off-the-shelf NER tools and a baseline thesaurus matching approach would yield results that are acceptable given archival production requirement with respect to quality, authority and service levels to external users. We conclude that results are acceptable in this context, with parameter settings that are optimized using a strict evaluation approach, allowing only terms when they are relevant as opposed to somewhat relevant. Precision given these parameter settings are sufficiently high to not disturb the archival quality requirements but the downside is that Recall is rather low: professional archivists will typically label content with more labels then the automatic approach. However, given the pressure on manual resources in the traditional work-flow, the current automated set-up is a useful starting point. Furthermore, having a stable production work-flow running allows us to

(i) monitor the longitudinal behavior of the approach, among others by asking for feedback from external users, allowing us to assess the effect of the change also from an end-user perspective, and (ii) work on incremental improvements, gratefully deploying the experimentation framework that was set-up during the research described here. We have seen that the NER modules used do not differ much so that considerations such as stability, speed and resource use may be the most important factors for choosing a module. However, we note that we only tested two modules and there are many others around such as the Stanford NLP toolkit [7] or GATE [8]. It is likely that NER modules that are trained specifically on the type of input (in our case speech transcriptions) would improve performance both in terms of recall and precision.

One of the first items on our future work list will be an analysis of results over different genres of programming. In the current experiment we took 'annotation priority' as a selection mechanism, but from a quality perspective it makes more sense to select input to the term-extraction pipeline based on expected performance. This will also allow us to investigate more effectively how improvements can be obtained. Based on observations in the field, we expect that there is room for improvement especially for non-common terms in the named-entity track and aiming for a better capturing of global document features to improve disambiguation and subject term assignment.

Other improvements in recall can be achieved through clustering of synonyms, using (external) structured vocabularies or by improving the named entity reconciliation (identifying the occurrence of the same entity in a text even though spelling variants are used). Finally, we will also look into the use of other collateral data sources such as program guides and scripts, and combinations of data sources, potentially also coming from multimodal analysis components such as speaker recognition and computer vision [9].

Acknowledgments. This research was funded by the MediaManagement Programme at the Netherlands Institute for Sound and Vision, the Dutch National Research Programme COMMIT/ and supported by NWO CATCH program (http://www.nwo.nl/catch) and the Dutch Ministry of Culture.

References

1. Gazendam, L., Wartena, C., Malaisé, V., Schreiber, G., de Jong, A., Brugman, H.: Automatic annotation suggestions for audiovisual archives: evaluation aspects. Interdisc. Sci. Rev. **34**(2–3), 172–188 (2009)
2. Ordelman, R., Heeren, W., Huijbregts, M., de Jong, F., Hiemstra, D.: Towards affordable disclosure of spoken heritage archives. J. Digital Inf. **10**(6), 17 (2009)
3. Declerck, T., Kuper, J., Saggion, H., Samiotou, A., Wittenburg, J.P., Contreras, J.: Contribution of NLP to the content indexing of multimedia documents. In: Enser, P.G.B., Kompatsiaris, Y., O'Connor, N.E., Smeaton, A.F., Smeulders, A.W.M. (eds.) CIVR 2004. LNCS, vol. 3115, pp. 610–618. Springer, Heidelberg (2004)
4. Iivonen, M.: Consistency in the selection of search concepts and search terms. Inf. Process. Manage. **31**(2), 173–190 (1995)

5. Bizer, C., Heath, T., Berners-Lee, T.: Linked data - the story so far. Int. J. Semantic Web Inf. Syst. **5**(3), 1–22 (2009)
6. Likert, R.: A technique for the measurement of attitudes. Arch. Psychol. **22**, 1–55 (1932)
7. Manning, C.D., Surdeanu, M., Bauer, J., Finkel, J., Bethard, S.J., McClosky, D.: The stanford corenlp natural language processing toolkit. In: Proceedings of 52nd Annual Meeting of the Association for Computational Linguistics: System Demonstrations, pp. 55–60 (2014)
8. Bontcheva, K., Tablan, V., Maynard, D., Cunningham, H.: Evolving gate to meet new challenges in language engineering. Nat. Lang. Eng. 10, 349–373 (9 2004)
9. Tommasi, T., Aly, R., McGuinness, K., Chatfield, K., Arandjelovic, R., Parkhi, O., Ordelman, R., Zisserman, A., Tuytelaars, T.: Beyond metadata: searching your archive based on its audio-visual content. In: IBC 2014, Amsterdam, The Netherlands (2014)

Measuring Quality in Metadata Repositories

Dimitris Gavrilis[1], Dimitra-Nefeli Makri[1], Leonidas Papachristopoulos[1],
Stavros Angelis[1], Konstantinos Kravvaritis[1], Christos Papatheodorou[1,2(✉)],
and Panos Constantopoulos[1,3]

[1] Digital Curation Unit, 'Athena' Research Centre,
Institute for the Management of Information Systems,
Athens, Greece
[2] Department of Archives, Library Science and Museology,
Ionian University, Corfu, Greece
[3] Department of Informatics, Athens University of Economics and Business,
Athens, Greece
{d.gavrilis,n.makri,l.papachristopoulos,s.angelis,k.kravvaritis,
c.papatheodorou,p.constantopoulos}@dcu.gr

Abstract. The need for good quality metadata records becomes a necessity given the large quantities of digital content that is available through digital repositories and the increasing number of web services that use this content. The context in which metadata are generated and used affects the problem in question and therefore a flexible metadata quality evaluation model that can be easily and widely used has yet to be presented. This paper proposes a robust multidimensional metadata quality evaluation model that measures metadata quality based on five metrics and by taking into account contextual parameters concerning metadata generation and use. An implementation of this metadata quality evaluation model is presented and tested against a large number of real metadata records from the humanities domain and for different applications.

Keywords: Information quality models · Metadata quality · Context-sensitive evaluation · Repositories · Research infrastructures

1 Introduction

The advent of large-scale information services and the appearance of information aggregators, thematic portals, institutional gateways and repositories has rendered metadata quality a key-issue for the development of research infrastructures as well as cultural heritage applications. Aggregators such as Europeana or the Digital Public Library of America harvest volumes of metadata and provide catalogues that link to millions of objects. Such organizations need mechanisms to assure the quality of metadata they provide. The concept of quality refers to the compliance to standards that make data appropriate for a specific use. Data quality is determined in terms of a set of specific criteria: completeness, validity, consistency, timeliness, appropriateness and accuracy constituents [1,2].

© Springer International Publishing Switzerland 2015
S. Kapidakis et al. (Eds.): TPDL 2015, LNCS 9316, pp. 56–67, 2015.
DOI: 10.1007/978-3-319-24592-8_5

However, the very nature of quality is subjective. For instance, a metadata record that is 30 % complete (according to its corresponding schema) might be of better quality for a specific application and domain than another record that is 70 % complete. One reason could be that the metadata schema used fits better to the domain of the described resources. Another reason could be that the structure of a schema and its mandatory elements might describe clearly the main features of the resources. Hence, although a record might not be complete, it enhances the accessibility of the resource by human or machine agents. The assessment of metadata quality is multidimensional and several parameters should be taken into account to specify the context in which metadata are generated and used. For instance the domain the metadata refer to, is one significant parameter, since the metadata schemas are not designed for just one domain. Furthermore the application and targeted use of the metadata is another parameter, because different parts of a record are used by different applications and the same elements are used differently.

This paper aims to develop a flexible metadata quality evaluation model (MQEM) that takes into account contextual elements. The proposed model provides a set of metadata quality criteria as well as contextual parameters such as the domain, the application and use. All these entities are combined in an evaluation model that allows curators, data custodians and metadata designers to assess the quality of their metadata and to run queries on existing datasets to explore the applicability of a metadata schema as utilized by an application. A major challenge that is also tackled is the implementation of MQEM so that it can allow the easy integration of new domains, schemas and applications on the basis of subjective as well as objective measurements of the defined metrics. The next section presents a review of the metadata quality models and approaches and sets the main requirements for the development of a context-based metadata quality assessment model. Section 3 introduces the main concepts that define the contextual parameters used by MQEM, while Sect. 4 defines the metadata quality metrics adopted by the proposed model. Section 5 presents some experimental results for its validation and concluding remarks are made in Sect. 6.

2 Background

The existing diversity of metadata quality definitions, facets and measures resulted in the need to establish common ground for communication within the researchers' community. Various initiatives attempted to create an unambiguous framework in which any assessment effort would be based having reliable indications about metadata quality. One of the first attempts in the field of metadata quality assessment followed content analysis - a time-consuming and labor-intensive method - in order to evaluate GILS aggregator records, establishing a narrow set of criteria such as accuracy, completeness and serviceability [3]. In fact the study highlighted the value of an "appropriateness or fit-for-use" metric. Other researchers [4] introduced a conceptual framework of metadata quality indicators, while in another work [5] they narrowed enough the perspective of the evaluator focusing only on the completeness dimension. The completeness of the metadata elements was also addressed by [6], which proposes a tertiary Metadata Quality Framework considering completeness as the base level of assessment. Built upon completeness results, accuracy and consistency metrics were applied on three Open Access Cultural Heritage Repositories. A quality assessment baseline was defined using a questionnaire to gather feedback on the importance of specific elements but not element groups.

The work of Bruce and Hillman [7] is considered as a benchmark in the pursuit of quality assessment as it specified a general set of metadata quality criteria such as accuracy, completeness, timeliness, accessibility provenance, etc., accompanied by a set of metrics. The domain-independence that the framework offered delegated the responsibility for a contextual wrap to posterior attempts. [8] provides particular metrics for each criterion of [7] and thus extends the scope of the Bruce and Hillman's framework towards an automatic evaluation of metadata quality. Moreover [11] adopted [7] to create a metric for metadata quality assessment at data element level.

Stvilia et al. [9] addresses the context-dependend character of metadata quality and claim that metadata quality issues follow four major concepts: mappings, changes to the information entity, changes to the underlying entity and context changes. The aforementioned sources correspond to a taxonomy of 22 information quality criteria which are clustered to three categories: intrinsic, relation and reputational and are measured via 41 metrics. Also [10] recognises the impact of the context in metadata quality and attempts a mapping of the dimensions of [9] to the model introduced in [7].

From the literature analysis it can be inferred that the existing approaches are either abstract [4] or extremely focused on one dimension e.g. completeness [5]; they are using various sets of metrics [3,6] to obtain the same objective and the majority of them are ignorant of the context of use [3,6,8]. MQEM assumes that metadata quality strongly depends on the viewpoint of the evaluator. The evaluation process should be in line with the domain, the use and the application of the metadata and that was the main motivation for defining weighting functions that depend on the aforementioned factors. Recent research suggests that a metadata quality framework doesn't have to "invent new dimensions in order to accommodate the needs of diverse communities of practice" [2] but to give the flexibility to each evaluator to assess the results within a specific context. Additionally the proposed framework fulfils the requirements of an evaluation framework as it (i) examines all the aspects of quality problems using a set of dimensions corresponding to them (ii) clarifies the use of each dimension and (iii) provides practical measures for the accomplishment of an evaluation process [12].

3   Defining the Contextual Framework

The Metadata Quality Evaluation Model (MQEM) provides the mechanism for incorporating the metadata quality criteria into a real-world quality measurement application which takes into account not only the various schemas, domains

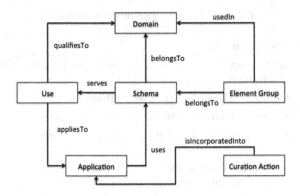

Fig. 1. The contextual framework

and uses, but also the fact that metadata change over time, e.g. through curation actions. In order to incorporate the notion of context into the proposed MQEM six primary classes are defined. The classes are related to each other through a set of properties. Hence a conceptual model is defined as Fig. 1 shows. The definitions of the six classes are as follows:

Schema: The schema represents the metadata schema of an object, dataset or collection. A metadata schema belongs to one or more domains, serves a specific use and consists of specific element groups. For example, the CARARE [13] metadata schema has been designed for the humanities domain and for the needs of the evaluation presented in this paper a set of six element groups was defined.

An *element group* is a group of elements that characterizes specific types of information (e.g. spatial information, thematic information, etc.). The introduction of the element groups is of major significance as these are mainly used to handle the complexities involved in the proposed model and they act as a glue between the applications and metadata schemas. Each schema is split into basic sets of elements that define the same type of information. Examples of such element groups are: Identity, Thematic, Spatial, Temporal, Agent, Event (see Table 1). Each element group can be split into sub-groups for denoting various detail levels. For example, the Spatial element group could be split into: Label, Address, Point and Area. It is easier to work with groups of elements rather than individual ones (especially when dealing with complex schemas). For simpler schemas, it is always feasible to define one element group per element.

Domain: Each evaluated dataset belongs to a domain. For example, the social sciences and humanities domain might be too general for certain applications and hence it should be split into sub-domains each one introducing different contextual parameters per element group.

Use: A use qualifies to a specific domain and applies to an application. It may range from a Generic use to more specific uses such as Research, Aggregation, etc. Use bind together the applications, the intended use and the element groups.

Table 1. Element groups descriptions

Element group	Description
Identity	Information used to identify a record. The id, title, collection information are such examples.
Thematic	Thematic information such as a subject.
Agent	Any agent related information such as a creator, contributor, etc.
Spatial	Any kind of spatial information from point coordinates to generic text place labels.
Temporal	Any kind of temporal information ranging from standard dates, scientific dates, periods, or simple place labels.
Activity	Information used to describe any kind of activity.

Application: An application at the instance level uses a specific schema and hence the element groups that belong to that schema. Modeling an application and incorporating it into an evaluation framework requires linking it to a schema and a use, and analyzing it into its primary functionalities. For example, a simple grey literature repository, from a user's point of view contains the following functionalities: Search, Browse and View. These functionalities are directly linked to element groups that contain the corresponding information. In this particular case, the browse functionality requires a thematic element group of a reasonable accuracy and a highly accurate Identity element group.

Curation Action: A curation action consists of operations/transformations performed on metadata records. Curation actions are incorporated into applications and thus ensure that adequate provenance and preservation information concerning the digital objects is available according to a curation/ preservation strategy.

4 Quality Criteria and Metrics

The main challenge of defining metadata quality involves the definition of a number of criteria along with appropriate metrics. The criteria introduced in this paper are the following: (i) *Completeness*, (ii) *Accuracy*, (iii) *Consistency*, (iv) *Appropriateness*, and (v) *Auditability*. These have already been introduced in the literature, yet this paper presents two main innovations: (i) it redefines them in a context-sensitive manner by using a weight function to regulate the sensitivity of each metric according to the context it is used in (domain, use, application) and (ii) it defines metrics not only on the basis of single elements but on element groups. In addition the computation of these metrics has been incorporated as a service into the MORe metadata aggregator [14].

Completeness: Completeness is a metric indicating the percentage of completion of the elements of a schema. This is a complex measure that can be analyzed into three partial measures: (i) completeness of the mandatory set of elements,

(ii) completeness of the 'recommended' element set and (iii) completeness of optional elements.

Although this definition may seem excessive for simple schemas such as Dublin Core, in the cases of more complex schemas like MODS, LIDO, CARARE, EAD this decomposition to element groups is necessary because: (a) complex schemas aim to minimize ambiguities for the captured information therefore are over-analytical by providing a large number of elements and (b) complex schemas usually incorporate elements used for different domains or types of objects in order to cover a broader range of needs. Therefore in such complex schemas, which are usually organized in sets of elements, not all the sets are required to properly describe an object and the absence of some in a record is not an indication of poor quality. In such cases it is essential to evaluate completeness on element groups rather than the schema as a whole. The metric for completeness is defined as follows:

$$COMP = \frac{\sum_{i=1}^{N} \sum_{j=1}^{Q} \gamma_j(d, u, a) comp(eg_{ij})}{\sum_{i=1}^{N} \sum_{j=1}^{Q} \gamma_j(d, u, a) max(comp(eg_{ij}))} \tag{1}$$

where $EG = \{eg_1, eg_2, ..., eg_N\}$, N are the available element groups; Q is the number of quantizations available. In particular for the completeness criterion there exist three quality quanta: completeness of mandatory elements, completeness of recommended elements and completeness of the total schema; $comp(eg)$ defines a function that computes the percentage of completeness for each element group that belongs to a quantum ($comp(eg) \in [0, 1]$); $\gamma(d, u, a)$ defines a weighting function that depends on the context classes *domain* (d), *use* (u) and *application* (a).

Accuracy: Indicates how accurate is the information provided to describe a certain element. For example, a thematic subject term encoded as text is less accurate than a subject term accompanied by a SKOS URI. Similarly, an address is more accurate than a place label and less accurate than a point (encoded in latitude/longitude). Although accuracy in general has an objective and a subjective aspect, we here only deal with what can be automatically measured, i.e. the objective aspect. Accuracy is measured per element group, and per accuracy quantum. More than one accuracy quantum may be applied to the subgroups of an element group ($eg_i = \cup_{j=1}^{Q_i} eg_{ij}$). The metric for accuracy is defined as follows:

$$ACCU = \frac{\sum_{i=1}^{N} \sum_{j=1}^{Q_i} \alpha_j(d, u, a) accu(eg_{ij})}{\sum_{i=1}^{N} \sum_{j=1}^{Q_i} \alpha_j(d, u, a) max(accu(eg_{ij}))} \tag{2}$$

where $EG = \{eg_1, eg_2, ..., eg_N\}$, N are the available element groups; Q_i is the number of quantizations available in each element group and $accu(eg)$ defines a function that assigns a boolean value of accuracy for each element group that belongs to a quantum ($accu(eg) \in \{0, 1\}$); $\alpha(d, u, a)$ defines a weighting function that depends on the context classes *domain* (d), *use* (u) and *application* (a).

Consistency: (i) Indicates whether the metadata values are consistent with the acceptable types of the metadata elements described by the metadata schema. For example, if a metadata schema defines that a language value has to be drawn from a specific list (e.g. ISO 639-2) then non-conformance indicates a lack of consistency. (ii) Indicates if the elements of a schema are used in a consistent manner across a metadata record. For example, in the scope of a specific academic repository if the contributor element (dc:contributor) of the schema is only used to define the committee of the reviewers of a thesis, then this element is used in a consistent manner. Notice that in the latter case (ii) manual assessment of the consistency of the value of an element is needed. Consistency is defined as follows:

$$CONS = k\frac{\sum_{i=1}^{M} cons(e_i)}{\sum_{i=1}^{M} \max(cons(e_i))} + l\frac{\sum_{i=1}^{M} \delta_i(d, u, a)}{\sum_{i=1}^{M} \max(\delta_i(d, u, a))} \tag{3}$$

where M the number of the metadata elements e_i; $cons(e)$ a function that evaluates to 0 or 1 according to case (i) above; $\delta_i(d, u, a)$ is a scoring function that depends on the context classes *domain* (d), *use* (u) and *application* (a), that correspond to the case (ii) consistency definition; k and l are non-negative weighting co-efficients such that $k + l = 1$.

Appropriateness: Indicates whether the values provided are appropriate for the targeted use. Hence this criterion is strongly affected by the contextual class *Use* and is defined as follows:

$$APPR = \frac{\sum_{i=1}^{M} \zeta_i(u)}{\sum_{i=1}^{M} \max(\zeta_i(u))} \tag{4}$$

where M the number of the metadata elements e_i and $\zeta_i(u)$ is a scoring function that depends on the context class *use* (u) with values in $[0, 1]$.

Auditability: Indicates whether the record can be tracked back to its original form. This is most useful especially when a curation service is present. For example, in the case of a metadata aggregator where metadata undergo transformations by various curation micro-services it is important that the record can be traced back to its previous form. Auditability is a Boolean metric ($AUDI \in \{0, 1\}$) indicating whether or not the curation/preservation model takes into account preservation and provenance allowing to trace back versions produced by various curation actions. Specific content models and schemas for audit logging, such as FoxML or PREMIS, can ensure auditability.

The mentioned individual metrics are combined by the following formula in order to measure the quality of metadata:

$$QUALITY = w_1 COMP + w_2 ACCU + w_3 CONS + w_4 APPR + w_5 AUDI \tag{5}$$

where w_1, w_2, w_3, w_4, w_5 are weights representing the importance if each metric in a metadata quality evaluation scenario.

5   Experimental Validation

In order to validate the MQEM model, the mentioned metrics have been implemented and incorporated in MORe metadata aggregator [14]. MORe aggregates content from data sources and the metadata quality service assists an agent to evaluate aggregated datasets before utilising them in applications and services. The prototype does not implement the appropriateness ($APPR$) metric, due to its exclusive dependency on the context of the agent which provides the metadata to be aggregated.

Two experiments were designed for the MQEM validation. The first one investigates the influence of the contextual classes *Application* and *Use* in measuring the quality of metadata of the same *Domain* and *Schema*. The second compares the quality of metadata of the same *Domain*, *Application* and *Use*, but of different *Schemas*.

5.1   Experimental Setup

Sixty seven (67) packages containing approximately two million metadata items (2113266) from the CARARE (Connecting ARchaeology and ARchitecture for Europeana) project were used for the two validation experiments. CARARE aggregated content from over 22 different providers related to the archaeology and architecture heritage and delivered to Europeana over 2 million records ensuring a high degree of homogeneity and quality [13,14]. Each item in these packages consists of two metadata records, one record in the CARARE schema and one in the Europeana Data Model[1] (EDM).

Three element groups were evaluated for each schema: (i) the Identity element group that includes elements for the monument, or collection, or digital object identification such as ids, titles, etc. (ii) the Spatial element group that incorporates geographical metadata elements such as place labels, addresses, point or area coordinates and (iii) the Temporal element group providing information about dates (either in textual or in date range form) as well as period names.

Regarding the computation of the values of the *Accuracy* metric ($ACCU$), three quanta were set. The *Low* concerns information of low quality including mostly textual information such as labels. The *Medium* corresponds to information that is structured but not highly accurate (such as an address or a geographical place) and can lead to a more accurate representation (enriched through some service). Finally the *High* concerns highly accurate information such as: Latitude/Longtitude points, scientific dates, URIs for SKOS concepts, URIs for links to collection information, etc. For each experiment different values for the weighting function $\alpha_j(d, u, a)$ were set, as analyzed in the next paragraphs.

For both the experiments the weighting values of the function $\gamma_j(d, u, a)$ of the *Completeness* metric ($COMP$) were set as follows: The weight of the *High* quantum is equal to 1.0 corresponding to the mandatory elements of a schema, the weight of the *Medium* quantum equals to 0.6 and corresponds to

---

[1] http://pro.europeana.eu/share-your-data/data-guidelines/edm-documentation.

the recommended elements of a schema, while the weight of the *Low* quantum equals to 0.2 and refers to the optional elements of a schema. If an element group contains metadata values, the function *comp(eg)* estimates the percentage of completeness of the element group.

The *Auditability* value (*AUDI*) was set to 1.0 because MORe aggregator uses PREMIS, a preservation metadata schema, to maintain a full log of the lifecycle of each object. When a record is changed a new PREMIS record is generated to keep track of the modification. The value of *Consistency* was also set to 1.0 because the metadata values are consistent with the types of the corresponding elements of the schemas. Note that the current experimental process investigates the influence of context only to *Completeness* and *Accuracy* metrics. Given that there are no studies concerning the impact that the weights for completeness w_1 and accuracy w_2 have on the total quality, they were indicatively set to 0.6 and 1.0 respectively for the needs of the current experiment.

5.2   Application Design Experiment

The first experiment aims at measuring the quality of metadata records in order to support the design of applications that will incorporate these metadata. In particular: (i) a Generic repository with search/browse functionality (PORTAL), (ii) a Virtual research environment (VRE) for humanities research and (iii) a Specialized geo-application (GEO) for viewing records on a map and time-line. Each application has different requirements implying respective contextual characteristics for the element groups that attribute different values to the weighting function $\alpha_j(d, u, a)$ for each accuracy quantum. PORTAL needs accurate identity and collection information and therefore the value of $\alpha_j(d, u, a)$ for the identity element group was set to 1.0 for the high accuracy quantum, 0.8 for the medium quantum and 0.5 for the low quantum. For the remaining element groups the weights of the high, medium and low accuracy quanta was set 1.0, 0.6. and 0.2 respectively. VRE requires the highest accuracy for the thematic and identity element groups and therefore the value of the $\alpha_j(d, u, a)$ for the identity and thematic element groups was set to 1.0, 0.8 and 0.5 respectively for each accuracy quantum, while for the remaining element groups its value was set to 1.0, 0.6. and 0.2 respectively. GEO requires the highest accuracy for spatial, temporal and identity information and therefore the value of the $\alpha_j(d, u, a)$ for all these element groups was set to 1.0, 0.8 and 0.5 respectively for each quantum, while for the remaining element groups was set to 1.0, 0.6 and 0.2 respectively for each quantum.

In this experiment only the quality of the CARARE metadata was assessed because the element groups of the EDM schema could not support the functionalities of the GEO application. Due to presentation reasons the results of only twelve packages are showed in Fig. 2[2]. The vertical axis shows the total quality score, while the horizontal axis shows the package id. Given that the

---

[2] The results for all the datasets are available at http://metaq.dcu.gr/.

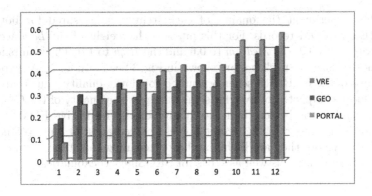

Fig. 2. Metadata quality for the application design experiment

PORTAL application has the less demanding applications, it exhibits the highest quality values. The GEO application provides good results meaning that they contain accurate spatial, temporal and thematic information in the corresponding element groups. The VRE application's quality measure falls behind by almost 10 %. This result is justified by the nature of the VRE applications, which require particular information about research Activities and Actors which are not provided by the element groups of the selected datasets.

5.3 Content Aggregation Experiment

Aggregation involves (i) the gathering of content from different content providers, mapped to an intermediate schema and (ii) its transformation to a target schema. As mentioned each item in a package comprises two metadata records, the source, which is in CARARE schema and the target which is in EDM. The CARARE schema contains over 50 elements in a multi-level nested structure and thus it is complex, rich and expressive. The EDM schema is less complex as it has been designed for representing cultural objects in general in the Europeana portal.

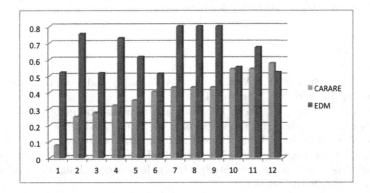

Fig. 3. Metadata quality for the CARARE and EDM metadata schemas

In this experiment the quality of each item was measured for both the CARARE and EDM records. For this purpose the weight of the *Low* accuracy quanta was set to 0.2, the *Medium* to 0.6 and the *High* to 1.0. The results for the twelve most indicative packages are shown in Fig. 3[See footnote 2]. As expected, the richness of the CARARE schema resulted to a lower quality value than EDM which is more compact. However for some packages the quality of the CARARE records is similar or better than the quality of the respective EDM records. Such records demonstrate high completeness percentages for the optional elements. Furthermore they are highly accurate, meaning that they provide exact geo-spatial coordinates as well as URIs for SKOSified vocabularies.

6   Conclusions

Existing metadata quality evaluation frameworks provide rather coarse grain results, while they do not provide mechanisms for the automated input from metadata sources. Such drawbacks prevent the development of flexible services that can incorporate contextual information, such as applications, schemas and uses. Moreover they do not allow curators or schema and application designers to evaluate quality scenaria allowing them to make informed decisions. This paper presented MQEM, a multidimensional, flexible model for evaluating the quality of metadata according to the context of their production and usage. The implementation of such a model presents a number of challenges as it should incorporate new schemas and model new applications. MQEM was validated in two different use cases, application design and content aggregation, each one having its own demands, represented by different values of contextual weighting functions. Further work includes performing more experiments on sensitivity analysis of the contextual weighting functions, as well as of the weights of each metric to estimate more precisely the metadata quality values.

Acknowledgment. This work has been funded by Greece and the European Regional Development Fund of the European Union under the O.P. Competitiveness and Entrepreneurship, NSRF 2007-2013 and the Regional Operational Program of ATTIKI in the frame of MEDA project, GSRT KRIPIS action.

References

1. Herzog, T., Scheuren, F., Winkler, W.: Data Quality and Record Linkage Techniques. Springer-Verlag, New York (2007)
2. Tani, A., Candela, L., Castelli, D.: Dealing with metadata quality: the legacy of digital library efforts. Inf. Process. Manag. **49**, 1194–1205 (2013)
3. Moen, W.E., Stewart, E.L., McClure, C.R.: The Role of Content Analysis in Evaluating Metadata for the U.S. Government Information Locator Service (GILS): Results from an Exploratory Study (1997). http://digital.library.unt.edu/ark:/67531/metadc36312/citation/

4. Margaritopoulos, T., Margaritopoulos, M., Mavridis, I., Manitsaris, A.: A conceptual framework for metadata quality assessment. In: International Conference on Dublin Core and Metadata Applications, Dublin Core Metadata Initiative, Singapore, pp. 104–113 (2008)
5. Margaritopoulos, M., Margaritopoulos, T., Mavridis, I., Manitsaris, A.: Quantifying and measuring metadata completeness. J. Am. Soc. Inf. Sci. Tech. **63**, 724–737 (2012)
6. Bellini, E., Nesi, P.: Metadata quality assessment tool for open access cultural heritage institutional repositories. In: Nesi, P., Santucci, R. (eds.) ECLAP 2013. LNCS, vol. 7990, pp. 90–103. Springer, Heidelberg (2013)
7. Bruce, T.R., Hillmann, D.I.: The continuum of metadata quality: defining, expressing, exploiting. In: Hillman, D., Westbrooks, E. (eds.) Metadata in practice. ALA Editions, Chicago (2004)
8. Ochoa, X., Duval, E.: Quality metrics for learning object metadata. In: World Conference on Educational Multimedia, Hypermedia and Telecommunications, pp. 1004–1011 (2006)
9. Stvilia, B., Gasser, L., Twidale, M.B., Smith, L.C.: A framework for information quality assessment. J. Am. Soc. Inf. Sci. Tech. **58**, 1720–1733 (2007)
10. Shreeves, S.L., Knutson, E.M., Stvilia, B., Palmer, C.L., Twidale, M.B., Cole, T.W.: Is "quality" metadata "shareable" metadata? the implications of local metadata practices for federated collections. In: 12th National Conference of the Association of College and Research Libraries (ACRL), Minneapolis, pp. 223–237 (2005)
11. Chen, Y.-N., Wen, C.-Y., Chen, H.-P., Lin, Y.-H., Sum, H.-C.: Metrics for metadata quality assurance and their implications for digital libraries. In: Airong, J. (ed.) ICADL 2011. LNCS, vol. 7008, pp. 138–147. Springer, Heidelberg (2011)
12. Lei, Y., Uren, V., Motta, E.: A framework for evaluating semantic metadata. In: 4th International Conference on Knowledge Capture, p. 135. ACM Press, New York (2007)
13. Papatheodorou, C., Dallas, C., Ertmann-Christiansen, C., Fernie, K., Gavrilis, D., Masci, M.E., Constantopoulos, P., Angelis, S.: A new architecture and approach to asset representation for europeana aggregation: the CARARE way. In: García-Barriocanal, E., Cebeci, Z., Okur, M.C., Öztürk, A. (eds.) MTSR 2011. CCIS, vol. 240, pp. 412–423. Springer, Heidelberg (2011)
14. Gavrilis, D., Dallas, C., Angelis, S.: A curation-oriented thematic aggregator. In: Aalberg, T., Papatheodorou, C., Dobreva, M., Tsakonas, G., Farrugia, C.J. (eds.) TPDL 2013. LNCS, vol. 8092, pp. 132–137. Springer, Heidelberg (2013)

Personal Information Management and Personal Digital Libraries

Memsy: Keeping Track of Personal Digital Resources Across Devices and Services

Matthias Geel$^{(\boxtimes)}$ and Moira C. Norrie

Department of Computer Science, ETH Zurich, Zürich, Switzerland
{geel,norrie}@inf.ethz.ch

Abstract. It is becoming increasingly difficult for users to keep track of their personal digital resources given the number of devices and hosting services used to create, process, manage and share them. As a result, personal resources are replicated at different locations and it is often not feasible to keep everything synchronised. In such a distributed setting, the types of questions that users want answers to are: Where is the latest version of this document located? How many versions of this image exist and where are they stored? We introduce the concept of global file histories that can provide users with a unified view of their personal information space across devices and services. As proof-of-concept, we present Memsy, an environment that helps users keep track of their resources. We discuss the technical challenges and present the results of a lab study used to evaluate Memsy's proposed workflow.

1 Introduction

It is now common for users to work with multiple computing devices in their daily lives, often using external storage devices or cloud services as a way of moving digital resources between devices and sharing them with others. In addition, web-based media services and social networking sites are used to not only publish resources, but also manage them and make them accessible on the move. As a result, while the problem of accessing resources from different places has been alleviated, the problem of keeping track of them has been exacerbated.

Since storage space has become extremely cheap, there is a tendency to copy, share and distribute more data than ever, unknowingly making it even harder to keep track of all the copies, versions and variants that we produce [12]. For example, the increased mobility offered by a new range of computing devices has resulted in a greater blurring between the private and working environments of users [5] and they now expect to be able to work on the same document in different locations. To do this, the users need to decide how they will move files around, possibly using a cloud storage service such as Dropbox or an external storage device such as a USB flash drive. In addition to being edited, a document may be copied, renamed or moved between folders and the user often ends with multiple versions in different places. Consequently, users often struggle to remember which is the latest version and where it is stored.

© Springer International Publishing Switzerland 2015
S. Kapidakis et al. (Eds.): TPDL 2015, LNCS 9316, pp. 71–83, 2015.
DOI: 10.1007/978-3-319-24592-8_6

Our goal is to provide users with a global view of their digital resources so they can get answers to questions such as: *"Where is the latest version of a document and on which device/service is it stored? Does a previous version still exist somewhere? Where is the original? How many copies exist and to which, if any, file hosting services have they been uploaded?"*. Importantly, we want to do this without requiring them to change their ways of working. The solution therefore has to be embeddable in standard desktop environments and work alongside existing applications and file management tools.

To achieve this, we propose the concept of a *file history graph* for managing metadata about a user's entire information space, possibly spanning several computing and storage devices as well as online file hosting services. As proof of concept, we have developed Memsy, an environment that helps users keep track of resources and reconcile versions based on the file history graph.

We provide an overview of related research findings and studies in Sect. 2 before proposing a general architecture for keeping track of resources in Sect. 3. We then discuss the technical challenges of implementing such a system in a highly distributed setting along with our Memsy solution in Sect. 4. A lab study carried out to evaluate the Memsy workflow is reported in Sect. 5 before presenting concluding remarks in Sect. 6.

2   Related Work

As the number of personal computing devices has increased, the requirements for information work to be carried out successfully have also changed [5,13,14]. In the study by Dearman et al. [5], they reported that participants found managing information across devices the worst aspect of using multiple devices. They argue that we need to be aware of a user's collection of devices and no longer assume a single personal computer.

Consumer-oriented cloud storage solutions have now become an integral part of many personal workflows as highlighted in recent interviews with 22 professionals from different industries carried out by Santosa and Wigdor [14]. While they enable resources to be readily available on multiple devices, none of their participants regarded the cloud as a complete data solution and data fragmentation remains an important topic even in the context of cloud solutions. Participants explicitly stated that they had difficulties in remembering where in the cloud they put their data. To remedy some of the issues, the authors recommend providing improved awareness and visibility of what is happening in the cloud.

Research into how human memory works has shown that users quickly forget important computing tasks [4]. Elsweiler et al. [7] point out similarities between everyday memory lapses and memory lapses in the context of personal information management and suggest that tools should help users recover from such lapses. Consequently, the ability to track and re-find information in a desktop environment is a highly desirable property for today's information management tools [9]. It has been shown that one effective way of providing memory cues is to use the idea of provenance information [1].

Previous research has mainly explored two strategies for helping users keep track of their resources. The first category of tools aim to provide better search and organisational facilities based on metadata/time [6] or tagging [3]. The second category focus on the associations between resources rather than the properties of the resources themselves and include navigation by associations [2] as well as the exploitation of semantic relations between resources [11]. However, most of these solutions have been designed with the underlying assumption of a single desktop computing environment and it is unclear how they could be adapted to multi-device environments. In fact, the authors of the *Stuff I've Seen* system [6] report from their user study: "The most requested new feature [in the user study] is unified access across multiple machines."

Studies have shown that users are actually very fond of their folder structures [10] as a means of not only categorising information but also decomposing problems/projects into smaller units. The fact that users have different work practices is reflected in how they set up their working environments. For example, some use the desktop as a temporary workspace for all kinds of documents based on how they use their physical desks [15] and cross-device work patterns can often be complex [14].

Since users often put a lot of effort into customising their work environments and file organisation, they are very reluctant to change their work habits. For example, Jaballah et al. [8] proposed a digital library system that enables documents to be browsed via various metadata axes, but, in their diary study, most users preferred the more limited folder view and this was in large part attributed to the time required to become familiar with a new application interface.

Despite their shortcomings, folder hierarchies remain the basis for file organisation. This apparent discrepancy between what has been proposed in research and how people actually work led us to conclude that user behaviour changes very slowly and new solutions should try to build upon existing conceptual models and incrementally refine them. Additionally, we argue that it is crucial to embrace rather than oppress the diversity in strategies and methods people employ to organise information, and take these into account when designing new tools for managing personal resources.

3   Version-Aware Environment

The main idea behind our approach is to build a global provenance index that spans the user's entire personal information space by collecting data from multiple devices and integrating third-party services. Because system access should be independent of any individual computing device, we opted for a classic client-server approach. Figure 1 illustrates the architecture of the Memsy environment. Our system is comprised of three main components which together realise a "version-aware" computing environment for end-users.

Global Resource Catalogue. This is the main service that orchestrates all components and communicates with various client services and tools. It provides a

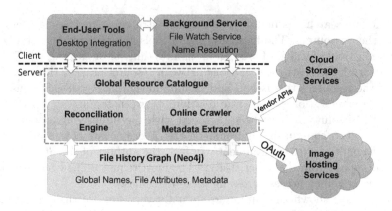

Fig. 1. Architecture for a Version-Aware Environment

global view of a single user's personal information space by ensuring that each resource is not only uniquely identifiable but can also be located in a unambiguous way. A file that is part of the global resource catalogue is called a **managed file**. All other files, including those that are not part of a user's personal information space, are **unmanaged**. The global resource catalogue is responsible for collecting updates from client devices as well as cloud storage and image hosting services (through online crawlers) and applying them to the file history graph.

File History Graph. The file history graph provides a lightweight, implicit versioning mechanism for files. It is essentially an index that stores the metadata required to identify and locate files across devices and services.

Reconciliation Engine. This component reconciles **unmanaged** files, which have not been tracked previously or cannot be tracked reliably (i.e. files on an cloud storage service beyond our control), with existing file histories. In both cases, the goal is to integrate those files into the global resource catalogue. Reconciliation is either done automatically based on file properties or user-driven. In the latter case, we use content-based similarity metrics to aid users in their decisions.

To build a global catalogue of a user's personal information space, each file needs to have a unique and non-ambiguous mapping between a global namespace and its actual location. Figure 2 shows the structure of the global address scheme and an example mapping. Our scheme follows the recommendations for Uniform Resource Identifiers[1]. A vital part of that scheme is the numerical identifier of the storage device on which the file is stored (`storageDeviceId`). In our environment, everything that has a mount point and is writeable by the user can be assigned a globally unique identifier that is then stored in the root directory of that particular storage device. This allows us to reliably recognise external storage devices such as USB flash drives or external harddisks, even when they are moved between computers.

---

[1] http://tools.ietf.org/html/rfc3986.

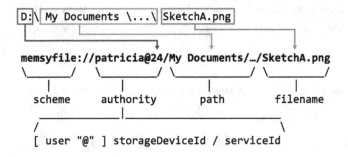

Fig. 2. Global namespace scheme. The example given uses a `storageDeviceId`.

To refer to file hosting services, we use pre-defined names, e.g. *dropbox* or *googledrive*, as identifiers (`serviceId`). The path component corresponds to the original path. In this address translation scheme, we assume that an absolute file path, including the file name, cannot be occupied by more than one file at the same time. This is true for all mainstream file systems (NTFS, ext3, HFS+).

On each computing device belonging to the user's personal information space, a background service is deployed that monitors connected storage devices and tracks the user's file activities. These computing devices also synchronise a list of connected storage devices upon startup and whenever the list changes. This information is used later to tell users about the last known location of "movable" storage devices (such as USB flash drives). Global name resolution at runtime is straightforward because the local service knows the identifiers and mount points of all locally connected storage devices. This is especially useful on the Windows platform, where the mount points (i.e. drive letters) of external storage devices usually depend on the order in which they were connected.

3.1 File History Graph

The file history graph is a data structure that records the metadata of each new version of a file and stores the last known location(s) for each file version. From a data model point of view, each file history is a tree of file version nodes that starts with a single root node. Each version node may point to one or more location nodes. The resulting structure is therefore a forest of directed trees, with each tree representing a separate history.

Figure 3 illustrates how that structure captures file operations. A *create* event triggers the creation of new file history with root V_1 and location L_1. If V_1 already exists, it represents a *copy* operation and we simply add a new location node to the existing node. When a *modify* event occurs, we first identify the (old) file version node that corresponds to the location of the modified file. If that version node is a leaf node, we derive a new version node and re-attach the location node to it. If it is not a leaf, a newer version must already exist somewhere else. In that case, we perform a fork. All other file operations (*rename*, *move* and *delete*) only affect the last known locations of file versions and update them accordingly.

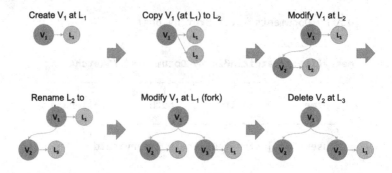

Fig. 3. Evolution of a single file history

As a consequence of the delete operation, file version nodes may end up with zero location nodes attached to them.

It is an important characteristic of our proposed index structure that we retain metadata about previous file versions, even if they no longer exist in the personal information space. With the information in these file histories, we can answer several of our initial questions such as locating copies, latest versions and file origins across devices.

File events are sent immediately to the server if the device is online, otherwise they are cached locally and stored on the storage device where the operations occurred. Thus, the cache travels with the storage device and if, for example, a USB drive is shared between computers that are both offline, the order of executed operations on that USB flash drive is always guaranteed, without the need for complicated time-synchronisation. We have implemented the cache as an event-queue backed by a transactional embedded database (HSQLDB). That queue is consumed by a background thread and entries are only removed if the event could be successfully submitted to server, otherwise processing is halted until Internet connectivity is restored. Since all events (update, move, rename, delete) are idempotent *if* executed in order, we not only achieve at-least-once semantics but can also guarantee that all locally observed transactions are eventually reflected correctly in the global file history graph.

3.2   Reconciliation of Related Resources

At some point, we are bound to encounter resources that either have not been tracked previously or originate from an environment that cannot be monitored directly. We call the process of turning such unmanaged resources into managed (tracked) ones *reconciliation*.

With file hosting services, reconciliation is required as changes to personal resources often cannot be tracked reliably from the outside. Even worse, some image hosting services re-size and re-compress uploaded images, thus changing the actual file content quite significantly. Furthermore, there will always be cases where a file enters a personal environment from *the outside*, for example as an e-mail attachment, downloaded file or copy from somebody else's USB stick.

To align such files with our global resource catalogue, we primarily rely on hashing and file path matching. To recognise copies, we need to be able to tell whether two files have the same content. Popular file hashing functions such as MD5 or SHA1 have proven to be very effective and we therefore use them to realise this functionality. Since we store these hashes for previously encountered versions as well as the current ones, we can reliably reintegrate files that correspond to older versions. For example, a user might have received a report from a co-worker by e-mail, saved it in a local, observed location and modified its content. Several weeks later, they might want to retrieve the revised version but cannot remember where they stored it. Luckily, they remember the e-mail that contained the original file. With the attachment from that e-mail serving as an entry point into the file history, they can learn about the location of the revised version. Having such provenance information allows users to discover newer versions of documents based on the copy of an older one.

To reconcile images that have been modified by online services upon upload (e.g. re-compression on Facebook), computer vision algorithms are used to produce a "perceptual hash". The main idea of such an image signature algorithm is to map an arbitrary image to an arbitrary *sparse image representation* which is significantly reduced in size but retains important perceptual information. An appropriate distance metric is then used to compare these sparse image representations in order to find near-duplicates. So far we have only implemented content-based reconciliation for images, but consider reconciliation to be a vital part of any environment for managing personal resources and plan to extend the service for other types of files.

3.3   End-User Experience

Users first install the Memsy client on their computer(s). Using these clients, they can add local folders to the observed environment. This allows very fine-grained control over which parts of the file system are monitored. Similarly, they can connect and add personal USB flash drives. Such storage devices can also be labelled to make it easier to recognise them when they show up in search results. After setting up the environment, users can continue to work as usual, freely creating, moving, renaming, copying and deleting files as well as creating new folder structures within the monitored folders.

If a user wants to find the latest version of a file, they follow a simple procedure shown in Fig. 4. (1) Small overlay icons in the file explorer view offer a quick glance of the current file status. From a user experience point of view, this works similarly to overlay icons in desktop integrations of popular cloud storage services (e.g. Dropbox) or version control systems (e.g. TortoiseSVN). A tick symbol indicates that the current file is up-to-date, while an exclamation mark warns the user that a more recent version is stored somewhere else. Such icon overlays give a simple and unobtrusive way of providing status information to users. They integrate well with the desktop experience, work in all standard file dialogues and are well-suited to handling directories with a large number of

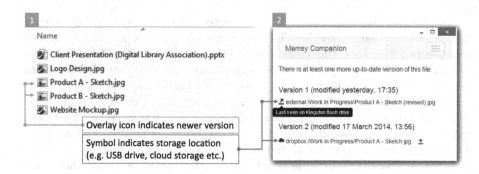

Fig. 4. A Version-Aware Environment: (1) Overlay icons inform user about file status [cropped screenshot from Windows file explorer]. Context menu entry invokes custom application (2) with further information about related versions.

files. When the user right-clicks one of those files, a few Memsy-specific commands appear in the contextual menu. (2) From there, the user can launch the Memsy Companion application, a small desktop tool that displays information retrieved from the global resource catalogue. In our example, the user learns that there are two newer versions located on other devices/services. We can show for each version whether that particular file can be accessed directly. Hovering over entries that are currently not available provides information about its last known location (a USB flash drive in this example).

4   Implementation

To demonstrate and evaluate our approach, we have developed a reference implementation of the Memsy environment in Scala, a JVM-based, multi-paradigm programming language that offers functional and object-oriented constructs. The global resource catalogue called *Memsy Global* runs on a dedicated server. It is built with a micro web framework called Scalatra[2]. As a backend for the file history graph, we use Neo4j[3], a graph database implemented in Java, which is a perfect fit for tree structure of our file history graph. On the client-side, we deploy a background service called *Memsy Local*, which is responsible for monitoring the users' file activities. It communicates with the server using its REST-style API over HTTP, with JSON as the data exchange format. To capture file system events, we make use of the Java 7 File API (NIO.2). To incorporate a number of popular online services, we have implemented crawlers for Dropbox, Google Drive, Facebook and Flickr. The former two cloud storage crawlers process all supported files whereas the latter two only retrieve and process images.

The user interface has been implemented for the Windows platform using a number of high- and low-level technologies. Both icon overlays and the context

---

[2] http://www.scalatra.org/.
[3] http://www.neo4j.org/.

menu have been realised as Windows shell extensions. To indicate a file's status, an icon overlay handler was created for the Windows file explorer that queries the local Memsy client, which in turn may have to ask Memsy Global for the current status. Because icon overlays are precious resources in Windows (only up to 16 handlers can be registered), our integration builds upon TortoiseOverlays, the icon handler of many popular desktop integrations of version control systems such as Subversion, Git or Mercurial. This makes the icons easily recognisable by users of any of these solutions.

Our dedicated end-user tool, the Memsy Companion app, is a Chrome app. These apps resemble native desktop applications but run on top of Google Chrome and can be built using web technologies. The tool communicates with the local background service and the global resource catalogue to answer the questions motivated earlier. We have chosen this web technology-centric approach because it allows us to seamlessly mix various technologies from high-level web interfaces down to low-level shell extensions. The consistent use of the HTTP protocol also facilitates similar integrations for other operating systems.

5 Evaluation

To evaluate our approach, we conducted a small user study with two primary goals. First, we wanted to gather valuable insight into work practices and strategies for keeping track of resources across different devices. Second, we wanted to find out whether the tool support would be sufficient when a user cannot remember what happened to a particular document.

The main study design challenges were: *How can we reliably create a condition that reflects the situation where a user has actually forgotten the location of the latest version? How can we eliminate non-controllable factors such as an individual's particular cognitive abilities?* For these reasons, we decided to perform a controlled lab experiment instead of an external field study.

Study Setup. We set up three work stations in a room to represent different work places. Each work station had a labelled USB stick. All 6 devices were configured to belong to the same user and all tools were pre-installed on the workstations. We prepared a set of 20 files for our hypothetical Memsy user: 5 presentations (PowerPoint), 5 reports (Word) and 10 images (JPEG). These files had been copied to a managed folder on each workstation.

Participants. Our participants were recruited from an interdisciplinary European research project where our group is a member. This allowed us to recruit not only computer science researchers (3 professors, 3 research assistants), but also people from media education/pedagogy (4), plus a designer and an architect. In total, we had 12 participants (half of them female), from industry as well as from academic institutions. Although we had a rather small group of participants, they were from 5 different organisations with quite different work environments and this matched our belief that a diverse set of study subjects would yield more interesting results than a large, but homogeneous, subject pool.

Procedure and Tasks. We divided our 12 participants into groups of 3 persons. All participants of a group were invited at the same time and each participant was randomly assigned to a designated work station. The study started with a pre-task questionnaire that collected background information about participants' experiences with file managers and their current techniques for version management of documents and copying files between devices.

The main study was a sequence of information management tasks, where each participant performed exactly the same steps but worked on a different subset of the files. In the first part of the study, participants were asked to perform three common file management and document editing tasks: update and rename a presentation, copy an image to a (newly created) subfolder and modify it, copy a report to the USB drive and modify it. Though we asked participants to follow our instructions closely, we did not tell them how to accomplish the tasks. It was up the individual user to decide whether they work with keyboard shortcuts, drag & drop or context menus. If they made a mistake (e.g. copied the wrong file), we asked them to revert the change how they saw fit. At that stage, the background service monitored the file management operations behind the scenes and propagated the changes to the global resource catalogue.

In the second part, participants were asked to gather information for an upcoming presentation which included files possibly modified by the other participants (1 presentation, 1 report, 2 images) and spread across the other two work stations and USB sticks. To complete the task, they had to indicate if there was a newer version of any of those files, and, if there was, where the newer versions were stored (location, drive and path). At any time, they could invoke the Memsy Companion app, shown in Fig. 4, for further information.

Results. In our pre-task questionnaire, we asked participants about their current work practices for managing versions and copying files between devices. In both questions, multiple answers were possible. By far the most popular method for managing multiple versions of office documents is a personal file naming scheme (10 participants). Although less popular, folder structures were still used by half of the participants. Although revision control systems are widely used to manage source code, our results indicate that they are considerably less popular for managing versions of documents (4), let alone to share files between devices (2). None of our participants used a dedicated document management system such as SharePoint.

Interestingly, even though all 12 participants use cloud storage services to move files between devices, almost none (2) use the implicit versioning provided by some cloud providers. This might be due to the fact that popular services such as Dropbox and Google Drive by default only store previous versions for up to 30 days. We conclude that cloud storage services have become a ubiquitous tool for transferring files between machines, but other means such as USB sticks (11 participants) and sending e-mail to yourself (11) still have their uses. These results affirm some of the assumptions underlying our initial use case and further inform our proposed approach.

All participants were able to use the Memsy Companion app to answer most of the questions from the second part correctly (overall completion rate was 46 out of 48 tasks). Figure 5 shows the results from the post-task questionnaire. For each of the statements, we asked participants to rate their level of agreement on a 7-point Likert scale, where 1 is *strongly disagree* and 7 *strongly agree*.

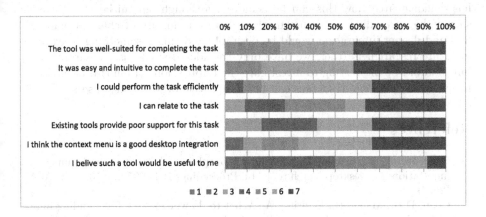

Fig. 5. Results from post-task questionnaire

The Memsy Companion app as well as the entire experience offered by our environment received fairly positive ratings. Most participants felt it was well-suited to the task at hand (median 6), that it was easy and intuitive to complete the task (median 6), and that they were efficient in doing so (median 6). Participants also agreed with our underlying assumption that existing tools [offered by the operating system] provide poor support for the task (median 6). However, while many participants could relate to the task (median 5), most users (mode 4) were undecided about whether such a tool would actually be useful for them. These general concerns were also reflected in some of the feedback we received, for example P3 noted: "*The general issue is what do I actually do when I realise that I don't have the most recent version in front of me.*"

We conclude that, while our prototype is successful in answering the kind of questions posed at the beginning, users may require more information to help them decide what to do when confronted with multiple versions on different devices. As P9 pointed out, "*[The tool should provide] some overview of what kind of change has been done*" and P10 agreed by saying "*[The tool should] provide the differences of versions (visual comparison of versions)*".

With regard to desktop integration, reactions from users were quite positive (median 6) but some expressed a preference for an even deeper integration with the regular desktop computing environment. To that end, we plan to experiment with other forms of shell extensions and consider providing more detailed information as part of the regular file explorer interface. However, further studies are required to steer development in the right direction.

6 Conclusion

We have shown how a global resource catalogue together with a file history graph can be used to help users keep track of personal resources across devices and services. Our aim was to provide a solution that could integrate with existing work practices and applications rather than replace them and our Memsy prototype has demonstrated how this can be achieved, although our initial studies have shown that providing more information about the nature of changes made to files and deeper integration would be desirable and this is something to explore in future research. Further, we have only considered single user environments, but it would be interesting to investigate the implications of multi-user environments where personal resources can be shared with other Memsy users.

References

1. Blanc-Brude, T., Scapin, D.L.: What do people recall about their documents?: implications for desktop search tools. In: Proceedings IUI 2007, pp. 102–111. ACM (2007)
2. Chau, D.H., Myers, B., Faulring, A.: What to do when search fails: finding information by association. In: Proceedings CHI 2008, pp. 999–1008. ACM (2008)
3. Cutrell, E., Robbins, D., Dumais, S., Sarin, R.: Fast, flexible filtering with phlat. In: Proceedings CHI 2006, pp. 261–270. ACM (2006)
4. Czerwinski, M., Horvitz, E.: An investigation of memory for daily computing events. In: Faulkner, B., Finlay, J., Détienne, F. (eds.) People and Computers XVI-Memorable Yet Invisible, pp. 229–245. Springer, London (2002)
5. Dearman, D., Pierce, J.S.: It's on my other computer!: computing with multiple devices. In: Proceedings CHI 2008, pp. 767–776. ACM (2008)
6. Dumais, S., Cutrell, E., Cadiz, J., Jancke, G., Sarin, R., Robbins, D.C.: Stuff I've seen: a system for personal information retrieval and re-use. In: Proceedings SIGIR 2003, pp. 72–79. ACM (2003)
7. Elsweiler, D., Ruthven, I., Jones, C.: Towards memory supporting personal information management tools. J. Am. Soc. Inf. Sci. Technol. **58**(7), 924–946 (2007)
8. Jaballah, I., Cunningham, S.J., Witten, I.H.: Managing personal documents with a digital library. In: Rauber, A., Christodoulakis, S., Tjoa, A.M. (eds.) ECDL 2005. LNCS, vol. 3652, pp. 195–206. Springer, Heidelberg (2005)
9. Jensen, C., Lonsdale, H., Wynn, E., Cao, J., Slater, M., Dietterich, T.G.: The life and times of files and information: a study of desktop provenance. In: Proceedings CHI 2010, pp. 767–776. ACM (2010)
10. Jones, W., Phuwanartnurak, A.J., Gill, R., Bruce, H.: Don't take my folders away!: organizing personal information to get things done. In: Proceedings CHI 2005 Extended Abstracts, pp. 1505–1508. ACM (2005)
11. Karger, D.R., Bakshi, K., Huynh, D., Quan, D., Sinha, V.: Haystack: a customizable general-purpose information management tool for end users of semistructured data. In: Proceedings CIDR (2005)
12. Karlson, A.K., Smith, G., Lee, B.: Which version is this?: improving the desktop experience within a copy-aware computing ecosystem. In: Proceedings CHI 2011. ACM (2011)

13. Oulasvirta, A., Sumari, L.: Mobile kits and laptop trays: managing multiple devices in mobile information work. In: Proceedings CHI 2007, pp. 1127–1136 (2007)

14. Santosa, S., Wigdor, D.: A field study of multi-device workflows in distributed workspaces. In: Proceedings UbiComp, pp. 63–72. ACM (2013)

15. Zacchi, A., Shipman, F.M.: Personal environment management. In: Kovács, L., Fuhr, N., Meghini, C. (eds.) ECDL 2007. LNCS, vol. 4675, pp. 345–356. Springer, Heidelberg (2007)

Digital News Resources: An Autoethnographic Study of News Encounters

Sally Jo Cunningham, David M. Nichols[✉], Annika Hinze,
and Judy Bowen

Department of Computer Science, University of Waikato, Private Bag 3105,
Hamilton, New Zealand
{sallyjo,daven,hinze,jbowen}@waikato.ac.nz

Abstract. We analyze a set of 35 autoethnographies of news encounters, created by students in New Zealand. These comprise rich descriptions of the news
sources, modalities, topics of interest, and news 'routines' by which the students
keep in touch with friends and maintain awareness of personal, local, national,
and international events. We explore the implications for these insights into
news behavior for further research to support digital news systems.

Keywords: News behavior · Qualitative research · Digital news resources ·
News encounter · Personal digital library

1 Introduction

The news landscape has changed considerably over the past decade with social media
platforms creating new dissemination channels for information. The inherently social
nature of systems such as Facebook has encouraged both recommendations of conventional news items and widened the conception of news itself. The resulting ecosystem of consumption, creation and sharing presents a complex landscape of news far
removed from just broadcast television news and print newspapers. Our understanding
of these news interactions informs both the creation of access tools for current users
(such as recommender systems and visualizations) and the design of news archive
collections for future users. In this present paper we add to this understanding of news
behavior by exploring the news practices of 35 New Zealand tertiary students in both
the digital and physical context. This research is qualitative; as such it inevitably
highlights areas for further research into news behavior and appropriate interface and
interaction design for more effective news provision systems.

2 Related Work

The growth of online content, and social media in particular, has been widely reported
as changing the news environment for users. Nielsen and Schrøder report that there is
limited understanding of how important social media is as a source of news [11].
Recent studies suggest that, despite the growth of social media, trusted legacy brands
and platforms remain important news channels [11, 15]. Social media has created a

S. Kapidakis et al. (Eds.): TPDL 2015, LNCS 9316, pp. 84–96, 2015.
DOI: 10.1007/978-3-319-24592-8_7

space where "consumers collaboratively create and curate news stories" rather than receiving news from a limited number of 'authoritative' sources [12]. Hermida et al. report that these social channels are valued by users for the alternative filtering they provide but that social media use by "traditional" news sources (such as newspapers) is an important source of information [6]. Meijer and Kormelink reflect the potential richness of social media-enhanced news interactions in the 16 types of news activity reported by their interviewees: reading, watching, viewing, listening, checking, snacking, monitoring, scanning, searching, clicking, linking, sharing, liking, recommending, commenting and voting [1].

These micro-activities are complemented by higher-level classifications of behavior. Marshall categorized study participants using a New York Times news reading application into three groups: "Reading primarily for relaxation and as a diversion; Reading as a newshound, following the narrative of specific breaking stories or particular recommendations; Reading broadly to stay informed or to keep up with events of the day" [10]. van Damme et al. categorize mobile news consumers as: *omnivores* (actively engaged using multiple channels), *traditionals* (intensive but loyal to established sources such as TV) and *serendips* (less of a news routine but digital when engaged) [15].

Characterizations of news consumers often use the intent of the user, whereas [16] note that *incidental* news exposure is common and, for digitally-connected citizens, increasingly difficult to avoid. Although sharing content is an important aspect of social media, [7] claim that it serves a personal as well a social function. Shared content persists in the platform as an archive that can be searched at a later date. Similarly, [10] notes that "a person's daily encounters with the news should become a fundamental part of ... a personal digital library." The diverse distributed cross-platform multimedia nature of users' news interactions suggests there are considerable technical and legal challenges to achieving that goal.

Although much has been written about social media, there is still uncertainty about the interaction between traditional news channels and social media [11]. Pentina and Tarafdar [12] claim that "no studies so far have attempted to explore or explain the mechanism of news consumption processes in the context of socially connected interactive participants". However, in the light of studies such as [8] this claim may reflect a literature gap between the journalism and user interaction communities.

Location is an important feature of news; both the location of the consumer and the places referred to in news reports. Systems such as *NewsStand* [14] and *NewsViews* [4] attempt to extract geographical references from news reports to enable map-based interactions such as queries and visualizations. Leino et al. [8] found that personal news recommendations are more effective, possibly because social media users have an implicit model of their friends' preferences which pre-filters suggestions.

In summary, timeliness and location are key elements of news-based information systems. Users have adapted the diverse tools of social media for sharing and recommending news items. Social recommendations are effective as they leverage existing models of users' preferences and are often innately timely. Studies of news behavior that predate the growth of social media (e.g., [3]) are of limited applicability and there may be a lack of interaction between journalism studies and those in the computing literature.

3 Methodology

Our study is based on a set of autoethnographies gathered from undergraduate students in New Zealand. In this section we describe the context in which the autoethnographies were created, our analysis method, and the limitations of this study.

3.1 Data Collection

The data collection for this study was performed using personal ethnography (incorporating the use of diaries, self-observation and self-interviews), gathered in a semester-long project in a third year university course on Human-Computer Interaction. The students were given the (deliberately broad and ambiguous) brief of designing and prototyping software to 'assist a person in accessing news'.

The first task for the students was to gather data on how people currently locate, manage, share or encounter news. To that end, they examined their own news-related behavior by completing a personal diary during the course of their everyday lives over a period of three days. For each encounter with a news item, the participants recorded the date and time, the number of news items they were exposed to during the encounter, the source, type (international, national, local or personal), the topic, how they encountered it and how believable it was (on a linear scale). The students then summarized and reflected on their diary entries as a post-diary 'debriefing'.

Next, the students reflected on how they managed their exposure to news items by creating autoethnographies [2] that identified the strategies, applications and resources they used and then investigated what types of media they were using, topics they were accessing and the activities they performed to actively locate news. They also observed any unexpected or chance encounters with news sources. Students were encouraged to reflect on their *actual* practices identified in the self-observation.

3.2 Data Analysis

The diary study summaries and autoethnographies for 35 students were retained for analysis, out of an enrolment of 103. As is typical of New Zealand IT students, these selected participants are predominantly young (under 30). Though the majority of students in the course were male (84 M, 19 F), we selected a higher proportion of the female students' work for analysis so that the female news experience would be better represented. The course also included a significant number of international students (33 of 103); accordingly, we also limited our selection to students who were New Zealand citizens or permanent residents, as the experiences of international students could be expected to both differ greatly by their country of origin. Table 1 presents the demographic details for the 35 students whose work is analyzed in this study. These students were assigned a unique label (i.e., P1, P2, ..., P35), and are referred to by that label in this paper.

Table 1. Student demographic details.

Gender	Count (%)	Age at time of study	Count (%)
Male	22 (63 %)	<30 years	29 (83 %)
Female	13 (37 %)	30–46 years	6 (17 %)

The diaries were retained by the students and so cannot be analyzed directly; instead, we view the recorded behavior through the diary study summaries and reflections. The entirety of the self-observation was available for analysis. These summaries and autoethnographies for the 35 students total over 200 printed pages; they were analyzed qualitatively using grounded theory methods [5], an iterative, inductive methodology that allows the participants' experiences, viewpoints, and conceptions to emerge naturally. Initial coding largely followed the categories included in the diary study summaries (Sect. 4), and further concepts emerged as these encounters were set in context by the autoethnographies (Sect. 5).

3.3   Limitations of Study

Participating in a study is known to have the potential to alter behavior. The students themselves recognized that undertaking the assignment introduced changes to their news behavior ("I was more aware entirely of the news around me... sub-consciously listening out for news to include" [P22]; "I found myself not wanting to look at the news online as much since I would have to record it down." [P31]). The assignment brief acknowledges these issues; in mitigation, the diaries included the option to indicate news events that occurred but were not noted, and the students were encouraged to explore deviations from their usual news behavior in the autoethnographies.

We recognize that the students likely felt a greater sense of commitment to completing the diaries and creating the autoethnographies than is usual with study participants, given that these activities were assessed. Participation in this present study, however, was not required; a student could opt out of inclusion by emailing a third party to indicate this desire. The assessors for the course were not informed of these decisions until after the semester's grades were finalized.

Given that this is an opportunistic study, we cannot claim to capture 'typical' behavior in searching/browsing/encountering news. As is characteristic of this style of study design, we instead build a rich picture of the news-related information behavior for these students, from their own perspective [13]. We demonstrate in Sect. 6 how this rich picture can suggest directions to explore both in developing software support for these behaviors and in further news information behavior studies.

4   Results

In this section we summarize the news sources consulted by the students, the news topics of interest to them, the characteristics of their common news behaviors ('routines'), and online and physical platforms that they use to access their news.

4.1   News Sources

The students encountered news from a wide variety of sources (Fig. 1), with an average of 4.4 significant news sources per student (Table 2). Only three students relied on a

Table 2. Number of significant news sources per student.

No. of sources	1	2	3	4	5	6	7	8
No. of students	3	2	10	13	4	1	1	1

Facebook is by far the most heavily relied upon news source both in *Social Media* and overall (Fig. 1a). Its use goes well beyond providing personal news from family and friends; it is also a source for 'breaking' news of all topics. Twitter can be a 'Facebook lite' for news ("Twitter is a more short form of Facebook where I look at various tweets posted from friends, random users and any company or news agency to find out the latest news." [P1]). The blogs, vlogs, and forums were often focused on a topic of interest to the student, where use of Reddit and Tumblr was described as being more exploratory or serendipitous ("I never visit [Reddit] intending to encounter news but sometimes find myself reading or watching news after clicking on a link that grabbed my attention…" [P19]).

Interpersonal news sources (Fig. 1b) include face-to-face conversations (both with individuals and

single news source; it was acknowledged in the other 32 autoethnographies that multiple sources were necessary for topic coverage and convenience of access ("The 3 main ways that I access news [Facebook, email, mobile news app] all have good and bad attributes and that's probably the reason why I don't just use one" [P17]).

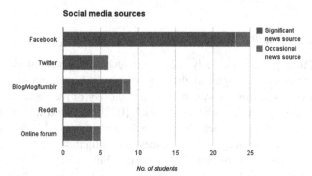

Fig. 1a. Social news sources

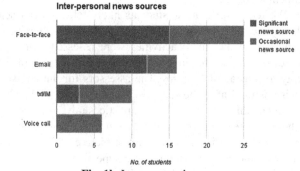

Fig. 1b. Inter-personal news sources

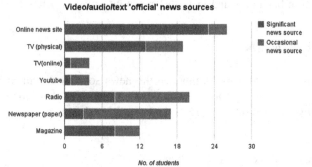

Fig. 1b. 'Official' news sources

Fig. 1. News sources used by students

groups), SMS or IM messages, voice calls (via mobiles or VOIP), and email. The latter source most frequently elicits contact by a commercial organization or from an interest-based mailing list, and more rarely messages from friends or relatives.

'Official' news sources such as television or radio broadcast news and newspapers are encountered in both digital and physical formats (Fig. 1c). The online news sites consulted were primarily websites for the major New Zealand newspapers and the local news aggregator stuff.co.nz. These provide primarily text and still images. The physical (paper) newspapers and magazines were New Zealand focused and were either free (local weekly free papers, the university's student-run weekly magazine) or were freely accessible (at work breakrooms, parents' houses). Only one student reported purchasing a magazine subscription, and three reported occasionally purchasing a single issue of a magazine or newspaper if a story attracted their interest.

4.2   News Topics

Figure 2 presents an overview of the news topics that students reported to be of significant interest (i.e., they frequently sought out information on these topics) and of occasional interest (i.e., they infrequently sought out or encountered information of interest). Each student held a significant interest in at least two topics (Table 3), with an average of five topics per individual.

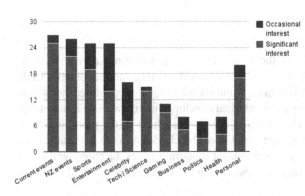

Fig. 2. News topics of significant and occasional interest

Table 3. Number of topics that students held a significant interest in following.

No. of topics of significant interest	2	3	4	5	6	7	8
No. of students	6	8	9	6	4	0	2

Current events is a broad category encompassing a range of categories typical of those covered by newspapers (e.g. wars, natural disasters, elections, etc.); *NZ current events* have a specifically New Zealand focus. *Sports* includes international, national, and local sporting news. *Entertainment* news covers the gamut of TV shows, music, movies, and other performances (but excluding games and sports), while *Celebrity* news is focused on a particular well-known person (including New Zealand and international celebrities). It is not surprising that students enrolled in an upper level

Computer Science course would have strong interests in *Technology and Science* (primarily 'popular science' events and new hardware and software releases) and in *Gaming*. While *Business* included limited interests in conventional topics in that category (finance, banking, commercial trends), students' primary concerns were with topics that directly impacted them—particularly notifications of upcoming sales.

We note that significant international or local events influence an individual's news interests. As the autoethnographies were performed during the 2012 Olympics, students reported a higher than usual exposure to sports news, sparking in some an interest specifically in Olympics-related news but not in sports news in general (P25: "Without these events [the Olympic games] happening, there is very little likelihood that sports would feature in any of my diary entries."). Similarly, three students pointed out that they normally have little interest in politics, but "when election time comes up I make sure to check out the people I can vote for" [P23]. Further, an individual story in an uninteresting topic can capture interest if the student sees a personal link: "… I just get onto the NZ Herald web site and I see … "Top lawyer guilty of misconduct". I do not like to read about politics because I find it boring, but I read about this because … my sister is a lawyer …" [P9].

4.3 News Routines

The overwhelming majority—31 of the 35 participants—reported having a news 'routine'. Some routines were simple: P14, for example, had arranged for a set of news feeds and email newsletters so that "The most common way that I encountered news was having it delivered to me." Other participants had developed more elaborate routines that spanned their entire day: "…I view news is usually first thing in the morning, check facebook see if any new news has appeared. Follow by checking to see the results from sports teams during the night. During the day I randomly check facebook and occasionally see new items in trending articles. At night I check stuff to see if anything interesting has occurred" [P5]).

The news activities in a particular routine could vary by:

Time of day: Generally the morning and/or the evening are important points in news routines,. News consumption can be helpful in waking up (on TV in the morning: "…it is a ritual I do when I wake up in the morning if I have a lot of time I'll watch it while eating breakfast, otherwise ill [sic] have it going in the background while I get ready." [P22]) and in relaxing after a day of study and work ([P20] reports "taking a good half hour to read through the news that has occurred over the course of the day while I wind down with a beer."). These news sessions tended to be longer and to involve active searching/browsing for news on the part of the participant. During the day, participants reported frequent news 'snacks' [1] to fill in time and avert boredom; these tended to be shorter (e.g., to fit in with work breaks or periods before a lecture started) and to involve checking newsfeeds.

Day of the week: Those students who reported having routines typically differentiated between routines for days involving scheduled work or study, and their free days. A free day might involve fewer news encounters ("…my Mondays this semester are… my lazy day at home. Because of this the amount of news I generally encounter on a

Monday is typically low." [P23]). However, if news encounters are motivated by relaxation or socialization, the number of encounters may increase (for face-to-face encounters, "on Saturday, the number of items each encounter yielded was greater. This could be related to the fact that on the weekends, my flatmates are all home, which allows us to have group conversation..." [P20]). A free day may also bring the student into contact with an additional source; for example, visiting the family home and finding "newspapers piled high in my parents' house" [P35]).

Availability of the source: Consumption of several news sources were tied to their availability. While none of the students subscribed to a print newspaper, 17 cited it as a significant (3) or occasional (14) news source; these students regularly read the newspapers provided at work during a break, looked out for the university's student-published weekly on its distribution day ("If I do read a magazine it is usually the nexus [student paper] and it's only once a week usually Tuesday afternoon during one of my lectures" [P5]), or watch TV news broadcasts only on visits to their parents ("As we don't have a TV at our flat this is just when I'm home at the weekend." [P26]). Radio news was most commonly serendipitously encountered while in a car, typically driving to/from university or work. Only one student incorporated radio news into daily routine, with a radio in his bedroom.

4.4   News Platforms

We identified three 'platforms' through which the students encountered news items: a standalone computer (desktop or laptop), a mobile device (tablet or handheld), and physical media (physical televisions, radio, newspapers, and magazines); see Fig. 3. Three of the participants did not differentiate between computer and mobile use, simply stating that they preferred to access news 'online'. Of the remaining 32 participants, 30 used computers as a significant news access platform, and 14 of those also described their

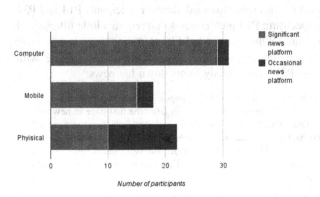

Fig. 3. Platforms for news encounters

mobile as a significant access platform. Only one participant described significant use of a mobile but not of a computer as well.

Physical news sources continue to see use, with approximately 60 % of the participants identifying them as significant or occasional sources for news. No student reported physical sources as their sole significant access platform.

Online access to news sources has obvious advantages: "...as long as there is an internet connection, they are easy and convenient to access, since I don't have to leave my bedroom to use them." [P17] Indeed, as students spend more time online, their access to physical or face-to-face news sources declines ("I spend approximately 6–10 h daily in front of my computer ... making the internet my only real source of information or news in any form" [P19]).

5 Discussion

The students' conception of what constitutes news goes beyond topics covered by traditional news media ("...the topics I find personally newsworthy are not necessarily the more traditional ideas of what news is defined to be." [P25]). Their news has a greater focus on the personal—activities of friends and family—and events or activities that impact them directly (e.g., grocery store sales, updates to their favorite game). News interests are more narrowly focused than the broad categories of traditional news media (e.g., specific genres of music, movies, and television). Further, broadcast media can be difficult or impossible to skim/scan to filter out irrelevant or 'boring' news. P8 points out that with TV news shows, even those viewable online, "It is hard to know if the news on TV will have anything I'm interested in. They don't describe every item of news that they will cover.... The news on TV is often only the important news at the start and less important news near the end.... This means I have to wait till nearly the end of the news to see something I'm interested in." The students' heavy use of non-traditional news sources (e.g., Facebook, blogs, forums) partly stems from this desire to create a more personalized news information feed than is possible with conventional broadcast news.

The desired degree of direct control over this feed varies. Participants P14 and P31 occupy opposite ends of this spectrum: P14 prefers news that requires little filtering ("I am someone who is very lazy in seeking out news. I like to utilize news sources that involve very little effort and are very easy to use. For example: TV, talking with friends, and Facebook."), while P31 meticulously hunts down his news:

> When I go looking for news at home I load the following pages in different tabs, Stuff.co.nz, Engadget and ESPN Soccernet. Then I'll open a few headlines from the first page in new tabs and close that news source.... Once I've found a few good news items on the next page I'll open them in new tabs then go back to the other tabs I opened since they should have loaded now. I'll then read those news items... Then once I've finished the news items from the first source, I'll move on to the ones from the second source.... I'll end up opening heaps of tabs then slowly work my way through them.

'Media multitasking' [9], the simultaneous information consumption from multiple sources, is recognized as a common behavior. Here, news consumption is interleaved with other entertainment activities (watching non-news TV shows) as well as serious activities (such as university assignments): "I get into bed... Then I turn my laptop on and start checking my emails, facebook, homework and I like to watch television episodes as well for background noise and something to watch. I like to flick between Facebook, NZ Herald web site and working on my homework..." [P9] Media multi-tasking can be deliberate (as with P9), or it can occur unintentionally ("Since I have a widget on my phone constantly updating me with news I simply 'ran across' a news item while checking a text message that caught my eye." [P1]).

While selected news items may be read or viewed carefully, students also engage in news satisficing behavior—getting the gist of a text news item by scanning news headings, summaries, or snippets, or by overhearing/viewing bits of news in passing. P12, for example, follows the New Zealand Herald newspaper and CNN on Twitter, and will "normally just read the headline on their tweet" rather than the entire tweet, much less follow a link to the full story. News applications that more readily support text news satisficing by prominently displaying headers and snippets are preferred to older-style interfaces that direct the reader through topic hierarchies before arriving at the news summaries; P11, for examples, resents having to "take time to read the headers and sub-headers of categories to find the information you want".

Searching for news related to a specific event (in contrast to encountering news on the event from existing feeds or browsing news resources) is often in reaction to exposure to a news snippet. When the snippet is encountered from face-to-face conversations or glancing at physical media, the most common response is to search online news sources for further details (rather than, for example, purchasing the magazine or newspaper). When the snippet is encountered online, the student may simply retrieve and read the text associated with the snippet, or may search additional sites for alternative viewpoints or updates. Social media such as Facebook or Twitter are more likely to be consulted when the event is in an early stage, to find the most recent reports ("… posts on the latest earthquakes in Christchurch came up on Facebook well before the news sites had any information on them" [P12]).

The students wrestled with issues of trust in news media and the believability of particular news items. There was no consensus on a set of sources that were more trustworthy than others; for example, P6 points to newspapers as "a great source of trustworthy news", where P3 prefers informal "online sources" because "online sources follow-up is often possible to find the original news source and establish if a news item is actually true". Personal news as encountered through text (via Facebook, SMS, Twitter, etc.) may not be accurate because "friends always boast or exaggerate" [P7]. It is easier to evaluate the believability of personal news delivered by "someone who is face-to-face, you can often tell with their body language and tone of voice whether or not what they are saying is true" [P11]. In general, the believability of news in most topics depends on the trustworthiness of the source, with the single exception of celebrity news–a topic panned as being "not very believable but still very entertaining" [P22].

Six students raised additional issues affecting believability of a news item: bias on the part of the author or sharer of that item, 'spin' or deliberate inaccuracies in the presentation of the item, and perceived manipulation of the reader to view or share particular news items. As noted above, personal news is particularly prone to biased presentation, as the people involved in a story may also be reporting it. However, the problem of bias on social media sharing runs deeper, as the external stories 'shared' in an individual's Facebook page are part of that person's social image–and so can skew their nature and topics ("… people tend to make an effort to post articles about things they feel are likely to provide "Likes" or discussion among their friends and are wary when posting controversial content." [P19]).

News can be slanted through its presentation or its content, and this 'spin' can be difficult to uncover. For example, P19 checked five different sources for a single breaking story, to identify potential bias in the reporting of the event ("I found that most articles had the same main points however some reporters attempted to put spins on these points..."). P11 points a cruder source of bias: the presence of "imitation style websites that make up stories or fake events that people can often mistake for reality". It can be difficult to differentiate between 'real' and 'imitation' news sources–hence the importance of identifying trustworthy, believable news sites and feeds. However, even those trusted sites may include dubious stories; of the 11 students who identified Stuff. co.nz as a site that they frequently used, none acknowledged that, as a news aggregator, the site does not verify the press releases posted to it.

The final issue affecting believability—perceived manipulation of the reader's attention—is an ongoing issue for both social media and commercial news sites. In social media sites such as Reddit, users can attempt to attract greater attention to a news item by "provoking inflammatory responses from other readers" [P20] in the comments threads or attempting to "blackmail" [P12] users into sharing a posted image by attaching an emotionally manipulative caption to it. Commercial online news sites manipulate news choices by introducing "advertiser links, pop up boxes and a plethora of tricks and techniques that divert my attention from where I was hoping to go to where someone wants me to go" [P35].

6 Summary

As is typical for qualitative work, the contribution of this study is to point to future directions in research and development for systems supporting the news behavior uncovered here. Specifically, we raise the following questions:

- Given that news consumption is a significant relaxation and entertainment activity, how can we make news encounters more enjoyable? Is it possible to make them more attractive, more pleasant to engage with, more ludic in nature?
- How can we support the 'newshound' [10] in tracking down the minute details of a story without getting lost in the process? (E.g., "I can sometimes find myself attempting to find out exactly how far the rabbit hole goes, and end up exactly where I started several hours later" [P20]).
- How can emerging, relevant sources be brought to the attention of users—and which existing sources will the new ones replace? Even the youngest of these students could reflect on sources that they once relied on but now rarely use.
- How may a personal Digital Library support up-to-date news encounters?
- How can we model a given user's news routines, and support those behaviors as they vary across times of day, days of the week, and the location of the user?
- As a user moves between different digital platforms (desktop/laptop, tablet, phone), how can we support a seamless news experience? Alternatively, should we tailor the news experience to the platform, given their different affordances? For example, small-screen mobiles are inherently well suited to 'news snacking', while desktops/laptops offer the screen real estate to support 'newshounds'.
- How can we model a given user's topics of interest, and how can we adapt that model to reflect change in those interests?

- What information presentation and organization designs can better support news satisficing? For example, how can we provide more informative news snippets, and how can we reduce the need for users to traverse topic hierarchies?
- What changes in news provision can support our users in identifying bias? For example, can we provide multiple versions of a story from different sources, or highlight relationships (such as corporate ownership) between publishers and the entities referred to in the news reports?

References

1. Costera Meijer, I., Groot Kormelink, T.: Checking, sharing, clicking and linking: changing patterns of news use between 2004 and 2014. Digital Journal. (2014). doi:10.1080/21670811.2014.937149
2. Cunningham, S.J., Jones, M.: Autoethnography: a tool for practice and education. In: Proceedings of CHINZ 2005, pp. 1–8. ACM (2005)
3. Diddi, A., LaRose, R.: Getting hooked on news: uses and gratifications and the formation of news habits among college students in an internet environment. J. Broadcast. Electron. Media **50**, 193–210 (2006)
4. Gao, T., Hullman, J.R., Adar, E., Hecht, B., Diakopoulos, N.: Newsviews: an automated pipeline for creating custom geovisualizations for news. In: Proceedings of the Conference on Human Factors in Computing Systems, pp. 3005–3014. ACM, April 2014
5. Glaser, B., Strauss, A.: The Discovery of Grounded Theory: Strategies for Qualitative Research, Chicago (1967)
6. Hermida, A., Fletcher, F., Korell, D., Logan, D.: Share, like, recommend: decoding the social media news consumer. Journal. Studies **13**(5–6), 815–824 (2012)
7. Lee, C.S., Ma, L.: News sharing in social media: the effect of gratifications and prior experience. Comput. Hum. Behav. **28**(2), 331–339 (2012)
8. Leino, J., Räihä, K.-J., Finnberg, S.: All the news that's fit to read: finding and recommending news online. In: Campos, P., Graham, N., Jorge, J., Nunes, N., Palanque, P., Winckler, M. (eds.) INTERACT 2011, Part III. LNCS, vol. 6948, pp. 169–186. Springer, Heidelberg (2011)
9. McDonald, D.G., Meng, J.: The multitasking of entertainment. In: Kleinman, S. (ed.) The Culture of Efficiency: Technology in Everyday Life, pp. 142–157. Peter Lang, New York (2009)
10. Marshall, C.C.: The gray lady gets a new dress: a field study of the Times News Reader. In: Proceedings of JCDL 2007, pp. 259–268. ACM, New York (2007)
11. Nielsen, R.K., Schrøder, K.C.: The relative importance of social media for accessing, finding, and engaging with news: an 8-country cross-media comparison. Digital Journal. **2**(4), 472–489 (2014)
12. Pentina, I., Tarafdar, M.: From "information" to "knowing": exploring the role of social media in contemporary news consumption. Comput. Hum. Behav. **35**, 211–223 (2014)
13. Resch, A., Berk, J., Akers, L.: Recognizing and Conducting Opportunistic Experiments in Education: A Guide for Policymakers and Researchers. REL 2014–037. National Center for Education Evaluation and Regional Assistance, USA (2014)

14. Samet, H., Sankaranarayanan, J., Lieberman, M.D., Adelfio, M.D., Fruin, B.C., et al.: Reading news with maps by exploiting spatial synonyms. Commun. ACM **57**(10), 64–77 (2014)
15. van Damme, K., Courtois, C., Verbrugge, K., Marez, L.: What's APPening to news? A mixed-method audience-centered study on mobile news consumption. Mobile Media Commun. **3**(2), 196–213 (2015)
16. Yadamsuren, B., Erdelez, S.: Incidental exposure to online news. Proc. Am. Soc. Inf. Sci. Technol. **47**(1), 1–8 (2010)

Exploring Semantic Web and Linked Data

On a Linked Data Platform for Irish Historical Vital Records

Christophe Debruyne[1,2], Oya Deniz Beyan[2(✉)], Rebecca Grant[1],
Sandra Collins[1], and Stefan Decker[2]

[1] Digital Repository of Ireland, Royal Irish Academy, Dublin, Ireland
{c.debruyne,s.collins,r.grant}@ria.ie
[2] Insight @ NUIG, National University of Ireland Galway, Galway, Ireland
{oya.beyan,stefan.decker}@insight-centre.org

Abstract. The Irish Record Linkage 1864–1913 is a multi-disciplinary
project aiming to create a platform for analyzing events captured in
historical birth, marriage and death records by applying semantic technologies for annotating, storing and inferring information from the data
contained in those records. This enables researchers to, for instance,
investigate to what extent maternal and infant mortality rates were
underreported. We report on the semantic architecture, provide motivation for the adoption of RDF and Linked Data principles, and elaborate
on the ontology construction process that was influenced by both the
requirements of the digital archivists and historians. Concerns of digital
archivists include the preservation of the archival record and following
best practices in preservation, cataloguing and data protection. The historians in this project wish to discover certain patterns in those vital
records. An important aspect of the semantic architecture is the clear
separation of concerns that reflects those requirements – the transcription and archival authenticity of the register pages and the interpretation
of the transcribed data – that led to the creation of two distinct ontologies and knowledge bases.

Keywords: Historical vital records · Cultural heritage · Linked data ·
Ontology engineering · RDF graph transformation

1 Introduction

We report on the semantic architecture and ontology creation of the multi-
disciplinary Irish Record Linkage (IRL) 1864–1913 project. The IRL project
aims to create a knowledge base containing historical birth-, marriage- and death
records translated into RDF and create a Linked Data [6] platform to analyze
those events. The project involves the expertise of three disciplines [3]: historians,
digital archivists and knowledge engineers. With the help of knowledge engineers
creating the ontologies and setting up the platform and the digital archivists
who curate, ingest and maintain the RDF, the historians will be able to analyze
reconstructed "virtual" families of Dublin in the 19th and early 20th centuries,

© Springer International Publishing Switzerland 2015
S. Kapidakis et al. (Eds.): TPDL 2015, LNCS 9316, pp. 99–110, 2015.
DOI: 10.1007/978-3-319-24592-8_8

allowing them to address questions about the accuracy of officially reported maternal mortality and infant mortality rates. To aid the historians in their data analysis, the knowledge engineers also contribute in linking people across records and the contextualization of the information with other datasets.

2   General Records Office

In Ireland, the General Register Office – GRO for short – is Ireland's civil registry responsible for recording information on births, deaths and marriages. In this project, the Registrar General of Ireland generously offered us records of 6,009,781 births (from 1864 to 1912), 4,314,963 deaths (from 1864 and 1912) and 1,443,110 marriages (from 1845 to 1912) under strict terms and conditions. It became compulsory to report and register births, deaths and marriages in 1864, but *non-Catholic* marriages were already being registered from 1845 onwards.[1] This explains the broader timespan for marriage records. *Records* of these events were captured on *register pages* (up to 10 per page for births and deaths, and up to 4 for marriages - see Fig. 1 for a redacted example) divided by district and sent to the GRO where volumes were then created and an *index* compiled. The information was provided to us as a database dump of the GRO's database with digitized versions of the register pages and indexes.[2]

The information system the GRO has built allowed one to search for vital records concerning a person based on a person's name, geographical area (to the level of district) and year; one of their core services to the public. Not only has the GRO spent resources in the construction of such a service, an enormous amount of effort also went into the digitization of register pages and indexes as accurately as the recording of a subset of the information in a relational database. A rational decision was made to only enter in the database the information sufficient to efficiently find records. While the system developed by the GRO works perfectly for finding historical records, information that is key in answering the IRL historians' questions were not captured by the database (such as the places of death, names of the informant, etc.). As such, we should call on the expertise of digital archivists – trained in processing, transcribing and curating the information – in preparation for the Linked Data platform to be developed.

The vital records and the goals of the IRL project lead to various challenges that need to be taken into account and those challenges reside at different levels: data protection, data transcription, historical evolution (medical knowledge, geographical, etc.) and, of course, the method for answering the historians' research questions. We will highlight some of the pertinent challenges below that will influence the design of the semantic architecture and the transcription workflow.

---

[1] http://www.irish-genealogy-toolkit.com/Irish-marriage-records.html.

[2] The terms and conditions of our data sharing agreement do not permit us to make public any data that would identify any individual [3]. One can access the historic records of the GRO at its dedicated research room in Dublin, but it is restricted per diem and there is an associated charge.

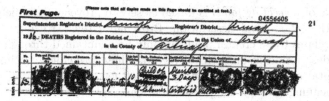

Fig. 1. Part of a register page containing death records (redacted as per our data sharing agreement). Copyright held by the General Register Office and reproduced with permission.

Data security and protection in terms of transfer, storage and use by authorized parties were covered by the data sharing agreement with the GRO. The goal of the IRL project is to build a platform that allows one to analyze the data captured in those records and not to replace the service already built by the GRO, although the new platform would support the queries typically executed by GRO as well. As per our data sharing agreement, the dataset in its entirety (that means data and digitized objects) should only be available to the members of the project team. With the help of the digital archivist, who is familiar with data protection legislation and best practices, we furthermore identified which guidelines to follow.

Records, knowledge and interpretation. Another challenge is the varying levels of detail in the records (seen in, for instance, the causes of death) and the variances in how subject names and places were recorded (initials, short hands, name of a building versus street name, etc.) [3]. These variances might imply something, which we are currently unaware of. Therefore, we should ensure that the transcription of the register pages transcribes exactly what was written down. In other words, the manipulation of the information should be kept to a minimum. This leads to another, yet related challenge, *clearly separate two concerns*: the exact transcription of what has been captured on the register pages as to have an authentic virtual account of historic events; and the interpretation, possibly with background knowledge, of certain aspects based on these interpretations. Examples of how interpretation can differ are the evolution of Ireland's geography (place names changing and streets disappearing, merging and even reappearing), evolution in knowledge (e.g., new insights in medicine) and even the adoptions of different theories (e.g., different classifications of social status).

Provenance and archival authenticity. Archival theory is based on two key principles, *respect de fonds* (original order) and *archival provenance*. Respect de fonds is the principle which guides archivists when exerting intellectual control over a collection, and ensures that the archival record is always described in relation to the context in which it is created as far as possible (for example a letter should only be described in terms of a set of correspondence where it is available). We follow this principle by transcribing not a line of data about an individual, which is meaningless in an archival context, but the entire register page that constitutes an archival record or object. The principle of *respect de fonds* is linked closely to provenance, which forms the foundation of archival description. Provenance refers to how the archival record relates to its creator, and can only be maintained through the appropriate description of an archival record. These principles are important in the digital sphere, and describing and authenticating records in this way gives meaning through the provision of context.

Other data challenges include the conversion to appropriate data formats as well as cataloguing of the digitized objects so as to ensure compliance with digital preservation best practices. These challenges, however, fall outside of the scope of this paper; work on the ingestion of the digitized objects in a suitable digital long-term preservation platform will be disseminated elsewhere.

3   IRL Semantic Architecture

This paper focuses on the semantic architecture on which the user interfaces for data analysis will be built. These interfaces are currently being developed and investigated, and will be reported elsewhere (see Sect. 8). The architecture is set up to cope with the requirements defined by the data challenges described in the previous section and the research questions the historians aim to address. Figure 2 depicts graphically our architecture in which the two aforementioned concerns – exact transcription on the left vs. interpretation on the right – are strictly separated. We will first motivate the adoption of RDF and semantic technologies and discuss some aspects of each concern. Details on the ontologies developed for this platform will be discussed in subsequent sections and build further upon the work reported in [3].

RDF and Linked Data principles were adopted for various reasons. RDF allows us to use a simple data model that facilitates the integration of internal and external data by creating links. Using RDF, the management of knowledge is scalable, and data access – for analysis, amongst others – is pushed closer

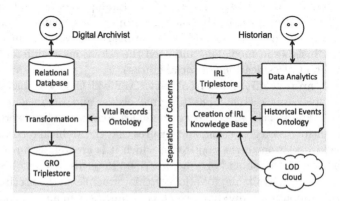

Fig. 2. The conceptual architecture of the IRL Linked Data platform. Transcription of register pages and the interpretation of the data are strictly separated.

to the user and application level by adopting the Linked Data principles (e.g., content negotiation) and the W3C SPARQL recommendation.

By reusing the existing HTTP infrastructure on which Linked Data is built, datasets that are behind firewalls can still link to other datasets in the Linked Data cloud. This allowed us to take a conservative approach by setting up our services behind a firewall and create (and exploit) outbound links; we thus benefit from all the Semantic Web technologies and the Linked Data cloud has to offer without violating our data sharing agreement and data protections legislations. Datasets relevant for this project that provide additional context include DBpedia [1] and Linked Logainm [13]. The latter is a Linked Data version of the authoritative bilingual database of Irish place names logainm.ie. Linked Logainm also provides links to places in DBpedia and geonames.org.

OWL 2 was adopted for the creation of the two ontologies allowing us to infer implicit information and rule languages were adopted to encode domain expert knowledge (historical, medical, etc.) to infer additional information that falls outside the capabilities of OWL.

There are four principles that Linked Data datasets should adhere to [2]: (1) use URIs as names for things; (2) use HTTP URIs so that people can look up those names; (3) provide information with standards (e.g., RDF) when URIs are looked up; and (4) include links to other URIs. Principles 1 to 3 are adhered to by both triplestores. The GRO triplestore provides links to other URIs within the same dataset to avoid interpretation and contextualization. The IRL triplestore links to external datasets to provide that contextualization. Since the datasets are behind a firewall, inbound links are not possible. Outbound links can be followed to discover more information. The authors are aware that the firewall can pose problems if one wishes to execute federated queries (across different datasets), but this has not yet been encountered within the context of this project.

For the platform, we adopted Jena TDB as triplestores and Jena Fuseki to provide the SPARQL endpoints.[3] Pubby is used to create a simple Linked Data frontend via those endpoints.[4] Details on the technologies adopted for the generation of RDF triples from the relational database and the transformation of triples for the interpretation of the data will be provided in the next sections.

4   Transcription of the Register Pages

We reiterate that the existing system the GRO has built took into account the attributes necessary to find records about individuals, thereby leaving out all fields on the register pages that were not relevant for this task. The digital archivists thus have the meticulous and laborious task of transcribing all that was captured on register pages, which is not merely transcribing those records, but also involves undertaking research and controlling the quality of what has been

---

[3] http://jena.apache.org/.
[4] http://wifo5-03.informatik.uni-mannheim.de/pubby/.

transcribed. Adopting Optical Character Recognition (OCR) was not possible as a very high level of precision in the transcription process was necessary.

In order to cope with the tension field of transcribing exactly what has been written down and the normalization of the data in some of these fields, a relational database has been set up that can capture in greater detail what can be observed on a register page. On death register pages, for instance, one can find a field "Certified Cause of Death and Duration of Illness". We observed variances in detail, which depended for instance on the registrar or on the informant (practitioner vs. relative). That field was sometimes used to indicate that the cause of death was uncertified. The database thus provided an additional field to indicate whether a death was explicitly certified, explicitly uncertified or neither. The duration of illness can be unknown or not applicable, e.g., in the case of drowning. The field can thus be NULL in case no information was provided.

Notes for each record and register page can be kept to capture anomalies or peculiarities such as signatures with a cross or crossed out information. As the project continues and the digital archivists transcribe register pages, these notes could be used as input for the creation of a controlled vocabulary for anomalies in register pages (see future work).

The database schema was developed in such a way that the data entered adheres to certain integrity constraints, thus effectively preventing certain errors. This relational database is then annotated with the Vital Records Ontology, presented in the next section, using D2RQ [5] and the generated triples are stored in a records triplestore.

5   Vital Records Ontology (VRO)

Births, deaths and marriages were captured per district (within a union, within a county) as single records on register pages. These pages can contain up to 10 records after which such a page is signed off by the registrar and sent to the superintendent registrar for inspection and validation. To create a first version of the Vital Records Ontology (VRO)[5], we "lifted" the information one could see on one such register page to an ontology.

To minimize interpretation, we choose to develop a "flat" ontology, which means that most information that can be found on such a register page was captured as literals. For example, instead of creating a concept Person that can have a forename and surname, we choose to relate the concept of a Record to these attributes. For the VRO, we thus defined a few concepts. A RegisterPage and a Record for representing the different types of records were declared. Each record must belong to a register page and each register page can have zero (which implies a blank pages) or more records. We make a distinction between a Certificate and a MarriageRecord, both of them being disjoint subclasses of the concept Record. The first has as a subject only one person and the latter two. The two concepts are disjoints, which makes that no instance of a certificate

---

[5] Available via http://purl.org/net/irish-record-linkage/records.

can be an instance of a marriage record and vice versa. Finally, we created two disjoint subclasses of the concept `Record`: `BirthRecord` and `DeathRecord`. The only object property, a relation between two concepts, we needed was to relate records to register pages. All other properties are datatype properties. Datatype properties are related to the greatest common denominator. For instance, all records are signed off by a registrar on a certain date. The date of registration as well as information on the registrar are therefore related to the concept of `Record` so that all subtypes of this class inherit this property.

One of the challenges is to capture the domain as well as possible, yet maintain a valid OWL 2 ontology. As explained by Motik and Horrocks in [14], it is difficult to reason about date and time intervals, and therefore only specific points in time (captured by both `xsd:dateTime` and `xsd:dateTimeStamp`) were "amenable for implementation" and those "can be handled by techniques similar to the ones for numbers." Together with the digital archivist, we choose not to capture dates mentioned in records as instances of `xsd:dateTime` as we do not know the exact times and we felt that encoding "default" times would not be in keeping with archival principles. We thus chose to declare the range of these properties as being `rdfs:Literal`, but provided transcription guidelines in which the use of `xsd:date` was to be highly encouraged.

One key requirement for Linked Data platforms in general is adequate identifiers. For our records knowledge base, we need to identify instances of records and register pages. Each register page and record is identified by a URI under the new subdomain http://irl.dri.ie/. Register pages are identified by a unique, physically stamped number provided by the GRO while digitizing. We use this stamp number for the creation of URIs identifying register pages. Individual records are identified by the combination of the stamp and entry-number. Figure 3 depicts the triples from a death record on a register page of a woman who died of paralysis in the year 1890.

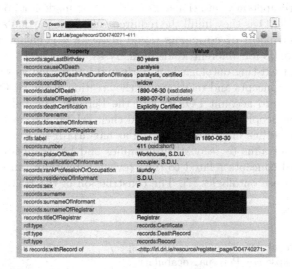

Fig. 3. Example of the triples from a death record in a register page.

6   Interpretation of the Register Pages and Records

We already described the importance of separating the information captured in the register pages and the interpretation thereof. The ontology that needs to support that kind of interpretation of the GRO data is more challenging given that the historians wishing to analyze the content are not necessarily familiar with ontology engineering and the knowledge base needs to support their activities, we adopted – reported in [3] – the approach proposed by Grüninger and Fox of having the stakeholders formulating *competency questions* [12]. The ontology must contain a necessary and sufficient set of axioms to represent and solve these questions [12]. These competency questions are not used to generate an ontology, but rather to evaluate it [11]. Using the types of queries the stakeholders wish to see answered, the knowledge engineers built an ontology, which was specifically tailored for the project, yet aimed to reuse existing, established vocabularies where possible. Competency questions formulated by historians included (paraphrased from [3]): "How many women died within n days after childbirth due to complications related to labor [...]?" and "What is the average sibship interval where the first child did not survive under various socio-economic conditions?" Those questions can be broken down in smaller competency questions such as: "Which infants died within the first 24 hours of their life?" and "What was the cause of death of a person?"

The questions were analyzed to identify the concepts and relations for the ontology, which were validated by the stakeholders. Graphical representations of the developed ontologies were used during discussions, e.g., as shown in Fig. 4. The VRO serves to reflect the historical records. Although it contains information about *events, people, places,* etc., the VRO does not capture these as *distinct entities.* However, to reconstitute families and analyze, we need distinct representations of events and persons involved. Therefore we developed the *Historical Events Ontology* (HEO) on top of the VRO as to provide a base ontology for answering the competency questions. The choice was made not to declare these concepts in the VRO as they fulfill the requirement of one particular set of tasks. This strict **separation of concerns** would allow for a **greater reuse of the historical records** for different kinds of analyses.

We looked at existing ontologies for reuse and integration as well as the creation of missing concepts and relations for the creation of the HEO. To describe people, we take into account FOAF[6] and the Persona Vocabulary[7]. Both are used to describe people, their activities and their relations to other people and objects. The latter has more relations such as hasChildren.

As the project aims to reconstitute families and health histories of people, we also included concepts related to time (events), relations, and reused available domain disease ontologies [7]. The construction of the HEO also included

---

[6] Friend-of-a-Friend: http://xmlns.com/foaf/spec/.

[7] http://wiki.eclipse.org/Persona_vocabulary.

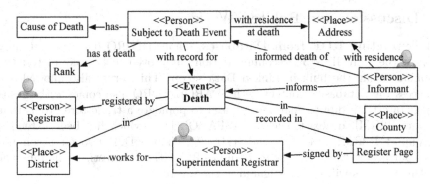

Fig. 4. Concepts and relations in the Historical Events Ontology for deaths.

formalizing information found in classification systems such as the International Statistical Classification of Diseases and Related Health Problems.[8]

Some of the concepts in the HEO are: `Person` for those involved during the event or registration; `Event` to capture the recorded births, deaths and marriages; `Place` for locations related to events or people; `CauseOfDeath` to facilitate reasoning and classifying causes of death; `Rank` for capturing the rank and occupation of involved persons; and `RegisterPage` to assure provenance. In a first instance, the data from the first triplestore is transformed to populate concepts and relations in the HEO by a series of SPARQL CONSTRUCT queries and SWRL rules. For instance, the following query allows us to create instances of the class `foaf:Person` from death records (prefixes omitted):

```
CONSTRUCT { ?new a foaf:Person; rdfs:seeAlso ?r;
                foaf:firstName ?f; foaf:familyName ?s.
} WHERE { ?r a rec:DeathRecord; rec:forename ?f; rec:surname ?s.
          BIND (URI(CONCAT(STR(?record),"/person")) AS ?new). }
```

Transforming graphs from the first knowledge base into the second leads to the creation of many persons. Matching techniques are adopted to identify the same persons across different vital records to assert `owl:sameAs` statements. This is an important as some names are very common and women adopted the name of their husband after marriage. Other fields (place, time) need to be taken into account to properly identify the same persons across records. When transforming the graphs from the first knowledge base into the second, many instances of persons are created. Another goal of the IRL platform is to add contextual information from other datasets [3]. We adopted Linked Logainm [13] for information on Irish place names and links with DBpedia resources.

---

[8] http://apps.who.int/classifications/icd10/browse/2010/en.

7   Discussion and Related Work

On Extracting RDF from Databases. Though D2RQ does not yet fully support the R2RML W3C recommendation[9], it proved to be easy to test the mappings using the built-in Linked Data server. This server also allowed one to access the database's content with SPARQL. D2RQ also comes with means to generate RDF dumps that can be used to populate a triplestore. Tools that support R2RML do exist, such as XSPARQL [4] – which has been extended to support R2RML, see [8] – and RML [9]. Though D2RQ so far accommodates our needs, "porting" the mapping to R2RML and investigate its different implementations will be investigated in the future.

On the Digitized Objects and the Transcriptions. We explained the reason why the Linked Data platform was placed behind a firewall in Sect. 3. Although not part of this project, one could investigate which subsets of the knowledge bases, and in particular the one containing historical events, do not violate the agreement and could be of benefit to the scientific community. The GRO also digitized the indexes for finding individual records. Indexes are currently not transcribed as they provide no additional information for our data analysis and individual records can be queried with SPARQL.

Important to consider in the future is the long-term preservation of the digitized objects and their RDF transcriptions. The Digital Repository of Ireland (DRI, http://www.dri.ie) is the national trusted digital repository for Ireland's social and cultural data. The DRI platform supports the ingestion of digitized objects and metadata, including Qualified Dublin Core (QDC) and Encoded Archival Description (EAD), and the configuration of access policies and licenses for these objects. Each object receives a Digital Object Identifier which will be referred to by the RDF transcription via, for instance, `rdfs:seeAlso` statements. We create an RDF file for each register page and related records by executing SPARQL DESCRIBE queries (an example with prefixes omitted is shown below). Those files are then used as input to create QDC files via an appropriate XSPARQL mapping.

```
DESCRIBE * WHERE { ?page r:stampNumber"4740271"; r:withRecord ?record. }
```

On Ontology Engineering. The digital archivists keep track of any anomalies or peculiarities in the register pages and individual records in a notes field in the database. Examples of anomalies include strikethroughs in fields or the occurrence of crosses where signatures are necessary. The first *could* indicate a correction or removal of information, the latter could indicate an illiterate person. We carefully chose to use the verb "could" as these are historical vital records and we should not give an interpretation to these anomalies when we are not sure. Depending on the nature of these anomalies and their frequency, we could consider using these for the creation of a controlled vocabulary; allowing one to look up these anomalies and decide how to interpret them. This vocabulary, captured as an ontology, would then reside next to the VRO.

---

[9] http://www.w3.org/TR/r2rml/.

8   Conclusions and Future Work

We reported on the creation of the semantic architecture, the ontologies and knowledge bases of the IRL Linked Data platform. Taking into account the requirements of both the digital archivists (archival authenticity, preservation, cataloguing and data protection) and the historians (answering their research questions), the Linked Data platform is comprised of two distinct knowledge bases, each supported by a different ontology, to separate those two concerns: the Vital Records Ontology for the exact transcription of the historical vital records and register pages, and the Historical Events Ontology for an interpretation of the register pages. The creation of the first was fairly straightforward and primarily the result of a collaboration between the knowledge engineers and digital archivists. The latter also involved the historians who were asked to formulate competency questions to identify concepts and relations. Reasoning provides one motivation for adopting semantic technologies. The second is the creation of links with other datasets providing additional context to interpret the data. As the transcription of register pages is a laborious process, the latter can only be meaningfully evaluated when we have an adequate number of transcriptions.

The lessons learned in this study arise from the value of the separation of concerns. Though digital archivists could have elicited facts from the register pages immediately and solely fit for answering the competency questions in this project, the resulting dataset would have had limited value for reuse and future research questions. We argue that the return in value justified the extra overhead in terms of transcription and platform complexity. Our approach is thus different from, for instance, the Dacura platform [10], which adopts crowdsourcing techniques to elicit facts from datasets such as newspaper articles according to a schema for a particular purpose.

Future work that we will prioritize will be the ingestion of the digitized images and their RDF in a long-term preservation platform according to best practices and standards and the investigation to what extent parts of the knowledge bases can be made available to the public without revealing the details of individuals. Finally, this paper focused on the semantic architecture upon which applications can be built. The user interfaces which will aid the historians in answering their research questions built on top of the semantic architecture are still being investigated and will be reported elsewhere.

Acknowledgements. We thank the Registrar General of Ireland for permitting us to use this rich digital content contained in the vital records for the purposes of this research project. This publication has emanated from research conducted within the Irish Record Linkage, 1864-1913 project supported by the RPG2013-3; Irish Research Council Interdisciplinary Research Project Grant, and within the Science Foundation Ireland Funded Insight Research Centre (SFI/12/RC/2289). The Digital Repository of Ireland (formerly NAVR) gratefully acknowledges funding from the Irish HEA PRTLI programme.

References

1. Auer, S., Bizer, C., Kobilarov, G., Lehmann, J., Cyganiak, R., Ives, Z.G.: DBpedia: a nucleus for a web of open data. In: Aberer, K., et al. (eds.) ASWC 2007 and ISWC 2007. LNCS, vol. 4825, pp. 722–735. Springer, Heidelberg (2007)
2. Berners-Lee, T.: Linked Data - Design Issues (2006). http://www.w3.org/DesignIssues/LinkedData.html. Accessed 7 June 2015
3. Beyan, O., Breathnach, C., Collins, S., Debruyne, C., Decker, S., Grant, D., Grant, R., Gurrin, B.: Towards linked vital registration data for reconstituting families and creating longitudinal health histories. In: KR4HC Workshop (in conjunction with KR 2014), pp. 181–187 (2014)
4. Bischof, S., Decker, S., Krennwallner, T., Lopes, N., Polleres, A.: Mapping between RDF and XML with XSPARQL. J. Data Semantics **1**(3), 147–185 (2012)
5. Bizer, C.: D2R MAP - a database to RDF mapping language. In: King, I., Máray, T. (eds.) Proceedings of the Twelfth International World Wide Web Conference - Posters, WWW 2003, Budapest, Hungary, May 20–24, 2003 (2003)
6. Bizer, C., Heath, T., Berners-Lee, T.: Linked data - the story so far. Int. J. Semantic Web Inf. Syst. **5**(3), 1–22 (2009)
7. Bodenreider, O.: Disease ontology. In: Dubitzky, W., Wolkenhauer, O., Cho, K., Yokota, H. (eds.) Encyclopedia of Systems Biology, pp. 578–581. Springer, New York (2013)
8. Dell'Aglio, D., Polleres, A., Lopes, N., Bischof, S.: Querying the web of data with XSPARQL 1.1. In: Verborgh, R., Mannens, E. (eds.) Proceedings of the ISWC Developers Workshop 2014, co-located with the 13th International Semantic Web Conference (ISWC 2014), Riva del Garda, Italy, October 19, 2014. CEUR Workshop Proceedings, vol. 1268, pp. 113–118. CEUR-WS.org (2014)
9. Dimou, A., Vander Sande, M., Colpaert, P., Verborgh, R., Mannens, E., Van de Walle, R.: RML: A generic language for integrated rdf mappings of heterogeneous data. In: Bizer, C., Heath, T., Auer, S., Berners-Lee, T. (eds.) Proceedings of the Workshop on Linked Data on the Web co-located with the 23rd International World Wide Web Conference (WWW 2014), Seoul, Korea, April 8, 2014. CEUR Workshop Proceedings, vol. 1184. CEUR-WS.org (2014)
10. Feeney, K.C., O'Sullivan, D., Tai, W., Brennan, R.: Improving curated web-data quality with structured harvesting and assessment. Int. J. Semantic Web Inf. Syst. **10**(2), 35–62 (2014). http://dx.doi.org/10.4018/ijswis.2014040103
11. Fox, M.S., Gruninger, M.: Enterprise modeling. AI magazine **19**(3), 109–121 (1998)
12. Grüninger, M., Fox, M.S.: The role of competency questions in enterprise engineering. In: Rolstadås, A. (ed.) Benchmarking Theory and Practice, pp. 22–31. Springer, New York (1995)
13. Lopes, N., Grant, R., Raghallaigh, B.Ó., Carragáin, E.Ó., Collins, S., Decker, S.: Linked logainm: enhancing library metadata using linked data of Irish place names. In: Bolikowski, Ł., Casarosa, V., Goodale, P., Houssos, N., Manghi, P., Schirrwagen, J. (eds.) TPDL 2013. CCIS, vol. 416, pp. 65–76. Springer, Heidelberg (2014)
14. Motik, B., Horrocks, I.: OWL datatypes: design and implementation. In: Sheth, A.P., Staab, S., Dean, M., Paolucci, M., Maynard, D., Finin, T., Thirunarayan, K. (eds.) ISWC 2008. LNCS, vol. 5318, pp. 307–322. Springer, Heidelberg (2008)

Keywords-To-SPARQL Translation for RDF Data Search and Exploration

Katerina Gkirtzou[1][(✉)], Kostis Karozos[2], Vasilis Vassalos[2],
and Theodore Dalamagas[1]

[1] "Athena" Research Center, GR, Maroussi, Greece
{kgkirtzou,dalamag}@imis.athena-innovation.gr
[2] Athens University of Economics and Business, GR, Athens, Greece

Abstract. Linked Data is the most common practice for publishing and sharing information in the Data Web. As new data become available, their exploration is a fundamental step towards integration and interoperability. However, typical search methods as SPARQL queries require knowing both the SPARQL syntax and the vocabulary used in the data. For this reason, keyword-based search has been proposed, allowing an intuitive way for searching an RDF dataset. In this paper, we present a novel approach for keyword search on graph-structured data, and in particular temporal RDF graph, i.e. RDF data that involve temporal properties. Our method, instead of providing answers directly from the RDF data graph, automatically generates a set of candidate SPARQL queries that try to capture users information need as expressed by the keywords used. To support temporal exploration, our method is enriched with temporal operators allowing the user to explore data within predefined time ranges. To evaluate our approach, we perform an effectiveness study using two real-world datasets.

Keywords: Keyword search · Graph data · RDF · Linked Data

1 Introduction

More and more corporate, scientific, governmental and user-generated datasets break the walls of traditional "private" management within their production site, are published and become available for potential data consumers. The Linked Data (LD) is the most common practice for publishing, sharing and managing information in the Data Web, offering new ways of data integration and interoperability. The main concept in LD is that all resources published on the Web are uniquely identified by a *Uniform Resource Identifier* (URI), and typed links between URIs, also identified by URIs, are used to semantically connect resources[1]. LD is implemented with the RDF[2] technology: (*a*) RDF is used for the representation and modeling of structured and semi-structured data on the Web and (*b*) RDF links are used to interlink data from different data sources.

---

[1] http://www.w3.org/TR/ld-bp/.
[2] http://www.w3.org/RDF/.

© Springer International Publishing Switzerland 2015
S. Kapidakis et al. (Eds.): TPDL 2015, LNCS 9316, pp. 111–123, 2015.
DOI: 10.1007/978-3-319-24592-8_9

The RDF representation is a set of statements about resources. Such statements are known as triples. A triple is an expression of the form *subject predicate object*. The subject refers to a resource to be described, the predicate is usually a term from existing vocabularies, while the object can either be a literal (i.e., string) or another resource. A set of RDF triples can also be represented by a directed labelled graph, known as the RDF data graph. The most common way for searching an RDF data graph is the SPARQL query language [9]. SPARQL queries involve conjunctions and disjunctions of triple patterns which are matched on the RDF data graph. However, the use of SPARQL requires, apart from knowing its syntax, also the knowledge of the RDF schema used to model the data. For this reason, keyword-based search has been proposed in bibliography, allowing an intuitive way for searching an RDF dataset.

In this paper, we present a novel approach for keyword search on graph-structured data, and in particular RDF graph. Our method, instead of providing answers directly from the RDF data graph, generates automatically a set of candidate SPARQL queries that try to capture users information needs as expressed by the keywords used. Our approached is tailored to temporal RDF data, i.e. RDF data that involve temporal properties. To this end, our method is enriched with temporal operators allowing the user to explore data within predefined time ranges. To evaluate our approach, we perform an effectiveness study using two real-world datasets.

2   Problem Definition

We assume that the user wants to explore of a dataset using a *user query language* $\mathcal{Q}_U$, while the system supports queries only in a specified *system query language* $\mathcal{Q}_S$ different from $\mathcal{Q}_U$. To deal with this case, a transformation from $\mathcal{Q}_U$ to $\mathcal{Q}_S$ is required. We focus on datasets modeled in RDF graph form, defined as follows:

Definition 1. *An RDF data graph G is a tuple $(V, E, \mathcal{L})$ where:*

- *V is a finite set of vertices. V is defined as the disjoint union $V_E \cup V_C \cup V_V$, where V_E is a set of vertices representing RDF entities, V_C is a set of vertices representing RDF classes and V_V is a set of vertices representing literals.*
- *$E \subseteq V \times V$ is a finite set of order pairs (v_1, v_2), called edges, where $v_1, v_2 \in V$.*
- *$\mathcal{L} : \{V \cup E\} \rightarrow L$ is a function assigning $\forall v \in V$ and $\forall (v_1, v_2) \in E$ a label from alphabet L. The alphabet L is defined as the disjoint union $L_E \cup L_C \cup L_V \cup L_R \cup L_A \cup \{type, subclass\}$, where L_E is a set of labels for the RDF entities, L_C is a set of labels for the RDF classes, L_V is a set of labels for the literals, L_R is a set of labels for the inter-entities properties, L_A is a set of labels for the entity-to-attribute properties. The following restrictions apply:*

  - *$\mathcal{L}(v) \in L_E$ if and only if $v \in V_E$,*
  - *$\mathcal{L}(v) \in L_C$ if and only if $v \in V_C$,*
  - *$\mathcal{L}(v) \in L_V$ if and only if $v \in V_V$,*
  - *$\mathcal{L}((v_1, v_2)) \in L_R$ if and only if $v_1, v_2 \in V_E$,*
  - *$\mathcal{L}((v_1, v_2)) \in L_A$ if and only if $v_1 \in V_E$ and $v_2 \in V_V$,*

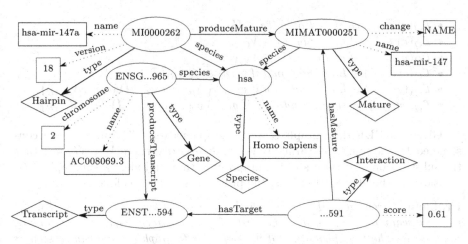

Fig. 1. An RDF subgraph from the DIANA dataset. If $v \in V_E$ then it has oval shape, if $v \in V_C$ a diamond shape and if $v \in V_V$ a rectangle shape. Also, dotted edges represent entity-to-attribute properties, while solid edges represent inter-entities properties.

- $\mathcal{L}((v_1, v_2)) = type$ *if and only if* $v_1 \in V_E$ *and* $v_2 \in V_C$,
- $\mathcal{L}((v_1, v_2)) = subclass$ *if and only if* $v_1, v_2 \in V_C$.

We consider two predefined types of edges, *type* and *subclass*, that have a special interpretation within RDF schema. The former states that an RDF entity is an instance of an RDF class, while the latter states class hierarchy [3]. The URIs of RDF entities define the set L_E, while the URIs of RDF class and RDF properties define the sets L_C and $L_R \cup L_A$, respectively. Finally, all literal data values define L_V. Figure 1 shows an RDF data subgraph for the DIANA dataset (see Sect. 6). The oval shape vertex MI0000262 represents an RDF entity of the RDF class Hairpin. The entity's name is hsa-mir-147a and it is related with the entity MIMAT0000251 of the RDF class Mature via the property producesMature.

In our scenario, the user query language $\mathcal{Q}_U$ is defined as a set of keywords $\mathcal{K} = \{k_1, k_2, \ldots, k_n\}$, while the system supports a query language $\mathcal{Q}_S$ that produces conjuctive SPARQL queries. These queries can also be viewed as graph patterns defined as follows:

Definition 2. *A graph pattern* G_Q *on an RDF data graph* $G = (V, E, \mathcal{L})$ *is a tuple* $(V_Q, E_Q, \mathcal{L}_Q)$ *where:*

- V_Q *is a finite set of vertices.* V_Q *is defined as the disjoint union* $V_{VAR} \cup V_{CL} \cup V_{VAL}$, *where* V_{VAR} *is a set of vertices representing variables,* $V_{CL} \subseteq V_C$ *is a set of vertices representing RDF classes and* $V_{VAL} \subseteq V_V$ *is a set of vertices representing literals.*
- $E_Q \subseteq V_Q \times V_Q$ *is a finite set of order pairs* (v_1, v_2), *where* $v_1, v_2 \in V_Q$.
- $\mathcal{L}_Q : \{V_Q \cup E_Q\} \rightarrow L_Q$ *is a function assigning* $\forall v \in V_Q$ *and* $\forall (v_1, v_2) \in E_Q$ *a label from alphabet* L_Q. *The alphabet* L_Q *is defined as the disjoint union* $L_{VAR} \cup L_C \cup L_V \cup L_R \cup L_A \cup \{type\}$, *where* L_{VAR} *is a set of labels for the variables. The following restrictions apply:*

- $\mathcal{L}_Q(v) \in L_{VAR}$ if and only if $v \in V_{VAR}$,
- $\mathcal{L}_Q(v) \in L_C$ if and only if $v \in V_{CL}$,
- $\mathcal{L}_Q(v) \in L_V$ if and only if $v \in V_{VAL}$,
- $\mathcal{L}_Q((v_1, v_2)) \in L_R$ if and only if $v_1, v_2 \in V_{VAR}$,
- $\mathcal{L}_Q((v_1, v_2)) \in L_A$ if and only if $v_1 \in V_{VAR}$ and $v_2 \in \{V_{VAL} \cup V_{VAR}\}$,
- $\mathcal{L}_Q((v_1, v_2)) = type$ if and only if $v_1 \in V_{VAR}$ and $v_2 \in \{V_{CL} \cup V_{VAR}\}$.

Given an RDF data graph G and a graph pattern G_Q, answers are constructed by mapping the variables $V_{VAR} \in G_Q$ to vertices $V \in G$ such that the substitution of variables in the graph pattern would yield a subgraph of G. Considering the above definitions, the problem is defined as follows:

Problem. *Given an RDF data graph G and a set of keywords $\mathcal{K} = \{k_1, k_2, \ldots, k_n\}$, we want to compute a ranked list of graph patterns G_Q given a ranking function R. The transformation of the keywords to graph pattern can be seen as a mapping fuction $\mu(\mathcal{K}) = G_Q$, where each keyword k_i will match to $L \setminus L_E$.*

3   Indexing RDF Graph Data

To assist the construction of graph patterns G_Q for a given set of keywords, we use two types of indices: (i) a term index and (ii) a schema-guide graph.

3.1   Term Index

A term index, given a keyword k_i, returns a set of matches from the RDF data graph G, $\mathcal{M}_i = \{m | m \in G = (V, E, \mathcal{L})\}$, along with other necessary information to assist query formation. More specifically, if $m \in L_A$, i.e. matches to a label of an entity-to-attribute property, then in the RDF data graph G exists at least one $e \in V_E, v \in V_V$ and $c \in V_C$ such that $(e, v) \in E$ with $\mathcal{L}((e, v)) = m$ and $(e, c) \in E$ with $\mathcal{L}((e, c)) = type$. In the index, we also keep the labels of the RDF classes c of the entities e that are subjects to the entity-to-attribute property with label m. Similarly, if $m \in L_R$, i.e. matches to a label of an inter-entities property, then in G exists at least one $e_s, e_o \in V_E, e_s \neq e_o$ and $c_s, c_o \in V_C$ such that $(e_s, e_o) \in E$ with $\mathcal{L}((e_s, e_o)) = m$, $(e_s, c_s) \in E$ with $\mathcal{L}((e_s, c_s)) = type$ and $(e_o, c_o) \in E$ with $\mathcal{L}((e_o, c_o)) = type$. In the index, we also keep the labels of the pairs of the RDF classes $\langle c_s, c_o \rangle$ of the entities e_s, e_o that are subjects and objects of the inter-entities property with label m, respectively. Finally, if $m \in V_V$, i.e. a literal, then in the RDF graph G exists at least one $e \in V_E, c \in V_C$ such that $a = (e, m) \in E$ and $(e, c) \in E$ with $\mathcal{L}((e, c)) = type$. It is possible that the same literal is met in multiple entities e of the same or even different classes c under both the same property or different properties a. To keep the index minimal, we keep one entry per property a for a given literal m, grouping only the labels of the RDF classes c. Table 1 shows analytically the information provided by the term index depending on the type of the matched element m.

Table 1. The information of the term index by the type of the matched element m.

Matched element	Term Index Structure	Notation
$m \in L_A$	$[m, \{\mathcal{L}(c_1), \ldots, \mathcal{L}(c_n)\}]$	$c_i \in V_C$ and $\mathcal{L}(c_i) \in L_C$
$m \in L_R$	$[m, \{\langle \mathcal{L}(c_{s_1}), \mathcal{L}(c_{o_1})\rangle, \ldots, \langle \mathcal{L}(c_{s_n}),$ $\mathcal{L}(c_{o_n})\rangle\}]$	$c_{s_i}, c_{o_i} \in V_C$ and $\mathcal{L}(c_{s_i})$, $\mathcal{L}(c_{o_i}) \in L_C$
$m \in V_V$	$[m, \mathcal{L}(a), \{\mathcal{L}(c_1), \ldots, \mathcal{L}(c_n)\}]$	$\mathcal{L}(a) \in L_A, c_i \in V_C$ and $\mathcal{L}(c_i) \in L_C$

3.2 Schema-Guide Graph

The schema-guide graph is an aggregated representation of the RDF data graph. It is used to guide the query computation process and it is defined as follows:

Definition 3. *A schema-guide graph G_{SCM} of an RDF graph $G = (V, E, \mathcal{L})$ is a tuple $(V_{SCM}, E_{SCM}, \mathcal{L}_{SCM})$ where:*

- *V_{SCM} is a finite set of vertices defined as the disjoint union $V_C \cup \{Thing\}$,*
- *$E_{SCM} \subseteq V_{SCM} \times V_{SCM}$ is a finite set of order pairs (v_1, v_2), where $v_1, v_2 \in V_{SCM}$ and*
- *$\mathcal{L}_{SCM} : \{V_{SCM} \cup E_{SCM}\} \to L_{SCM}$ is a function assigning $\forall v \in V_{SCM}$ and $\forall e \in E_{SCM}$ a label from alphabet L_{SCM}, which is defined as the disjoint union $L_C \cup L_R \cup \{Thing, subclass\}$. The following restrictions apply:*
  - *$\mathcal{L}_{SCM}(v) \in L_C$ if and only if $v \in V_C$,*
  - *$\mathcal{L}_{SCM}(v) = Thing$ if and only if $v = Thing$,*
  - *$\mathcal{L}_{SCM}((v_1, v_2)) \in \{L_R, subclass\}$ if and only if $v_1, v_2 \in V_C$.*

Every vertex $c \in V_C$ of the schema-guide graph G_{SCM} represents all the entities vertices $e \in V_E \subset V$ of the RDF data graph G that have RDF type c and $\mathcal{L}_{SCM}(c) \in L_C$. On the other hand, vertex $Thing \in V_{SCM}$ with $\mathcal{L}_{SCM}(Thing) = Thing$ represents all entities vertices $e \in V_E \subset V$ of the RDF data graph G that have no given type. Similarly, an edge $(c_1, c_2) \in E_{SCM}$ of the schema-guide

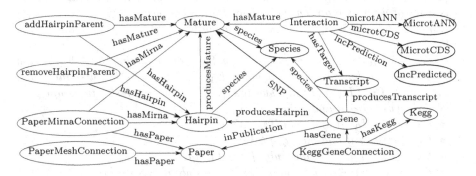

Fig. 2. The schema-guide graph of DIANA dataset.

graph G_{SCM} represents all edges $(e_1, e_2) \in E$ where $e_1, e_2 \in V_E$ of the data graph G if and only if $(e_1, c_1) \in E$ with $\mathcal{L}((e_1, c_1)) = type$ and $(e_2, c_2) \in E$ with $\mathcal{L}((e_2, c_2)) = type$. In this case, $\mathcal{L}_{SCM}((c_1, c_2) = \mathcal{L}((e_1, e_2)) \in L_R$. Finally, the edge $(c_1, c_2) \in E_{SCM}$ with $\mathcal{L}((c_2, c_2)) = subclass$ represents class hierarchy.

Figure 2 shows an example of the schema-guide graph for the DIANA dataset. Note that vertices here represent all RDF entities from the RDF graph of a specific type rather than specific RDF entity. For example vertex MI0000262 in Fig. 1 is represented by vertex `Hairpin` in Fig. 2. Similarly, the property `producesMature` between the RDF entities MI0000262 and MIMAT0000251 in Fig. 1 is represented by the abstract edge `producesMature` between vertex `Hairpin` and vertex `Mature` in Fig. 2.

4   Query Pattern Graph

To compute the graph patterns as a response to use keyword-based queries, we perform the following steps: (1) for each keyword $k_i \in \mathcal{K}$ we retrieve all its matches $\mathcal{M}_i$ on the RDF data graph, (2) we calculate all possible combinations of the matched elements $\mathcal{C} = \mathcal{M}_1 \times \ldots \times \mathcal{M}_n = \{c = (m_1, \ldots, m_n) | m_i \in \mathcal{M}_i, \forall i = 1, \ldots, n\}$, ($3$) for each combination $c = (m_1, \ldots, m_n) \in \mathcal{C}$ that contains one matched element m_i per keyword k_i, we create an augmented schema-guide graph G_{AUG}, (4) for each augmented schema-guide graph G_{AUG}, we generate the graph query pattern G_{QP} which is used to form a SPARQL query and (5) we rank each query pattern graph G_{QP} based on a ranking function $\mathcal{R}$.

4.1   Augmented Schema-Guide Graph

The augmented schema-guide graph is used as a data guide for the query pattern formation and it is defined as follows:

Definition 4. *An augmented schema-guide graph G_{AUG} of an RDF data graph $G = (V, E, \mathcal{L})$ is a tuple $(V_{AUG}, E_{AUG}, \mathcal{L}_{AUG})$ where:*

- *V_{AUG} is a finite set of vertices defined as the disjoint union $V_C \cup V_{VAL} \cup V_U \cup \{Thing\}$, where $V_{VAL} \subseteq V_V$ is a set of vertices representing literals and V_U is a set of vertices representing unknown literal values,*
- *$E_{AUG} \subseteq V_{AUG} \times V_{AUG}$ is a finite set of order pairs (v_1, v_2), where $v_1, v_2 \in V_{AUG}$ and*
- *$\mathcal{L}_{AUG} : \{V_{AUG} \cup E_{AUG}\} \to L_{AUG}$ is a function assigning $\forall v \in V_{AUG}$ and $\forall e \in E_{AUG}$ a label from an alphabet L_{AUG}, defined as the disjoint union $L_C \cup L_V \cup L_R \cup L_A \cup \{Thing, subclass\}$. The following restrictions apply:*
  - *$\mathcal{L}_{AUG}(v) \in L_C$ if and only if $v \in V_C$,*
  - *$\mathcal{L}_{AUG}(v) \in L_V$ if and only if $v \in V_{VAL}$,*
  - *$\mathcal{L}_{AUG}(v) = \emptyset$ if and only if $v \in V_U$,*
  - *$\mathcal{L}_{AUG}(v) = Thing$ if and only if $v = Thing$,*
  - *$\mathcal{L}_{AUG}((v_1, v_2)) \in \{L_R \cup subclass\}$ if and only if $v_1, v_2 \in V_C$,*
  - *$\mathcal{L}_{AUG}((v_1, v_2)) \in L_A$ if and only if $v_1 \in V_C$ and $v_2 \in \{V_{VAL} \cup V_U\}$.*

Given a set $(m_1, \ldots, m_n) \in \mathcal{C}$, we construct the augmented schema-guide graph G_{AUG} from the schema-guide graph G_{SCM} as follows:

- If $m_i \in L_V$, add an edge (c, v_{m_i}) and a vertex v_{m_i} with $\mathcal{L}_{AUG}(v_{m_i}) = m_i$. The newly inserted edge (c, m_i) will be attached to the proper $c \in V_C \cup Thing$ vertex and labelled $\mathcal{L}_{AUG}(c, v_{m_i}) = \mathcal{L}(a)$, according to the information provided by term index for m_i.
- If $m_i \in L_A$, add an edge (c, v) and a vertex v. The newly inserted edge (c, v) will be attached to the proper $c \in V_C \cup \{Thing\}$ vertex and will be labelled as $\mathcal{L}_{AUG}(c, v) = \mathcal{L}(c) \in L_A$, while the newly inserted vertex $v \in V_U$ will not be labelled.
- If $m_i \in L_R$ or $m_i \in L_C$, do nothing as they are already part of the schema-guide graph G_{SCM}.

4.2   Query Pattern Graph Formulation

Definition 5. *The Query Pattern Graph G_{QP} is the minimal connected subgraph of the augmented schema-guide graph G_{AUG} such that it includes all the matched elements $m_i \in c$ and there exists no other query pattern graph G'_{QP} such that $C(G'_{QP}) < C(G_{QP})$ for a given cost function C.*

In our framework, we define the cost function as the average pairwise distance between the matched elements m_i, in other words $C(G_{QP}) = \sum \text{dist}(m_i, m_j)$ $\forall i, j = 1, \ldots, n$ and $i \neq j$. In order to minimize the cost function for every pair of matched elements (m_i, m_j) $i \neq j$, we calculate their shortest path in G_{AUG}. Note that during the shortest path calculations we ignore the directionality of the edges. Moreover, since a matched element m_i, i.e. a source or sink of the shortest path algorithm, can also be an edge, then the distance between two matched elements counts the number of both vertices and edges that needs to traverse across the augmented summary graph G_{AUG}. Finally, we combine all the pairwise shortest path forming a connected subgraph, the Graph Query Pattern G_{QP}. Note that the G_{QP} is a "compressed" form of graph pattern G_Q. Each node $v \in \{V_C \cup \{Thing\}\}$ from G_{QP} is a hypernode that corresponds to a single subtree $G'_Q \subset G_Q$ which contains two vertices $v_1, v_2 \in G'_Q$, where $v_1 \in V_{VAR}$ and $v_2 \in V_C$ connected with an edge (v_1, v_2) with label *type*. Any edge e initially attached to hypernode v can be attached to vertex v_1.

4.3   Query Mapping

The next step is to translate the query pattern graph G_{QP} into a SPARQL query. Remember that $G_{QP} = (V_{QP}, E_{QP}, \mathcal{L}_{QP})$ is a subgraph of the augmented schema-guide graph G_{AUG} and it is also a compressed form of a graph pattern G_Q, thus some of its elements have known fixed values and some need to be associated with variables. More specifically, vertices $v \in V_C \cup V_U$ represent generic RDF elements and need to be associated with variables. Note also that the labels of the vertices can be used as constants in the triple patterns, while the labels of the edges as predicates. Given these observations, to produce conjunctive SPARQL queries for every graph element $\in G_{QP}$ we perform:

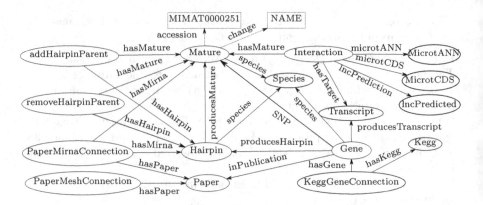

Fig. 3. One of the 6 possible Augmented schema-guide graph for the keywords *MIMAT0000251, name* and *hasTarget* in the DIANA dataset.

- if $v \in \{V_U \cup V_{VAL}\}$, then associate the vertex v with a new variable $\mathbf{var}(v)$,
- if $v \in V_C$, then associate the vertex v with a new variable $\mathbf{var}(v)$ and produce the triple pattern $\mathbf{var}(v)$ $\mathbf{rdf:type}$ $\mathcal{L}_{QP}(v)$, where $\mathcal{L}_{QP}(v) \in L_C$.
- if $(s,o) \in E$ from vertex $s \in V_C$ to vertex $o \in V_C$ represents an interentities property where $\mathcal{L}_{QP}((s,o)) \in L_R$, then produce the triple pattern $\mathbf{var}(s)$ $\mathcal{L}_{QP}(e)$ $\mathbf{var}(o)$,
- if $(s,o) \in E$ from vertex $s \in V_C$ to vertex $o \in V_{VAL}$ represents an entity-to-attribute property where $\mathcal{L}_{QP}(o) \in L_V$ and $\mathcal{L}_{QP}((s,o)) \in L_A$, then produce the triple pattern $\mathbf{var}(s)$ $\mathcal{L}_{QP}((s,o))$ $\mathbf{var}(o)$. $\mathbf{FILTER}(\mathbf{var}(o) = \mathcal{L}_{QP}(o))$ and
- if $(s,o) \in E$ from vertex $s \in V_C$ to vertex $o \in V_U$ represents an entity-to-attribute property, such that $\mathcal{L}_{QP}((s,o)) \in L_A$, then produce the triple pattern $\mathbf{var}(s)$ $\mathcal{L}_{QP}((s,o))$ $\mathbf{var}(o)$.

Finally, all produced queries are ranked based on a given ranking function $\mathcal{R}$. In our framework, we provide three different ranking functions (a) the number of triplet patterns, (b) the average shortest path distance and (c) the longest shortest path distance. The former works on the SPARQL form of the generated query, while the latter two work on the query pattern graph G_{QP}. Note that the smaller the score of the ranking function $\mathcal{R}$ for a given query, the higher the ranking position. The idea behind this is based on the assumption known as "Locality of Information", meaning that the information required by the user can be modelled in terms of entities which are closely related [11].

4.4   Example

Let's assume that the user is interested in exploring the DIANA dataset and has provided the following as keywords: *MIMAT0000251, name, hasTarget*. Given the subgraph depicted in Fig. 1, the keywords match to the following elements: (a) *MIMAT0000251* to the literal MIMAT0000251 that is connected via the property accession met only with RDF entities of Mature type, (b) *name* to the literal NAME that is connected via the property change met with RDF entities of type either Mature and to the entity-to-attribute property name met with entities of type Hairpin, Mature, Species and Gene, resulting in 5 possible matches, (c) hasTarget to the inter-entities property hasTarget met with subject of type Interaction and object of type Transcript.

Overall we have $1 \times 5 \times 1 = 5$ possible combinations of the keywords to matched elements, that would result to 5 possible queries. Let's consider one combination, where the *name* keyword matches to the literal "NAME". Figure 3 shows the augmented schema-guide graph G_{AUG} for this combination. From this G_{AUG} we calculate the shortest paths between all pairs of matched elements and since we have 3 keywords, we need to calculate of $\binom{3}{2} = 3$ shortest paths. We then combine them into a single connected component, generating the Query Pattern graph G_{QP} shown in Fig. 4. Note that the extra node Transcript is attached to the property hasTarget in order to form a complete triple pattern, although it is not part of neither of the previous calculated shortest paths. Finally, the G_{QP} is mapped to the SPARQL query shown in Fig. 5.

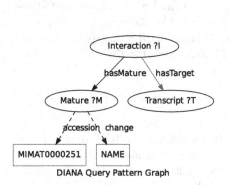

DIANA Query Pattern Graph

```
SELECT ?I ?M ?T WHERE
{
    ?I a diana:Interaction.
    ?M a diana:Mature.
    ?T a diana:Transcript.
    ?I diana:hasMature ?M.
    ?I diana:hasTarget ?T.
    ?M diana:accession ?a.
    FILTER( ?a = "MIMAT0000251").
    ?M diana:change ?n.
    FILTER(?n = "NAME").
}
```

Fig. 4. The Query Pattern Graph extrapolated from the Augmented Schema-Guide Graph of Fig. 3. In red we depict the matched elements (Color figure online).

Fig. 5. The generated SPARQL query from the Query Pattern Graph of Fig. 4.

5 Temporal Operators

When working with diachronic data, querying should also involve temporal constraints. In our method, we support the following three temporal operators: (a) at (b) before and (c) after. The first one can be used to retrieve data at a specific time point, while the other two can be used to define a time window constraint. The temporal operators are used as follows: property operator:value, where property is the temporal entity-to-attribute property that the selected

Table 2. Aggregated statitics for the datasets of AI4B and DIANA.

Dataset	# Triples	# Classes	# Properties	# Unique String Values
AI4B	$2,7 \times 10^6$	15	148	6.350
DIANA	$4,6 \times 10^9$	16	76	613.408

temporal operator will be applied and value is a desired value. For example, if the user provides as input *"paper let7a year after:2006"*, she wants to retrieve publications related with miRNAs named "let7a" published after 2006.

When a temporal operator is used in a query, we consider the selected temporal property as an extra keyword and proceed to the construction of the query pattern graph G_{QP} and its mapping to SPARQL as described in Sect. 4. There are only two small differences. The first one is that when we construct the augmented schema-guide graph G_{AUG}, we add, apart from the temporal property itself, an extra node representing the desired value provided by the user. The second difference is that when we translate the temporal entity-to-attribute property (recall that it will be represented as an edge (s, o) within G_{QP} from vertex $s \in V_C$ to vertex o that represents the selected value) we produce the following triplet var(s) $\mathcal{L}_{QP}(e)$ var(o). followed by (a) FILTER(var(o) $= \mathcal{L}_{QP}(o)$), when the at temporal operator is used, (b) FILTER(var(o) $\leq \mathcal{L}_{QP}(o)$) when the before temporal operator is used and (c) FILTER(var(o) $\geq \mathcal{L}_{QP}(o)$), when the after temporal operator is used.

6   Evaluation

We evaluated our proposed keyword search method using two RDF datasets, the AI4B[3] and the DIANA[4]. The AI4B dataset contains information about biomass products, while the diachronic DIANA dataset contains aggregated information of the miRNA world from well-known biology databases that change and evolve throughout their lifespan. Table 2 shows detailed information about the characteristics of the two datasets. The implementation of the presented method for the DIANA dataset is available at http://snf-624527.vm.okeanos.grnet.gr:8080/KeywordSearchDiana/. Finally, our approach has also been incorporated in the collaborative platform LinkZoo [7].

To evaluate our approach we perform an effectiveness study. We have asked our collaborators to provide keyword queries along with a natural language description of the required information. We have aggregated 20 queries in total, 15 for the DIANA (Q1-Q15) and 5 for the AI4B (Q16-Q20). An example query is *"'Alzheimer's disease' mature version at:18"* and the corresponding description is *"Retrieve all mature miRNAs of miRBase version 18 that are related with Alzheirmer's disease"*. All queries used in the evaluation can be found at https://web.imis.athena-innovation.gr/redmine/projects/lodgov/wiki/Deliverable2_4_Evaluation.

---

[3] http://snf-629975.vm.okeanos.grnet.gr:8897/ai4b/sparql.

[4] http://leonardo.imis.athena-innovation.gr:8891/diana/sparql.

The effectiveness is calculated by the *Reciprocal Rank* metric defined as $RR = 1/r$, where r is the ranking position of the query that corresponds to the provided natural language description. To further assist our collaborators in the evaluation process, we provide also a natural language description of the generated SPARQL queries by incorporating the verbalization system SPARQL2NL [8]. Figure 6 shows the Reciprocal Rank we have calculated for our three ranking functions. In the 17 out of 20 queries, we got an RR of 1 meaning that we were able to get the information requested by the users.

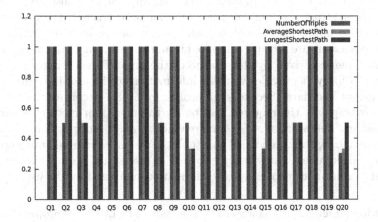

Fig. 6. Reciprocal Rank of different ranking functions on DIANA and AI4B datasets.

7 Related Work

The keyword search problem over structured data, either tree structured [4,6] or graph structured [1,2,5], is a problem that has widely been explored. These works involve the following basic steps: (a) mapping the keyword elements to structured data elements (b) connect the keyword elments by searching for substructures on the data, and (c) return as output the retrieved substructures given a scoring function. Contrary to the previous approaches, Tran et al. [10] proposed a different solution for the keyword search problem. Instead of computing the answers directly on the data, they compute structured queries allowing the user to choose the appropriate one. The advantages of this process are the valuable information provided by the queries allows better comprehension of the retrieved results and the exploitation of the existing query optimization techniques. Our approach keywords-to-sparql queries follows [10] approach, but enriches the information stored within indices and also uses a different exploratory method. More specifically, in comparison with Tran et al.'s keyword index our term index maintains also information about RDF classes and inter-entities properties allowing to efficiently track possible matches of keywords under a uniform space, while our augmented schema-guide graph allows also the encoding of temporal properties. Furthermore, we create multiple augmented schema-guide graphs one per keywords combination and use the notion of shortest paths to create a query pattern graph.

8   Conclusion

In this paper, we have presented a novel method for keyword search on data modeled under the RDF graph representation. Our approach can also be applied to generic graph-structured data if a schema can be extracted. Contrary to the most common approaches found in bibliography for the keyword search problem, where answers are directly computed from the data, our algorithm generates structured queries driven by the schema of the data. This leverages a two-fold advantage. Firstly, it provides valuable information to the user in terms of comprehension of the data. Secondly, it profits from the system optimization in order to extrapolate the required results. Furthermore, our method is enriched with temporal operators allowing a more efficient and deeper exploration of the diachronic data within predefined time points. We have evaluated our approach under an effectiveness study using two real datasets and we have achieved an excellent performance capturing the information requested by the users.

Acknowledgements. This study has been supported by LODGOV project, Research Programme ARISTEIA (EXCELLENCE), General Secretariat for Research and Technology, Ministry of Education, Greece and the European Regional Development Fund.

References

1. Bhalotia, G., Hulgeri, A., Nakhe, C., Chakrabarti, S., Sudarshan, S.: Keyword searching and browsing in databases using BANKS. In: ICDE, pp. 431–440 (2002)
2. Bikakis, N., Giannopoulos, G., Liagouris, J., Skoutas, D., Dalamagas, T., Sellis, T.: RDivF: diversifying keyword search on RDF graphs. In: Aalberg, T., Papatheodorou, C., Dobreva, M., Tsakonas, G., Farrugia, C.J. (eds.) TPDL 2013. LNCS, vol. 8092, pp. 413–416. Springer, Heidelberg (2013)
3. Brickley, D., Guha, R.V.: RDF Schema 1.1 W3C Recommendation, 25 February 2014. www.w3.org/TR/rdf-schema/
4. Cohen, S., Mamou, J., Kanza, Y., Sagiv, Y.: XSEarch: a semantic search engine for XML. In: VLDB, pp. 45–56 (2003)
5. He, H., Wang, H., Yang, J., Yu, P.S.: BLINKS: ranked keyword searches on graphs. In: SIGMOD, pp. 305–316 (2007)
6. Kimelfeld, B., Sagiv, Y.: Finding and approximating Top-k answers in keyword proximity search. In: PODS, pp. 173–182. ACM (2006)
7. Meimaris, M., Alexiou, G., Gkirtzou, K., Papastefanatos, G., Dalamagas, T.: RDF resource search and exploration with LinkZoo. In: DATA. p. (2015) (to appear)
8. Ngonga Ngomo, A.C., Bühmann, L., Unger, C., Lehmann, J., Gerber, D.: Sorry, I Don'T Speak SPARQL: Translating SPARQL Queries into Natural Language. In: WWW, pp. 977–988 (2013)

9. Prud'hommeaux, E., Seaborne, A.: SPARQL Query Language for RDF. W3C Recommendation (2008). http://www.w3.org/TR/rdf-sparql-query/
10. Tran, D.T., Wang, H., Rudolph, S., Cimiano, P.: Top-k exploration of query candidates for efficient keyword search on graph-shaped (RDF) data. In: ICDE (2009)
11. Tran, T., Cimiano, P., Rudolph, S., Studer, R.: Ontology-based interpretation of keywords for semantic search. In: Aberer, K., et al. (eds.) ASWC 2007 and ISWC 2007. LNCS, vol. 4825, pp. 523–536. Springer, Heidelberg (2007)

Author Profile Enrichment for Cross-Linking Digital Libraries

Arben Hajra[1(✉)], Vladimir Radevski[1], and Klaus Tochtermann[2]

[1] South East European University (SEEU), Tetovo, Republic of Macedonia
{a.hajra,v.radevski}@seeu.edu.mk
[2] Leibniz Information Centre for Economics (ZBW), Kiel, Germany
k.tochtermann@zbw.eu

Abstract. This work aims at enriching author profiles with additional information to better support search and retrieval of publications across different digital libraries. To achieve this objective we exploit concepts for cross-linking data to identify correlations between one author and other authors, publications or other related information. We will introduce a profile enrichment approach which adds additional information (e.g. biographic information) from different sources to existing author profiles. Within this context, the linked open data repository DBpedia serves a valuable source for our profile enrichment approach. Still, one of several challenges in this context is the identification of the same author in different sources. To address this challenge we will exploit VIAF (virtual authority file) for author identification. Technically we apply data mining and clustering techniques to uniquely identify authors.

Keywords: Digital libraries · VIAF · Author disambiguation · Data mining · Profile enrichment · Linked open data

1 Introduction

It is widely accepted that Digital Libraries (DL) play an important role in modern scholarly communication. Typically DLs hold domain specific information (e.g. economics) which makes it difficult to search across different domains. For example, would a scholar need literature from economics and agriculture he or she would have to access two different DLs. To overcome this limitation, we will explore options for achieving interoperability by cross-linking authors and/or publications from different DLs with one another.

The main aim of our work is enriching the content of a DL with additional information from other DLs especially regarding information which is somehow related to the authors. Our primary objective is as follows: Assume we have found publications and bibliographic information from an author in one DL, we want to harvest other DLs for correlations to other publications of the same authors, of his or her co-authors and for additional bibliographic information of the initial author.

Our approach suggests creating an author profile, based on the information we have collected from one DL. This profile will continuously be enriched with additional information found in other DLs. To enrich the search results from one DL with

© Springer International Publishing Switzerland 2015
S. Kapidakis et al. (Eds.): TPDL 2015, LNCS 9316, pp. 124–136, 2015.
DOI: 10.1007/978-3-319-24592-8_10

additional results from other DLs we apply author name disambiguation, author identification and false authorship prevention.

To uniquely identify authors and to create correlations between them, we consider bibliographic repositories offered by several libraries and institutions. Very promising is data which are presented in the form of Linked Open Data (LOD), as part of the LOD cloud [1, 3, 4]. As a test case, we will leverage the following repositories: German National Library - DNB, Library of Congress - LC, National Library of France - BNF, National Library of Sweden - KB / LIBRIS.

Finally, we put the Virtual International Authority File - VIAF[1] in the center of our work and utilize it as a "bridge" to those DLs we want to cross-link with each other.

The reminder of this paper is structured as follows. In Sect. 2 we put our work in context with related work. As a contribution to theory and practice of digital libraries, Sects. 3 and 4 introduce formally our concepts for profile enrichment, i.e. we present how we collect information for author profiles, how we model them and how we correlate them with one another using VIAF as a bridge. Section 5 shows the practical implementation of our work. It is followed by Sect. 6 which highlights the most important evaluation results. The paper closes with an outlook on our future work.

2 Related Work

In general, author disambiguation includes two main steps, measuring the similarity and clustering similar records [7]. The main challenge is the identification of whether two authors in the same or different DLs have the same identity or not. The most explored strategies consider the string processing approach which measures the similarity of authors' names [8, 9]. The comparisons are one-to-many and many-to-many, by applying iterative methods [10]. The explored disambiguation process is generally divided using the following approaches: supervised with heuristic similarity functions, unsupervised and hybrid [6, 11]. In our approach, the similarity measurement is not only based on the author's names. We also consider the semantic distance between publication titles, co-authors correlations and co-authors publications. As a result, we suggest a completely automated unsupervised clustering technique.

The most explored strategies in the center of the process apply similarity measurements by employing data mining algorithms for text based distances. The data are represented as vector space model where the distance between vectors represents the similarity. Such algorithms include the Cosine Similarity (CS) with TF-IDF, Jaccard Similarity, Jaro Winkler, and Levenshtein algorithms [7, 9, 12–14].

In almost all these strategies, the author disambiguation process is primarily based on relationships among co-authors and similarity of publications, by discovering other relationships in other DLs [15]. The approach presented in [12], gathers information from citations and submits queries to a Web search engine with the aim to find relevant information about authors. That is, the possibility of user feedback is emphasized on

---

[1] http://viaf.org/.

ambiguous references across iterations in which the feedback in combination with the hybrid supervised process is applied for assigning references to authors [13].

Additionally, there are several efforts for generating authority profiles for uniquely identifying resources and researchers [16]. We emphasize: ORCID, VIAF, VIVO, RESERCHERID and OPENID as most appropriate approaches facilitating the disambiguation of authors and which are used as a "bridge" for retrieving accurate information from different repositories.

ORCID - Open Researcher and Contributor Identifier create and maintain a registry of unique researcher identifiers and a method of linking research activities. Main contributors are several publishing houses, scientific communities and universities. It has available APIs under an open source license [18].

VIAF - Virtual International Authority File hosted by OCLC (Online Computer Library Center, Inc.) is a service that virtually integrates multiple authority files from several national libraries into a single OCLC name authority service. VIAF began as a common project with the LC, DNB BNF and OCLC [19].

VIVO - enables the discovery of researchers across institutions. It is an open source semantic web application where through it, institutions such as Cornell, Harvard, and Indiana University, manage and publish information about researchers and their activities [20].

RESEARCHID – to identify potential collaborators and avoid author misidentification, each member is assigned a unique identifier to enable researchers to manage their publication lists. The ResearcherID information integrates with the Web of Science of Thomson Reuters Company [21].

OPENID – is a foundation that promotes OpenID technologies. OpenID Foundation members include leading companies and individuals in the digital identity industry such as Google, Microsoft and Yahoo [22]. Even though this currently has no direct application in the scholarly communication, there is a promising potential.

In our work, we consider VIAF with the highest usage relevance. The main idea of VIAF is to link authority files from several national libraries into a "super" virtual authority record, i.e., cluster. Currently, the most known national libraries maintain their own authority files, which brings a distinctive way of preserving them [19]. The VIAF API can be used by anyone without the need of authentication. In addition, there is also the option of VIAF LOD repositories. However, VIAF strongly recommends the usage of API because of the frequency of updates of the VIAF content.

VIAF links disparate names for the same person by integrating authority files from 35 national libraries from 30 countries into a particular cluster. Each cluster is assigned with a unique number, a VIAF ID. However, there are cases when the VIAF clustering algorithm shows deficiencies, such as: several clusters for the same person, different people into the same cluster, incorrect bibliographic data and clusters with poor content [23]. Based on the results from [17] in a search of 283,114 names, 59 % were not ambiguous, meaning that only one heading was found, 26 % matched two headings, 10 % matched three headings, 3 % matched four and 2 % more than four.

3 Basic Principles for Profile Enrichment

Our primary goal is enriching the content of a Digital Library with content from other repositories by cross-linking information related to authors. Our research is based on the EconStor[2] repository, the leading German Open Access repository for economics which is maintained by ZBW. EconStor content has also been published in the LOD.

For each EconStor author, we harvest several other repositories for correlations with other authors, publications or other relevant information about the initial author. As a result, we create a wider author profile enriched with additional information. This profile serves two purposes, to enrich the search result and to solve author ambiguities by global identification of the same author written in different ways or same name referring to different authors.

The process of correct author identification in different repositories is related to the challenge of author's name ambiguity, when determining if two or more references correspond to the same person [2, 5, 6]. For example, an author can be represented with different spellings in several bibliographic repositories or different authors can share the same name, which increases the complexity to the data cross-linking process.

Considering the fact that EconStor content is represented as RDF statements, i.e., linked open data, we extend our interest to other bibliographic repositories in the LOD cloud. Still, the author name ambiguity remains to be the major obstacle for direct information retrieval about a given author from these repositories.

As an example, we would like to find as much information as possible about an EconStor author by harvesting other repositories. We often encounter cases in which the same author is presented with different spelling variations, such as: **Adam Smith; Smith, Adam; A. Smith-; Smith, Adam, 1723–1790; Смит, Адам, 1723; Smith. A.; Smith, Adam T.; and Smith, Adam, 1930.** In addition, there could be different authors all with the name **Adam Smith**. In principles, a similar problem concerns the metadata about titles of publications which can vary across different repositories.

3.1 EconStor Metadata

The process for data cross-linking is based and initiated from the *metadata* that are used to describe the authors and publications in EconStor. The most basic metadata for describing an author are Name and Surname. An author $\mathbf{a}(a_{name}, a_{surname})$ is represented by the vector $\mathbf{a} = (t_1, t_2)$. Given this, the set of publications where $\mathbf{a}$ is author is represented as $P_a = \{p_1^a, p_2^a, p_3^a, \ldots, p_k^a\}$. Consequently, every certain publication will be composed by the set of terms (strings) found in the title, such: $p_i^a = \{t_1^{pi}, t_2^{pi}, t_3^{pi}, \ldots, t_m^{pi}\}$.

Accordingly, for each publication from P_a, other authors are considered to be co-authors of $\mathbf{a}$. The union of authors from all P_a publications, will represent the set of co-authors, which are denoted as $A_a = \{a_1^a, a_2^a, a_3^a, \ldots, a_n^a,\}$.

---

[2] http://www.econstor.eu/.

The set of co-authors' publications is of particular importance for determining the co-authorships at the initial repository. With $\bar{P}_a$ we will represent the set of publications of co-authors of **a**, where $\bar{P}_a = \{\bar{p}_1^{a1},..,\bar{p}_k^{a1},\bar{p}_1^{a2},..,\bar{p}_k^{a2},\bar{p}_1^{a3},..,\bar{p}_k^{an}\}$. Thus, $\bar{P}_a = \{\bar{P}_j^{ai}; i = 1,n; j = 1,k\}$.

Table 1a represents the set of these metadata. A detailed picture of the relationships is shown in Fig. 1, where can be seen that p_1^a and p_2^a have a common author.

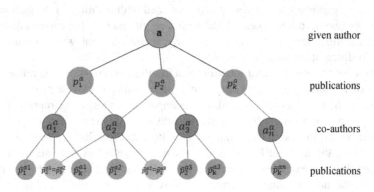

Fig. 1. Relationship among authors, co-authors, publications and co-authors publications for a given author **a**

3.2   VIAF Metadata

VIAF clusters are considered as the target repository in which metadata are analyzed. The similarity measurement will be performed between the metadata from the VIAF clusters and the metadata from our repository. For an input author in VIAF the output is delivered by a set of clusters for that author, denoted as c_j, where $j = 1, k$. Inside each of these VIAF clusters different forms of authors' name presentations can be found for a particular author, obtained from the native libraries. In this paper, the set of variations is denoted $A_{cj} = \{a_1^{cj}, a_2^{cj}, a_3^{cj},\ldots, a_l^{cj}\}$, where each $a_i^{cj} = (t_1, t_2)$, similarly as in the initial repository. Except this information, in any cluster c_j, a possible list of publications can be found in addition to the list of co-authors assigned to that author. The set of publications found in a particular cluster is notated with $P_{cj} = \{p_1^{cj}, p_2^{cj}, p_3^{cj},\ldots, p_k^{cj}\}$, while the set of co-authors inside a cluster will be $\hat{A}_{cj} = \{\hat{a}_1^{cj}, \hat{a}_2^{cj}, \hat{a}_3^{cj},\ldots, \hat{a}_n^{cj}\}$.

Besides these data, the set of publications retrieved directly from the libraries or institutions that are contributing in that cluster can be of a particular importance. These publications can be retrieved by referring the identification number of each library for that cluster. Thus, the set of publications extracted from all the sources like this, are presented with the set $\breve{P}_{cj} = \{\breve{p}_1^{cj}, \breve{p}_2^{cj}, \breve{p}_3^{cj},\ldots, \breve{p}_k^{cj}\}$. Table 1b represents the set of metadata from a particular VIAF cluster that we are considering.

Table 1a. Notation table - metadata from the initial repository

$\mathbf{a}, \mathbf{a} = (t_1, t_2)$.	the author to be disambiguated
$P_a = \{p_1^a, p_2^a, p_3^a, \ldots, p_k^a\}$	publications of author $\mathbf{a}$
$p_i^a = \{t_1^{pi}, t_2^{pi}, t_3^{pi}, \ldots, t_m^{pi}\}$	title's terms from the publication
$A_a = \{a_1^a, a_2^a, a_3^a, \ldots, a_n^a\}$	co-authors of the author $\mathbf{a}$
$\bar{P}_a = \{\bar{p}_1^{a1}, \ldots, \bar{p}_k^{a1}, \ldots, \bar{p}_1^{a3} \ldots, \bar{p}_k^{an}\}$	publications of co-authors of $\mathbf{a}$

Table 1b. Notation table - metadata from a VIAF cluster

c_j	clusters to be checked at VIAF
$A_{cj} = \{a_1^{cj}, a_2^{cj}, a_3^{cj} \ldots, a_l^{cj}\}$	author's names variations in a VIAF cluster c_j, j = 1, k
$P_{cj} = \{p_1^{cj}, p_2^{cj}, p_3^{cj}, \ldots, p_k^{cj}\}$	publications in a VIAF cluster c_j
$\hat{A}_{cj} = \{\hat{a}_1^{cj}, \hat{a}_2^{cj}, \hat{a}_3^{cj}, \ldots, \hat{a}_n^{cj}\}$	co-authors in a VIAF cluster c_j
$\check{P}_{cj} = \{\check{p}_1^{cj}, \check{p}_2^{cj}, \check{p}_3^{cj}, \ldots, \check{p}_k^{cj}\}$	publications from other sources in the VIAF cluster

4 Application of the Profile Enrichment

In our work we consider VIAF as a "bridge" to cross-link different bibliographic repositories. It is a challenge to detect accurately a particular author from a repository, i.e., EconStor, and to connect this author with the corresponding author in VIAF. Achieving the right identification will facilitate the process of retrieving information from other repositories, especially from libraries that contribute to VIAF records, such as, DNB, LC, BNF and LIBRIS. We also consider other publications from a given author, correlations with co-authors, biographical data, publishers, etc. (Fig. 2).

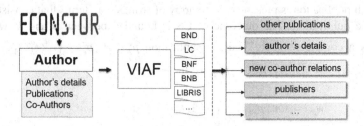

Fig. 2. The overview for enriching process with additional information about authors

4.1 Identifying Authors in VIAF

In Sect. 3.2 we highlighted that a search in VIAF results in several records which match the name of an author. In a second step, we assess the accuracy for each retrieved record.

For this purpose, we implement data mining techniques, by adopting different vector space algorithms. With highest priority, we use the Cosine Similarity (CS) in combination with TF-IDF for the distance between publications, while we apply Levenshtein distance and Jaro distances for similarity author names. The algorithm we propose follows ideas from the process of name deduplication and address information [24].

We start by defining the metadata for the publications in our native repository. These metadata are described in detail in Sect. 3. In the very beginning, the process starts by using the VIAF API for identifying a particular author. Each retrieved cluster is analyzed in iterative fashion according to these steps:

i. **Similarity among author's name with the alternatives within a cluster.**

In cases when at least one full match is found, a particular weight is assigned to the variable, denoted as w_{ac}. In detail, the similarity check is done only in the context of the authors name and surname as terms in a vector, i.e. $\mathbf{a} = (t_1, t_2)$ and $a_i^{cj} = (t_1, t_2)$. Thus, iteratively for each name alternative a_i^{cj} within a cluster, similarity measurement is calculated with the author $\mathbf{a}$.

$$w_{ac} = sim(a, a_i^{cj}), \ i = 1, n; \ j = 1, k; \tag{1}$$

The similarity among names in this step is calculated with CS and TF-IDF where only the perfect match among names is considered. We take this simplified approach to avoid any unreliable results that could be infiltrated when otherwise.

ii. **Similarity between publications that an author has in our repository with the publications found in the VIAF cluster.**

With $\mathbf{P}_a$ is assigned the set of all publications that this author has in our repository, while with $\mathbf{P}_{cj}$ the set of publications found in a particular cluster. Each publication from our repository is compared with each publication found in the cluster. The similarity between publications can be measured based on Cosine Similarity with TF-IDF, where each publication is presented as an array of strings, i.e., terms that consist of the title of the publication. The outcome of CS is bounded between 0 and 1, where 1 represents a complete match. Thus, a publication $p_e^a \in P_a, p_e^a = \left\{ t_1^{pi}, t_2^{pi}, t_3^{pi}, \ldots, t_k^{pi} \right\}$ and $p_f^{cj} \in P_{cj}, p_f^{cj} = \{ t_1^{cj}, t_2^{cj}, t_3^{cj}, \ldots, t_m^{cj} \}$ we have:

$$w_{pc} = sim\left(p_e^a, p_f^{cj} \right), e = 1, k; \quad f = 1, m; \quad k, m \geq 3; \tag{2}$$

In this case for each comparison a specific weight w_{pc} is assigned. Its value is determined if the similarity among the compared titles is above the defined threshold, which is 0.6 for publications that have more than three terms in the title. This value is set based on our preliminary analysis, which showed that lower thresholds and less that three terms in the title, resulted in inaccurate matching.

Before performing the similarity algorithm, the cleaning and formatting of the data is conducted, such as: removing punctuation, eliminating "stopwords", lowercase and encoding the data to Unicode character encoding (UTF-8).

iii. **Comparing the list of co-authors for an author with co-authors found in the cluster.**

Let us consider $A_a = \{a_1^a, a_2^a, a_3^a, \ldots, a_n^a,\}$ the set of co-authors with whom the author a has at least one common publication, while $\hat{A}_{cj} = \{\hat{a}_1^{cj}, \hat{a}_2^{cj}, \hat{a}_3^{cj}, \ldots, \hat{a}_n^{cj}\}$ is the set of co-authors in a particular VIAF cluster c_j. In this case, as it is explained in (ii), each co-author from A_a is compared with each co-author from $\hat{A}_{cj}$.

$$w\hat{a}_c = sim(a_e^a, \hat{a}_f^{cj}), \ e = 1, k; \quad f = 1, m; \tag{3}$$

At least one match, $A_a \cap \hat{A}_{cj} \neq \emptyset$, can be a significant proof that our repository and the cluster have a common co-author. In that case variable $w\hat{a}_c$ will get a weight for each iteration in this step. Having more than one match increases the evidence that it is the required cluster. A more suitable similarity metric for names is applied based on the Jaro-Winkler similarity metric. In this case the similarity is calculated according to the characters. The threshold for names calculated by CS remains 1.0, while for Jaro-Winkler it will be above 0.9.

iv. **Checking the list of publications directly from the sources (libraries) that belong to the cluster.**

The set of publications retrieved from the libraries that belong to the cluster c_j, is denoted with $\check{P}_{cj}$. For example, if DBN has its records in that cluster, we are measuring the similarity between them and publications from our repository, $p^a \in P_a$ with $\check{p}^{cj} \in \check{P}_{cj}$. For each check, a particular weight is assigned to the variable $w\check{P}_c$, absolutely in the same manner as in the step (ii).

$$w\check{p}_c = sim\left(p_e^a, \check{p}_f^{cj}\right), e = 1, k; \quad f = 1, m; \quad k, m \geq 3; \tag{4}$$

4.2 Determining the Matching Degree

The key factors for determining the matching degree between an author from our repository with a particular VIAF cluster, are precisely the components presented above. At each of these components, under (i), (ii), (iii) and (iv) the weight is calculated iteratively with Eqs. (1), (2), (3) and (4). The overall "*weight*" is calculated in accumulative way such as $sum(w_{ac}, w_{pc}, w\,\hat{a}_c, w\,\check{p}_c)$, by respecting the threshold.

5 Sample Implementation for Profile Enrichment

In this section we describe the prototype used for the evaluation of the developed algorithms. This prototype automatically checks VIAF for a particular author and automatically determines the appropriate clusters according to the principles presented in the previous sections. For each cluster found, the VIAF ID is taken and assigned to the corresponding author in the initial repository (EconStor in our case). As a result, an author's profile is enriched with additional information found in the cluster.

For the implementation we use EconStor and an RDF dump file of Econstor. EasyRdf PHP library and rdf4j Sesame are applied for processing and storing the RDF data. The current version of dump file contains 1.635.599 RDF statements, 36.490 publications and 27.580 authors.

To give an example, we select a particular author, i.e. "*Kubler, Felix*", in EconStor. As a result a list of all publications, co-authors and co-author's publications from our repository will be created and returned to the user of our prototype. Considering this author, the prototype found six clusters in VIAF, of which the third one is depicted in Fig. 3. In this cluster, similarities are found related to the author's name, publications, co-authors and publications from the libraries that belong to it. From the list of six publications, the prototype has highlighted three with 100 % match to the EconStor publication. Also, four co-authors of "*Kubler, Felix*" with 100 % match were found.

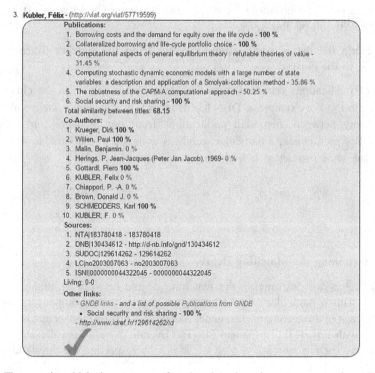

Fig. 3. The case in which the prototype found and evaluated as correct match an EconStor author with a VIAF cluster

Additionally, there are in total five libraries ("Sources" in Fig. 3) or institutions which contain this cluster, thus a possible exploration in these resources would endorse the match. For example, in the German National Library, a publication is found with 100 % similarity (see "Other links" in Fig. 3). However this result is excluded from the calculation because the same publication appears in the cluster's publications, $p^a = \check{p}^{cj}$ (publication 6 in Fig. 3). Overall, all these elements provide evidence that this cluster is correct for the author "*Kubler, Felix*".

For a performed search the number of retrieved results can vary from zero to some hundreds. The above example had only six clusters, with only one correct cluster. However, there are several cases in which for one author the number of correct clusters can be zero, one or more than one cluster that really represents him. In case that at least one cluster is found, the VIAF ID is saved in our local database, for each author.

6 Evaluation

We have randomly analyzed 991 authors from EconStor to VIAF and generated the evaluation metrics of recall, precision and F1 score. In our case, precision represents the fraction of the clusters that are retrieved as correct match. In fact it is the fraction among the truly correct clusters (true positive) with all clusters that the system has retrieved as correct, including clusters that are retrieved as correct but are not (false negative).

$$\text{Precision} = \frac{\text{true positive}}{\text{true positive} + \text{false positive}} = \frac{|\{\text{truly correct clusters }\}|}{|\{\text{all retrieved clusters as correct}\}|}$$

The recall represents the fraction between the truly correct clusters with all correct clusters, including the clusters that are correct but the system has not identified them as such (false negative).

$$\text{Recall} = \frac{\text{true positive}}{\text{true positive} + \text{false negative}} = \frac{|\{\text{truly correct clusters }\}|}{|\{\text{all correct clusters}\}|}$$

Based on the manually checked evaluations the system gives an overall precision of 98.1 % and the recall of 95.9 %. Thus, the efficiency of our system is measured with 0.970 as F1 score.

The results in Table 2 represent only the clusters that are marked as positive and the prototype has marked them as correct clusters. However, there are cases in which for an author only one or more than one clusters are retrieved as correct match.

Table 2. The number of found VIAF clusters for EconStor authors.

Number of checked authors from EconStor	Number of truly found clusters in VIAF	%	Precision	Recall	F1
for 598	1	60.3 %	0.988	0.957	0.972
for 125	2	12.6 %	0.957	0.972	0.964
for 18	3	1.8 %	0.952	0.976	0.964
for 9	> 3	0.9 %	0.951	0.978	0.964
for 241	0	24.3 %	/	/	/

Each of these found clusters are manually evaluated for accuracy of matches. Based on these evaluations, very satisfactory results are generated. In the cases when an author is matched with only one VIAF cluster, we gain 98.8 % precision, 95.7 % recall and F1 score of 0.972. Thus the possibility for it to be the correct cluster is almost absolute. In the cases when two clusters are retrieved as correct match for one author, the precision is 95.7 % and 97.2 % recall, with F1 score of 0.964.

For each checked author from our repository, the corresponding VIAF ID is stored locally. Grouping authors like this can be a huge benefit for clustering them inside a local repository and for creating a local authority profile. Beyond this, the found VIAF ID offers a permanent link to that cluster in VIAF. This avoids to repeat the process of identification again. With the right VIAF ID, all the relevant information found in the cluster are instantly retrieved, such as new publications and new co-authorship correlations. Figure 3 shows an example of this.

In addition, each cluster keeps in it the identification number of libraries or institutions that are contributing with content. We are considering these IDs as valuable information for extending the enrichment of an author profile. Therefore, by having that id, such as *13043612* for DNB, *129614262* for SUDOC, we can refer directly to these repositories to search this author. This can be done by different Web Services and APIs which these libraries offer, or by querying the LOD repositories. Most know libraries including DNB, LC, BNE, BNB, BNF, and LIBRIS offer their data or metadata as LOD in LOD cloud. Consequently, by performing a SPARQL query in these repositories, direct information retrieval is possible.

In several cases, a particular VIAF cluster offers alignment to DBpedia for the corresponding author. We consider this as a possibility to extend an author profile with several other information. The prototype automatically realizes a SPARQL query in DBpedia and retrieves information such as: a short bio, an author picture, a link to Wikipedia page and a downloadable list of works. Figure 4 depicts details from the output of this process.

Fig. 4. Finding and extracting author's information from DBpedia

7 Conclusion and Future Work

Relying on the initial idea of creating enriched author profiles in a digital library by extracting data from several other repositories, the process of author disambiguation is inevitable. We referred to VIAF for avoiding ambiguity and uniquely identifying each author from our repository. Note, that our algorithm is not limited to EconStor only; it should work for any repository given that the following input data are provided: author name, list of publications, co-author names and their publications.

Using our promising results, author profiles as part of a digital library can be enriched by useful information such as new publications which are not part of the initial repository, new co-authorship correlations, publications of co-authors, possibility to cluster authors in the initial repository, biographic information, and DBpedia content.

As future work, improvements in the process of similarity measurements will be performed. This will be done by incorporating and combining several metadata elements and by performing other analyses for similarity calculations. Such analyses will impact the process of threshold calculations and consequently improve the determinations of a cluster's accuracy.

References

1. Bizer, C., Heath, T., Idehen, K., Berners-Lee, T.: Linked data on the web. In: Proceedings of the 17th International Conference on World Wide Web, pp. 1265–1266. ACM (2008)
2. Elmagarmid, A.K., Ipeirotis, P.G., Verykios, V.S.: Duplicate record detection: a survey. IEEE Trans. Knowl. Data Eng. 19(1), 1–16 (2007)
3. Hajra, A., Latif, A., Tochtermann, K.: Retrieving and ranking scientific publications from linked open data repositories. In: Proceedings of the 14th International Conference on Knowledge Technologies and Data-Driven Business (I-Know), p. 29. ACM (2014)
4. Latif, A., Borst, T., Tochtermann, K.: Exposing data from an open access repository for economics as linked data. D-Lib Magazine 20(9/10) (2014)
5. Laender, A.H., et al.: Keeping a digital library clean: new solutions to old problems. In: Proceedings of the Eighth ACM Symposium on Document Engineering, Sao Paulo, Brazil, pp. 257–262. ACM (2008)
6. Santana, A.F., Goncalves, M.A., Laender, A.H., Ferreira, A.: Combining domain-specific heuristics for author name disambiguation. In: Proceedings of the IEEE/ACM Joint Conference on Digital Libraries, pp. 173–182. IEEE (2014)
7. Chin, W.S., et al.: Effective string processing and matching for author disambiguation. J. Mach. Learn. Res. 15(1), 3037–3064 (2014)
8. Torvik, V.I., Smalheiser, N.R.: Author name disambiguation in MEDLINE. ACM Trans. Knowl. Discov. Data (TKDD) 3(3), 11 (2009)
9. Bilenko, M., Mooney, R., Cohen, W., Ravikumar, P., Fienberg, S.: Adaptive name matching in information integration. IEEE Intell. Syst. 18(5), 16–23 (2003)
10. Bhattacharya, I., Getoor, L.: Iterative record linkage for cleaning and integration. In: Proceedings of the 9th ACM SIGMOD Workshop on Research Issues in Data Mining and Knowledge Discovery, pp. 11–18. ACM (2004)

11. Tang, J., Fong, A.C.M., Wang, B., Zhang, J.: A unified probabilistic framework for name disambiguation in digital library. IEEE Trans. Knowl. Data Eng. **24**(6), 975–987 (2012)
12. Pereira, D.A., et al.: Using web information for author name disambiguation. In: Proceedings of the 9th ACM/IEEE-CS Joint Conference on Digital Libraries, pp. 49–58. ACM (2009)
13. Godoi, T.A., et al.: A relevance feedback approach for the author name disambiguation problem. In: Proceedings of the 13th ACM/IEEE-CS Joint Conference on Digital Libraries, pp. 209–218. ACM (2013)
14. Fan, X., Wang, J., Pu, X., Zhou, L., Lv, B.: On graph-based name disambiguation. J. Data Inf. Qual. (JDIQ) **2**(2), 10 (2011)
15. De Nies, T., et al.: Towards named-entity-based similarity measures: challenges and opportunities. In: Proceedings of the 7th International Workshop on Exploiting Semantic Annotations in Information Retrieval, pp. 9–11. ACM (2014)
16. Mazov, N.A., Gureev, V.N.: The role of unique identifiers in bibliographic information systems. Sci. Tech. Inf. Process. **41**(3), 206–210 (2014)
17. Freire, N., et al.: Author consolidation across european national bibliographies and academic digital repositories. In: Proceedings of the 11th International Conference on Current Research Information System (2012)
18. What is ORCID?, http://orcid.org/content/about-orcid
19. Virtual International Authority File, http://www.oclc.org/viaf.en.html
20. What is VIVO?, http://www.vivoweb.org/about
21. What is ResearcherID?, http://www.researcherid.com/
22. OpenID Foundation, http://openid.net/foundation/
23. DNB-Virtual International Authority File (VIAF), http://www.dnb.de/viaf
24. Bilenko, M., Mooney, R.J.: Adaptive duplicate detection using learnable string similarity measures. In: Proceedings of the Ninth ACM SIGKDD International Conference on Knowledge Discovery and Data Mining pp. 39–48. ACM (2003)

User Studies for and Evaluation of Digital Library Systems and Applications

On the Impact of Academic Factors on Scholar Popularity: A Cross-Area Study

Pablo Figueira[✉], Gabriel Pacheco, Jussara M. Almeida,
and Marcos A. Gonçalves

Computer Science Department, Universidade Federal de Minas Gerais,
Belo Horizonte, Brazil
{pabfigueira,gabriel.pacheco,jussara,mgoncalv}@dcc.ufmg.br

Abstract. In this paper we assess the relative importance of key academic factors – conference papers, journal articles and student supervisions – on the popularity of scholars in various knowledge areas, including areas of exact and biological sciences. To that end, we rely on curriculum vitae data of almost 700 scholars affiliated to 17 top quality graduate programs of two of the largest universities in Brazil, as well as popularity measures crawled from a large digital library, covering a 16-year period. We use correlation analysis to assess the relative importance of each factor to the popularity of individual scholars and groups of scholars affiliated to the same program. We contrast our results with those of two top programs of a major international institution, namely, the Computer Science and Medicine departments of the Stanford University.

Keywords: Citation analysis · Scholar popularity · Academic factors

1 Introduction

Scholar productivity and success is a topic that has attracted a lot of attention from researchers. While many have proposed alternative metrics to measure scholar popularity, prestige or influence [7,15,17], others have quantified the influence of individual articles or publication venues [4,8]. We here focus on one particular metric, scholar popularity, estimated by the total citation count [7], which is one of the most widely used metrics of research performance [9,19].

Our present goal is to assess the relative importance of different academic factors, that is, factors directly related to academic productivity, to the scholar popularity. Such knowledge is valuable to individual researchers, as it can help supporting their decisions to build the career in research, as well as to research institutions, guiding the design of policies to incentivize the productivity growth of their research group members. Understanding how different factors impact scholar popularity can also bring insights into the design of more effective popularity prediction models [1,6], which in turn can be used to improve various services (e.g., expert or collaboration recommendation services).

Our work complements a recent characterization of scholar popularity in the computer science community [10]. However, unlike [10] and most prior studies,

© Springer International Publishing Switzerland 2015
S. Kapidakis et al. (Eds.): TPDL 2015, LNCS 9316, pp. 139–152, 2015.
DOI: 10.1007/978-3-319-24592-8_11

which focus on scholars from the same knowledge area, we here analyze and contrast the impact of academic factors on the popularity of scholars in *different knowledge areas*, including exact (e.g., Computer Science, Physics) and biological sciences (e.g., Medical Sciences, Neurosciences, and Animal Sciences). Moreover, unlike [10], we look into the role of student supervisions, and contrast the importance of publications in scientific events (e.g., conferences, workshops) and in journals. We aim at addressing questions such as: *what is the relative importance of each academic factor to scholar popularity?*, and *how does such importance vary for scholars across different knowledge areas?*

Specifically, we quantify the correlations between three core indices of academic productivity – number of student supervisions, number of conference papers and number of journal articles – and popularity for almost 700 scholars affiliated to 17 graduate programs in two major public universities in Brazil. The selected programs are ranked among the best ones in the country (in their particular knowledge areas). Our analyses are performed on a per-scholar basis and aggregated for individual programs, but always separately for different areas. Moreover, we compare our findings with the patterns observed for two top programs of a major international institution, namely the Computer Science and Medicine departments of the Stanford University in the United States.

The rest of this paper is organized as follows. Section 2 discusses prior work. Our datasets are described in Sect. 3. Sections 4 and 5 discuss the impact of the productivity indices on the popularity of individual scholars and graduate programs. Finally, Sect. 6 offers conclusions and directions for future work.

2   Related Work

The literature is rich in efforts to assess influence, popularity or productivity in scientific research. One of the earliest studies relied on citation analysis to measure dependences among journals [4]. More recently, there have been several proposals of metrics of research performance. Ding and Cronin [7] distinguished between weighted (e.g., h-index [11]) and unweighted citation counts, using the former to assess scholar prestige and the latter to estimate scholar popularity. More sophisticated metrics, relying on machine learning techniques [2] and customized indices sensitive to the productivity in specific research fields [5,15], have also been proposed. Others have used centrality metrics (e.g., PageRank and its variants), applied to the co-authorship graph, to assess the influence of a scholar, publication, or publication venue within a community [3,16].

As argued in [13], there is not a clear winner metric, as each one has its own bias. We here are not interested in proposing a new metric of research performance but rather assessing the relative importance of core indices of scholar productivity to this performance. While any particular metric could be used in this study, we choose one that is widely used, namely scholar popularity estimated by total number of citations (as defined in [7]). Though simple and easy to compute, this metric has been shown to be very important for various types of analyses. For instance, after analyzing the productivity of more than 700 scholars, Riikonen et al. [19]

concluded that actual publication and citation counts are better indicators of the scientific contribution of researchers than impact factors. Similarly, citation count has also been used to study the impact of a scholar experience and prestige on the number of references used in her publications [9].

Our present work extends a recent effort to study the impact of various academic factors to the popularity of scholars in the Computer Science area [10]. Unlike this prior study and all prior analyses of scholar performance, we here consider scholars in different knowledge areas, aiming at identifying similarities and differences across them. Indeed, to our knowledge, the only few prior efforts covering multiple knowledge areas are focused on the popularity of *publication venues* [3,13], not on scholars. Moreover, we here also analyze the role of student supervision to scholar popularity, a factor that was not tackled by most prior work. Indeed, we are aware of only two prior analyses that exploit student supervisions [5,14]. However, they are restricted to computer scientists and do not directly correlate student supervision with scholar popularity, as we do here.

Our work is also orthogonal to prior efforts to develop popularity prediction models for publications [6] and individual scholars [1], as it aims at assessing the relative importance of various features to scholar popularity, and thus draws insights that might help the design of future prediction models.

3   Datasets

Our study is focused on scholars affiliated to different graduate programs of two of the largest and most important public universities in Brazil, namely Federal University of Minas Gerais (UFMG) and State University of Campinas (Unicamp). These institutions have some of the best graduate programs of the country in different knowledge areas according to CAPES, the Brazilian Ministry of Education's organization for graduate courses and curricula. CAPES periodically evaluates all graduate programs of the country by assigning a score which varies from 3 (lowest) to 7 (highest). This score, referred to as CAPES level, takes into account the scientific production of students and faculty, the curriculum of the program, the institution's research infrastructure, among other factors. A score of 5 is given to programs that reached the national level of excellence, whereas scores 6 and 7 are assigned to courses with international-level high quality. In this study, we selected graduate programs that have received scores of 5 or higher in the most recent evaluation by CAPES. These programs are shown in Table 1.

We collected the names of all scholars affiliated to each selected program from the SOMOS web portal of each university[1]. The SOMOS portal, developed by the UFMG's Office of Technology Transfer and Innovation, contains detailed information about each department and graduate program of the university. For example, for each scholar affiliated to each program, SOMOS provides a list of keywords, areas of expertise, research groups, and the time series of various indices of academic productivity. The goal of SOMOS is to ease the identification of expertises available in the university, fostering a better interaction with public

---

[1] http://somos.ufmg.br and http://somos.unicamp.br (both in portuguese).

Table 1. Selected graduate programs.

University	Program	CAPES level (3-7)	Number (fraction) of scholars listed in MS-AS	Total citations
UFMG	Computer Science	7	39 (65 %)	18,284
	Electrical Engineering	7	24 (43 %)	1,976
	Mathematics	5	13 (20 %)	348
	Physics	7	33 (36 %)	6,839
	Animal Science	6	27 (100 %)	1,903
	Biochemistry and Immunology	7	51 (69 %)	21,569
	Bioinformatics	6	17 (100 %)	6,644
	Cellular Biology	6	16 (100 %)	2,658
	Ecology	5	16 (100 %)	2,545
	Neurosciences	5	14 (100 %)	2,468
	Molecular Medicine	5	15 (100 %)	2,733
Total UFMG			**265**	**67,967**
UNICAMP	Computer Science	7	41 (80 %)	8,347
	Chemical Engineering	6	35 (70 %)	2,138
	Electrical Engineering	7	60 (65 %)	12,137
	Mathematics	7	36 (33 %)	4,678
	Physics	7	61 (62 %)	7,495
	Medical Sciences	5	188 (55 %)	31,440
Total UNICAMP			**421**	**66,235**
Stanford	Computer Science	–	80 (77 %)	393,883
	Medicine	–	301 (66 %)	360,486
Total Stanford			**381**	**754,369**

and private institutions. Originally deployed only at UFMG, SOMOS is currently also available at Unicamp and other smaller Brazilian universities.

The main source of data behind SOMOS is the Lattes platform[2], a Web-based application created by CNPq (the Brazilian National Council for Scientific and Technological Development) to collect and integrate curriculum vitae information from the academic community at large. Scholars and students are required to have and keep up-to-date their vitae in Lattes as a precondition to applying for grants and other forms of financing. Thus, Lattes is considered a very reliable source of curriculum vitae data, as it is maintained by the scholars themselves with the most accurate and up-to-date information. All information in Lattes is publicly available, and currently covers practically all active Brazilian scholars.

We here focus on three key indices of academic productivity, namely, number of student supervisions (including Master and PhD students who have already graduated), number of publications in events (workshops, conferences, symposiums), and number of publications in journals, book chapters and books. We grouped journal publications, book chapters and books into a single index, referring to it as simply journal articles, as books and book chapters are usually in much smaller number. We also refer to publications in events as simply conference papers. For

---

[2] http://lattes.cnpq.br.

each scholar affiliated to each program, we collected the time series of each index covering the period from 1995 to 2010. To support our analysis of the importance of each index to whole programs, we also collected the year when the scholar joined the program as a faculty from her vitae on Lattes.

Recall that our goal is to correlate each productivity index with scholar popularity (estimated by citation count). Finding a good source of citation counts for such heterogeneous group of scholars is quite a challenge. We experimented with various online digital library services, including ArnetMiner, Google Scholar, and Microsoft Academic Search (MS-AS)[3], choosing MS-AS, as it has a better coverage of the scholars in our dataset. Each scholar name was submitted as a query to MS-AS to retrieve the popularity time series. In case the query returned multiple entries, we used the scholar's affiliation and area of expertise to select the correct one. We focused on the same aforementioned 16 year-period (1995-2010), which was chosen based on recent reports that MS-AS performance in indexing scholarly documents and tracking their citations was considered to be very competitive until 2010[4]. Yet, some selected programs (e.g., Mathematics in both universities) are underrepresented, as shown in the 4th column of Table 1, which presents the numbers and fractions of scholars who are also listed in MS-AS. Rather than disregarding such programs, we chose to disregard only individual scholars whose names were not found in MS-AS. Table 1 also shows the total citation count for all scholars considered in each program (5th column).

In order to put our results into perspective, we also analyze scholars of two top departments of a high quality university in international standards. We focus on the Computer Science and Medicine departments of the Stanford University (USA), covering two knowledge areas in our dataset of Brazilian scholars. In the absence of platforms like SOMOS and Lattes for Stanford, we had to use alternative data sources. We collected the names of affiliated scholars from each department's webpage[5], and searched for their lists of publications in public digital libraries, notably DBLP for Computer Science and Scopus for Medicine[6]. We focus only on publications for Stanford scholars, as we could not find a reliable source of data about their student supervisions. As done for the Brazilian scholars, we collected citation counts from MS-AS, and focused on the same 16-year period (1995-2010). Table 1 also shows, for each Stanford department, the number of scholars covered in our dataset and their citation counts.

4   Popularity of Individual Scholars

We start our study by analyzing the relative importance of each productivity index to the popularity of individual scholars. To that end, we take the total

---

[3] http://arnetminer.org/citation, http://scholar.google.com, http://academic.research.microsoft.com.

[4] http://blogs.nature.com/news/2014/05/the-decline-and-fall-of-microsoft-academic-search.html.

[5] http://www-cs.stanford.edu, http://medicine.stanford.edu.

[6] http://dblp.uni-trier.de/ and http://www.scopus.com, respectively.

Table 2. Popularity and productivity indices of individual scholars in each program (averages and coefficients of variation (CV) across scholars).

University	Program	Popularity average (CV)	Conference papers average (CV)	Journal articles average (CV)	Student supervisions average (CV)
UFMG	Computer Science	468.9 (1.9)	66.5 (0.9)	23.1 (1.3)	38.2 (1.1)
	Electrical Engineering	82.3 (1.5)	64.9 (0.6)	21.7 (1.0)	40.5 (0.9)
	Mathematics	26.8 (1.2)	5.4 (1.0)	15.9 (0.6)	20.6 (0.6)
	Physics	207.2 (2.4)	48.8 (0.9)	50.5 (0.6)	25.3 (0.7)
	Animal Science	70.5 (2.2)	102.9 (0.7)	67.3 (0.6)	47.6 (0.6)
	Biochemistry and Immunology	422.9 (1.4)	83.5 (0.7)	64.0 (0.8)	43.7 (0.7)
	Bioinformatics	390.8 (1.0)	101.8 (0.6)	52.9 (0.9)	44.7 (0.6)
	Cellular Biology	166.1 (0.7)	73.2 (0.8)	48.2 (0.7)	38.2 (0.7)
	Ecology	159.0 (2.2)	83.8 (0.7)	57.0 (0.9)	52.9 (0.7)
	Neurosciences	176.3 (1.2)	69.3 (1.1)	68.9 (0.9)	33.8 (0.7)
	Molecular Medicine	182.2 (1.1)	42.7 (1.1)	107.7 (1.0)	34.4 (0.8)
UNICAMP	Computer Science	203.6 (1.5)	49.2 (0.9)	15.3 (0.9)	26.4 (0.8)
	Chemical Engineering	61.1 (1.3)	107.7 (1.4)	48.1 (1.4)	45.2 (0.7)
	Electrical Engineering	202.3 (2.0)	73.3 (0.7)	27.6 (0.8)	29.9 (1.3)
	Mathematics	129.9 (1.3)	23.7 (1.4)	28.0 (0.7)	18.3 (0.9)
	Physics	122.9 (3.2)	38.3 (1.2)	50.8 (0.8)	16.1 (0.9)
	Medical Sciences	167.2 (2.1)	88.8 (1.1)	75.9 (0.9)	22.2 (1.1)
Stanford	Computer Science	4,923.5 (1.1)	42.0 (0.9)	19.5 (0.8)	–
	Medicine	1,197.6 (1.5)	2.9 (2.0)	42.9 (1.4)	–

numbers of conference papers, journal articles and student supervisions, and the total citation count of each scholar from 1995 to 2010. We then quantify the correlation of each index with popularity for scholars of each program.

Note some scholars may have joined the programs *after* 1995, and may even have published papers or supervised students before joining the program, either during their studies or as a faculty in another program. Since our focus in this section is on individual scholars, we choose to take all publications, student supervisions and citations of the scholars during the whole 16-year period, regardless of when the scholar joined the program under analysis. Our assumption is that the scholar's academic productivity has been on the same knowledge area throughout the period, even though she might have been affiliated to other programs. Thus, the scholar's current affiliation is used as a token of her main area of expertise, and we group scholars by their current affiliation.

We start our discussion by first briefly analyzing how the four metrics, namely popularity, number of conference papers, number of journal articles and number of student supervisions, vary across individual scholars. Table 2 provides average values along with corresponding coefficients of variation (CV) – ratio of the standard deviation to the average value – of each metric computed across all scholars of each program, considering the whole 16-year period.

We make the following observations. Technological programs related to Computer Science (UFMG, Unicamp), Bioinformatics (UFMG), and Electrical Engineering (Unicamp) tend to have the most "popular" scholars of the analyzed programs in the respective universities. The same holds for Stanford: the Computer Science scholars have an average popularity 4 times higher than those in

Medicine. One exception is Biochemistry and Immunology (UFMG), a CAPES level 7 program (highest) whose affiliated scholars also have very high popularity (on average). Despite these similarities, the average popularity of scholars in the two Stanford programs are almost 10 times higher than the respective programs in Brazil. We conjecture that this may be due to a greater focus of Stanford's scholars on publishing in higher quality venues. Evidence of this argument can been seen in some of the publication patterns in the period. If we consider only Computer Science, the per scholar average number of conference papers (for UFMG and Unicamp) and journal articles (for UFMG) are higher than for Stanford's scholars. The same is true for the Medical related programs in both Brazilian universities. In other words, the Brazilian scholars publish many more papers but are much less cited than Stanford's scholars. These findings are consistent with previous observations [18] that the Brazilian productivity has been increasing steadily but its impact is not following the same pattern.

Regarding student supervisions, the differences within each university, in general, are not very large, although there are exceptions. Note the smaller average number of students per scholar in the Physics and Mathematics courses (both universities), which might reflect some idiosyncrasies of the area. On the other hand, the UFMG numbers. when compared to the Unicamp's for similar programs, are a bit higher, which might be due to the size of the respective universities: UFMG has 15 % more students than Unicamp, on average[7].

Table 2 also shows great variability (large CVs) across scholars of the same program in all three universities, particularly in their popularity values. This reflects a natural heterogeneity of scholars and their particular areas of research.

We now turn to our correlation analysis. Figure 1 illustrates some of our findings by presenting scatter plots[8] of each productivity index (x-axis) versus popularity (y-axis, logarithm scale) for scholars in Computer Science (Fig. 1(a–c)) and Neurosciences (Fig. 1(d–f)) at UFMG, and Medical Sciences at Unicamp (Fig. 1(g–h))[9]. All three indices seem (visually) correlated with scholar popularity, though with variable strength. For Computer Science and Medical Sciences, the correlations between popularity and number of publications, particularly in journals, seem higher than the correlation with number of student supervisions, although the latter does not seem negligible. The same is not as clear for Neurosciences: the number of student supervisions exhibits no clear relationship with scholar popularity, possibly due to the small population (14 scholars) and great variability across individual scholars (Fig. 1(f)).

We summarize these results by computing the Spearman's rank correlation coefficient [12] between each productivity index and scholar popularity, for all programs. The Spearman coefficient is a nonparametric measure of statistical

---

[7] http://www.ufmg.br/conheca/nu_index.shtml, and http://pt.m.wikipedia.org/wiki/Universidade_Estadual_de_Campinas.

[8] Each point represents one scholar.

[9] The scatter plot for conference papers for Medical Sciences is omitted due to space constraints.

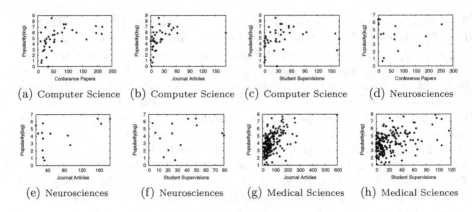

Fig. 1. Productivity indices (x-axis) vs. popularity (y-axis, logarithm scale) for individual scholars (Computer and Neurosciences at UFMG, Medical Sciences at Unicamp).

dependence between two variables that does not assume linear relationships[10]. Table 3 shows the correlations for all analyzed programs[11]. The strongest correlations for each program are shown in bold[12], and statistically significant results (p-value under 0.1) are indicated by a "*" mark[13].

Although the correlations in Table 3 vary greatly across areas, and even across programs of the same area, there are some common patterns. The number of journal articles is the index with strongest correlation with popularity in most programs, including all Unicamp and Stanford programs, and 8 out of 11 programs in UFMG (being statistically significant in 6). Thus, out of the three indices, it is the most important one to scholar popularity in most areas.

Interestingly, we find that, for Computer Science in both UFMG and Stanford, conference papers are as important as journal articles to popularity, being only slightly less important (somewhat weaker correlation) for the program in Unicamp. This observation reflects the importance of scientific events to publicize recent results in such dynamic area. Yet, for the three Computer Science programs, the average numbers of conference papers per scholar are around three times higher than the corresponding average numbers of journal articles (Table 2). The same holds for scholars in both Electrical Engineering programs: despite the much larger average number of conference papers per scholar, the number of journal articles is much more strongly correlated with scholar popularity. This suggests that, although scholars in these two areas do find it necessary to often publish in scientific events, more consolidated journal publications, even if in much smaller number, still generate similar (or higher) popularity.

We also note that, unlike observed for other programs of Biological sciences, conference papers do seem to be important for scholars in Bioinformatics

---

[10] The Pearson linear correlation coefficients are slightly smaller but lead to similar conclusions.

[11] Recall that we do not have data on student supervision for Stanford scholars.

[12] Differences under 5 % are consideredties.

[13] We used the t-test available in the SciPy library – http://scipy.org.

Table 3. Spearman's correlation between each productivity index and scholar popularity for individual scholars ("*" indicates statistical significance with p-value < 0.1).

University	Program	Conference papers	Journal articles	Student supervisions
UFMG	Computer Science	**0.616***	**0.612***	0.437*
	Electrical Engineering	0.450*	**0.670***	0.481*
	Mathematics	**0.575***	0.234	0.138
	Physics	0.174	**0.645***	0.308*
	Animal Science	**0.312**	0.314	0.318
	Biochemistry and Immunology	0.124	**0.526***	0.403*
	Bioinformatics	**0.462***	0.241	0.229
	Cellular Biology	0.037	**0.521***	0.467*
	Ecology	0.614*	**0.728***	0.496*
	Neurosciences	-0.370	**0.263**	0.147
	Molecular Medicine	-0.234	0.304	**0.428**
UNICAMP	Computer Science	0.500*	**0.645***	0.525*
	Chemical Engineering	0.417*	**0.752***	0.403*
	Electrical Engineering	0.197	**0.452***	0.076
	Mathematics	0.347*	**0.717***	0.377*
	Physics	0.069	**0.462***	0.173
	Medical Sciences	0.185*	**0.506***	0.401*
Stanford	Computer Science	**0.396***	**0.384***	–
	Medicine	0.480*	**0.672***	–

(UFMG), possibly due to the presence of many computer scientists as faculty members. Other results that are worth mentioning are the negative correlations between conference papers and popularity for Neurosciences and Molecular Medicine (UFMG), and the strong correlation between conference papers and popularity for Mathematics at UFMG (unlike Unicamp). Although such results might reflect some peculiarities of scholars affiliated to these programs, they might also be due to the small coverage of such programs (e.g., Mathematics) or small population sizes (e.g., NeuroSciences and Molecular Medicine). Indeed, we note that the negative correlations are not significant with 90 % confidence. We leave for future work to further analyze these peculiar results.

Finally, student supervision is also important to scholar popularity in specific areas, particularly of Biological sciences, such as Cellular Biology and Biochemistry and Immunology (UFMG), and Medical Sciences and Chemical Engineering (Unicamp). It is also important, to a lesser extent, to Computer Science.

5   Aggregated Popularity of Graduate Programs

Having analyzed the relative importance of each productivity index to the popularity of individual scholars, we now assess their importance to each graduate program. To that end, we aggregate the values for each index and for the popularity of all scholars affiliated to each program in each year, over the 16-year period. Note that by summing up the citations and publications of all scholars of a program, publications coauthored by multiple scholars of the program are counted multiple times. Thus, we estimate the (aggregated) productivity and popularity of each program in each year by taking the average values of each metric across all scholars of the program. Given the great variability of all metrics across different scholars (Table 2) considering such average values may smooth out individual extremes, reflecting a better picture of each program as a whole.

Yet, as mentioned, scholars may have joined the selected programs after the beginning of the analyzed period. To be fair to all programs, a scholar should *not* be included in the computation of the program's productivity and popularity in the years that precede the scholar's admission to that program. Unfortunately we are able to do so only for the Brazilian scholars, as the years of their admissions are extracted from the Lattes platform (see Sect. 3). Since we could not find this information for Stanford scholars, we make a simplifying assumption that they were affiliated of the selected programs throughout the 16-year period.

Figure 2 shows the time series of average productivity and average popularity (in logarithm scale for better visualization) for six example programs: Computer Science in all three universities, Biochemistry and Immunology (UFMG), Medical Sciences (Unicamp), and Medicine (Stanford). For all six programs, and for the others analyzed, we observe an increase in the average scholar popularity over time. This may reflect an increase of visibility, thanks to a larger number of scientific digital libraries and other mechanisms for publicizing scientific work. A general increasing trend is also observed for the average number of student supervisions, possibly reflecting the growth of incentives (e.g., scholarships) from

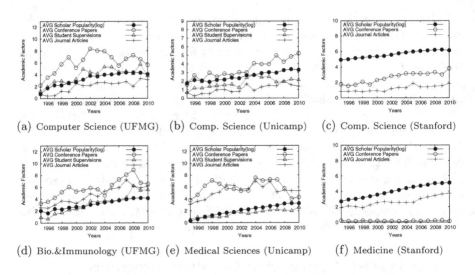

(a) Computer Science (UFMG)   (b) Comp. Science (Unicamp)   (c) Comp. Science (Stanford)

(d) Bio.&Immunology (UFMG)   (e) Medical Sciences (Unicamp)   (f) Medicine (Stanford)

Fig. 2. Time series of the average productivity and popularity of example programs.

the Brazilian research supporting agencies during the period, particularly in the 2000s. Regarding publications, whereas the average number of journal articles increased over the years for most programs, the trend in the number of conference papers vary across programs. For some, e.g., Computer Science (UNICAMP and Stanford) and Bioinformatics (UFMG), the numbers increased over the years, but for most programs, no clear trend is observed (e.g., Fig. 2(e,f)).

We assess the relative importance of each productivity index to the popularity of each program over the years by computing the correlations between the values of each pair of metrics in each year[14]. Since the impact of publications and student supervisions on popularity may be observed with some delay in time, we compute the correlations between the (average) productivity of the program in year i and its (average) popularity in year $i+\delta$. Despite some variability, we found that the results are qualitatively very similar for δ equal to 0, 1 and 2, possibly reflecting some degree of maturity of most programs such that their productivity and popularity remain roughly stable over short time periods (up to 2 years). Thus, due to space constraints, we show results only for $\delta=2$ in Table 4. Once again, the strongest correlations for each program are shown in bold[15], and statistically significant correlations (p-value < 0.1) are marked with "*".

In general, the per-program correlations are much stronger than those for individual scholars (Table 3). By averaging the metrics over all scholars of a program, we are less susceptible to the impact that individual differences have on the programs (especially on the small ones). For example, the negative correlations between conference papers and popularity observed for individual scholars in Neurosciences and Molecular Medicine (UFMG) appear reasonably strong (above 0.5 and statistically significant) at the program level. This suggests that despite the great variability at the individual level, both programs tend to benefit from a larger number of conference papers of their scholars (on average).

Also, unlike observed for individual scholars, student supervisions is the most important factor for most programs in both Brazilian institutions. In the few cases where it is not the most important index, its correlation with popularity is still quite strong. For example, the correlations for Electrical Engineering, Physics, Biochemistry and Immunology, and Cellular Biology (UFMG), as well as for Medical Sciences (Unicamp) are above 0.9. Some of these strong correlations are also observed at the individual level (see previous section). Others, such as for Electrical Engineering, Physics and Ecology (UFMG), emerge only at the program level. As mentioned, this might simply be due to the averaging procedure. Yet, collaborations among professors within or cross research groups, supported by sharing (formally or informally) student supervisions, may also help to increase the importance of this factor in a collective perspective.

Finally, once again journal articles are as important as (and often more important than) conference papers to most programs. As observed for individual scholars, conference papers are also very important to the Computer Science programs, particularly at Unicamp and Stanford, and to Bioinformatics (UFMG).

---

[14] Each point represents one year in this computation.

[15] Differences under 5 % are considered ties.

Table 4. Spearman's correlation between each productivity index and popularity for programs over the years (productivity in year i and popularity in year $i + 2$, "*" indicates statistical significance with p-value < 0.1).

University	Program	Conference papers	Journal articles	Student supervisions
UFMG	Computer Science	0.578*	**0.842***	**0.824***
	Electrical Engineering	0.723*	0.722*	**0.920***
	Mathematics	-0.179	**0.662***	0.552*
	Physics	0.727*	0.689*	**0.886***
	Animal Science	0.519*	0.771*	**0.812***
	Biochemistry and Immunology	0.842*	**0.974***	**0.978***
	Bioinformatics	**0.837***	0.667*	**0.846***
	Cellular Biology	0.679*	0.717*	**0.902***
	Ecology	0.095	0.361	**0.745***
	Neuroscience	0.510*	**0.757***	0.458*
	Molecular Medicine	0.640*	0.835*	**0.943***
UNICAMP	Computer Science	**0.913***	0.829*	0.824*
	Chemical Engineering	0.257	**0.916***	0.855*
	Electrical Engineering	0.317	**0.825***	0.581*
	Mathematics	0.009	0.759*	**0.836***
	Physics	-0.095	0.705*	**0.871***
	Medical Sciences	0.530*	0.626*	**0.970***
Stanford	Computer Science	**0.929***	**0.896***	–
	Medicine	0.228	**0.930***	–

However, for many programs, such as Mathematics (both universities), Ecology (UFMG), Physics (UNICAMP), and Medicine (Stanford), the correlations between conference papers and popularity are not significant. In those cases, despite the trend of increasing scholar popularity, the average number of conference papers per scholar remained roughly stable over the years, suggesting that this type of publication has little impact on the popularity of such programs.

6   Conclusions and Future Work

We have presented a cross-area study on the impact of key academic factors on scholar popularity, considering a diverse set of top quality graduate programs in Brazil and internationally, with more than 1,000 scholars and 880,000 citations, over a 16-year period. Some of our key findings are: (i) scholars in technological programs (e.g., Computer Science, Electrical Engineering, Bioinformatics) tend to be the most "popular" ones in their universities (though exceptions exist); (ii) international popularity in still much higher than that obtained by Brazilian scholars; (iii) when taken individually, the most important factor to popularity is the number of journal articles, even in areas such as Computer Science and Electrical Engineering, for which conferences are important venues for communicating research results; (iv) while student supervisions are only moderately correlated with popularity at the individual level, it is one of the most important factors to explain the popularity of programs, possibly due to the increasing motivation for collaborations among scholars and students. Our findings are original and thought-provoking, as cross-area analyses are rarely performed, and provide insights into how to increase research impact for individual scholars and graduate programs. Future work includes covering new universities and knowledge areas, tackling self-citations, and designing popularity prediction models.

Acknowledgements. This work is partially funded by the Brazilian National Institute of Science and Technology for the Web, CNPq, CAPES and FAPEMIG.

References

1. Acuna, D.E., Allesina, S., Kording, K.P.: Future impact: predicting scientific success. Nature **489**(7415), 201–202 (2012)
2. Bergsma, S., Mandryk, R.L., McCalla, G.: Learning to measure influence in a scientific social network. In: Sokolova, M., van Beek, P. (eds.) Canadian AI. LNCS, vol. 8436, pp. 35–46. Springer, Heidelberg (2014)
3. Bollen, J., Rodriquez, M.A., de Sompel, H.V.: Journal status. Scientometrics **69**(2), 669–687 (2006)
4. Cason, H., Lubotsky, M.: The influence and dependence of psychological journals on each other. Psychol. Bull. **33**(2), 95–103 (1936)
5. Cervi, C., Galante, R., Oliveira, J.P.d.: Comparing the reputation of researchers using a profile model and scientific metrics. In: CSE, pp. 353–359 (2013)
6. Chakraborty, T., Kumar, S., Goyal, P., Ganguly, N., Mukherjee, A.: Towards a stratified learning approach to predict future citation counts. In: JCDL, pp. 351–360 (2014)
7. Ding, Y., Cronin, B.: Popular and/or prestigious? measures of scholarly esteem. IP&M **47**(1), 80–96 (2011)
8. Fersht, A.: The most influential journals: impact factor and eigenfactor. PNAS **106**(17), 6883–6884 (2009)
9. Frandsen, T.F., Nicolaisen, J.: Effects of academic experience and prestige on researchers' citing behavior. JASIST **63**(1), 64–71 (2012)
10. Gonçalves, G., Figueiredo, F., Almeida, J.M., Gonçalves, M.A.: Characterizing scholar popularity: a case study in the computer science research community. In: JCDL, pp. 57–66 (2014)
11. Hirsch, J.E.: An index to quantify an individual's scientific research output. PNAS **102**(46), 16569–16572 (2005)
12. Jain, R.: The Art of Computer Systems Performance Analysis: Techniques for Experimental Design, Measurement Simulation and Modeling. Wiley, New York (1991)
13. Leydesdorff, L.: How are new citation-based journal indicators adding to the bibliometric toolbox? JASIST **60**(7), 1327–1336 (2009)

14. Lima, H., Silva, T., Moro, M., Santos, R., Meira Jr., W., Laender, A.H.F.: Assessing the profile of top Brazilian computer science researchers. Scientometrics **103**(3), 879–896 (2015)
15. Lima, H., Silva, T., Moro, M., Santos, R., Meira, Jr., W., Laender, A.: Aggregating productivity indices for ranking researchers across multiple areas. In: JCDL, pp. 97–106 (2013)
16. Ma, N., Guan, J., Zhao, Y.: Bringing pagerank to the citation analysis. IP&M **44**(2), 800–810 (2008)
17. Martin, T., Ball, B., Karrer, B., Newman, M.: Coauthorship and citation in scientific publishing. CoRR, 1304.0473 (2013)
18. Noorden, R.: The impact gap: South America by the numbers. Nature **510**(7504), 202–203 (2014)
19. Riikonen, P., Vihinen, M.: National research contributions: a case study on finnish biomedical research. Scientometrics **77**(2), 207–222 (2008)

A Comparison of Offline Evaluations, Online Evaluations, and User Studies in the Context of Research-Paper Recommender Systems

Joeran Beel[(⊠)] and Stefan Langer

Docear, Magdeburg, Germany
{beel,langer}@docear.org

Abstract. The evaluation of recommender systems is key to the successful application of recommender systems in practice. However, recommender-systems evaluation has received too little attention in the recommender-system community, in particular in the community of *research-paper* recommender systems. In this paper, we examine and discuss the appropriateness of different evaluation methods, i.e. offline evaluations, online evaluations, and user studies, in the context of research-paper recommender systems. We implemented different content-based filtering approaches in the research-paper recommender system of *Docear*. The approaches differed by the features to utilize (terms or citations), by user model size, whether stop-words were removed, and several other factors. The evaluations show that results from offline evaluations sometimes contradict results from online evaluations and user studies. We discuss potential reasons for the non-predictive power of offline evaluations, and discuss whether results of offline evaluations might have some inherent value. In the latter case, results of offline evaluations were worth to be published, even if they contradict results of user studies and online evaluations. However, although offline evaluations theoretically might have some inherent value, we conclude that in practice, offline evaluations are probably not suitable to evaluate recommender systems, particularly in the domain of research paper recommendations. We further analyze and discuss the appropriateness of several online evaluation metrics such as click-through rate, link-through rate, and cite-through rate.

Keywords: Recommender systems · Evaluations · Offline evaluation · User study

1 Introduction

Thorough evaluations are paramount to assess the effectiveness of research-paper recommender systems, and judge the value of recommendation approaches to be applied in practice or as baseline in other evaluations. The most common evaluation methods are user studies, offline evaluations, and online evaluations [1].

User studies typically measure user satisfaction through explicit ratings. Users receive recommendations generated by different recommendation approaches, rate the recommendations, and the community considers the approach with the highest average

S. Kapidakis et al. (Eds.): TPDL 2015, LNCS 9316, pp. 153–168, 2015.
DOI: 10.1007/978-3-319-24592-8_12

rating most effective [1]. Study participants are typically asked to quantify their overall satisfaction with the recommendations. However, they might also be asked to rate individual aspects of a recommender system, for instance, how novel or authoritative the recommendations are [2] or how suitable they are for non-experts [3]. A user study can also collect qualitative feedback, but this is rarely done in the field of (research paper) recommender systems. Therefore, we will ignore qualitative studies in this paper. We distinguish further between "lab" and "real-world" user studies. In lab studies, participants are aware that they are part of a user study, which, as well several other factors, might affect their behavior and thereby the evaluation's results [4, 5]. In real-world studies, participants are not aware of the study and rate recommendations for their own benefit, for instance because the recommender system improves recommendations based on the ratings (i.e. relevance feedback), or ratings are required to generate recommendations (i.e. collaborative filtering).

Online evaluations originated from online advertisement and measure acceptance rates of recommendations in real-world recommender systems. Acceptance rates are typically measured by click-through rate (CTR), i.e. the ratio of clicked recommendations to displayed recommendations. For instance, if a recommender system displays 10,000 recommendations and 120 are clicked, CTR is 1.2 %. Other metrics include the ratio of downloaded or bought items. Acceptance rate is typically interpreted as an implicit measure for user satisfaction. The assumption is that when a user clicks, downloads, or buys a recommended item, the user liked the recommendation. Of course, this assumption is not always reliable because users, for example, might buy a book but after reading it yet rate it negatively. If the recommender system's objective is revenue, metrics such as CTR can be explicit measures of effectiveness, namely when the operator receives money, e.g. for clicks on recommendations.

Offline evaluations typically measure the *accuracy* of a recommender system based on a ground-truth, but also *novelty* or *serendipity* of recommendations can be measured [6]. Offline evaluations were originally meant to identify a number of promising recommendation approaches [1]. These approaches should then be evaluated in detail with a user study or online evaluation to identify the most effective approaches. However, criticism has been raised on the assumption that offline evaluation could predict an algorithm's effectiveness in online evaluations or user studies. More precisely, several researchers have shown that results from offline evaluations do not necessarily correlate with results from user studies or online evaluations [7, 8]. This means that approaches that are effective in offline evaluations are not necessarily effective in real-world recommender systems. As a consequence, *McNee et al.* criticised that "the research community's dependence on offline experiments [has] created a disconnect between algorithms that score well on accuracy metrics and algorithms that users will find useful" [9]. Several more researchers voiced criticism of offline evaluations. *Jannach et al.* stated that "the results of offline [evaluations] may remain inconclusive or even misleading" and "real-world evaluations and, to some extent, lab studies represent probably the best methods to evaluate systems" [10]. *Knijnenburg et al.* reported that "the presumed link between algorithm accuracy [...] and user experience [...] is all but evident" [11]. Others believe that "on-line evaluation is the only technique able to measure the true user satisfaction" [12]. The main reason for the criticism in the literature is that offline evaluations ignore human factors, yet human factors strongly affect overall user satisfaction with recommendations. For instance, users may be dissatisfied with recommender systems, if they must wait for too long to receive recommendations [13], or the presentation is unappealing [1].

Despite the criticism, offline evaluations are the predominant evaluation method in the recommender community [14]. This is also true in the field of research-paper recommender systems, where the majority of recommendation approaches are evaluated offline, and only 34 % of the approaches are evaluated with user studies and only 7 % with online evaluations [15, 16]. However, online evaluations and user studies are also not without criticism. For instance, results of user studies may vary, depending on the questions [17]. *Zheng et al.* showed that CTR and relevance do not always correlate and concluded that "CTR may not be the optimal metric for online evaluation of recommender systems" and "CTR should be used with precaution" [18]. In addition, both user studies and online evaluations require significantly more time than offline evaluations, and can only be conducted by researchers who have access to a recommender system and real users, or at least some participants (e.g. students) to participate in a user study.

2 Research Objective and Methodology

In the field of research-paper recommender systems, there is no research or discussion about how to evaluate recommender systems. In addition, the existing comparisons in other recommender disciplines focus on offline evaluations and user studies [19, 20], *or* offline evaluations and online evaluations [18], but not on all three methods. Our research goal hence was to explore the adequacy of online evaluations, user studies, *and* offline evaluations. To the best of our knowledge, we are first to compare the results of all three evaluation methods, and to discuss the adequacy of the methods and metrics in detail. In addition, we are first to discuss the appropriateness of the evaluation methods in the context of research-paper recommender systems, aside from our previous paper on recommender system evaluation [21]. Compared to our previous paper, the current paper is more comprehensive, covers three instead of two evaluation methods, considers more metrics, is based on more data, and provides a deeper discussion.

To achieve the research objective, we implemented different recommendation approaches and variations in the recommender system of our literature management software Docear. We evaluated the effectiveness of the approaches and variations with an offline evaluation, online evaluation, and user study, and compared the results.

Docear is a free and open-source literature suite, used to organize references and PDFs [22]. It has approximately 20,000 registered users and uses mind-maps to manage PDFs and references. Since 2012, Docear has been offering a recommender system for 1.8 million publically available research papers on the web [23]. Recommendations are displayed as a list of ten research papers, showing the title of the recommended papers. Clicking a recommendation opens the paper in the user's web browser. Figure 1 shows an example mind-map that shows how to manage PDFs and references in Docear. We created categories reflecting our research interests ("Academic Search Engines"), subcategories ("Google Scholar"), and sorted PDFs by category and subcategory. Docear imported annotations (comments, highlighted text, and bookmarks) made in the PDFs, and clicking a PDF icon opens the linked PDF file. Docear also extracts metadata from PDF files (e.g. title and journal name) [24, 25], and displays metadata when the mouse hovers over a PDF icon. A circle at the end of a node indicates that the node has child nodes, which are hidden ("folded"). Clicking the circle would unfold the node, i.e. make its child nodes visible again.

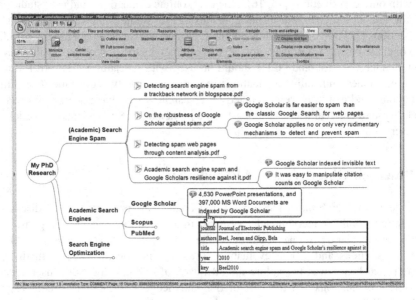

Fig. 1. A screenshot of Docear, showing the management of research articles and references

Docear users, who agree to receive recommendations, automatically receive recommendations every five days upon starting the program. In addition, users can request recommendations at any time. To create recommendations, the recommender system randomly chooses one of three recommendation approaches [23, 26]. The first approach is classic content-based filtering, which utilizes terms from the users' mind-maps [27–29]. The most frequently occurring terms are extracted from the nodes, and research papers that contain these terms are recommended. For instance, if the mind-map in Fig. 1 was used, terms such as *Google*, *Scholar*, *Academic*, and *Search* would be used for the user modeling because these terms occur frequently in the mind-map. Terms are weighted with TF-IDF, stop words are removed, and recommendations are generated based on cosine similarity in the vector space. The second approach utilizes citations in the same way that the first approach utilizes terms. In the example mind-map in Fig. 1, the four PDF links and annotations would be interpreted as citations. The citations would also be weighted with TF-IDF and documents from the corpus that contain the same citations would be recommended. Both content-based filtering approaches are automatically assembled by a number of random variables. Details about this process are provided later in the paper. The third approach implements the stereotype concept, introduced by Rich in 1979 [30]. Based on this approach, Docear generalizes that all users are researchers, which is not strictly true since some use Docear only for its mind-mapping functionality. However, the very nature of stereotyping is to generalize, and the majority of Docear's users are researchers. To give recommendations based on the stereotype approach, we manually compiled a list of research articles that we assumed were relevant for researchers, namely articles and books about academic writing. If the stereotype approach is randomly chosen, the pre-compiled list of articles is recommended. We mainly chose the stereotype approach as a baseline and to have an approach that is fundamentally different from content based filtering. For more details on the architecture of the recommender system refer to [23].

For the offline evaluation, we considered papers that users cited in their mind-maps to be the ground-truth. For each Docear user, we created a copy of their mind-maps, and removed the paper that was most recently added to the mind-map. We then applied a randomly selected recommendation approach to the modified mind-map. Overall, we calculated 118,291 recommendation sets. To measure the accuracy of the algorithm, we analyzed whether the removed paper was within the top10 (P@10) or top3 (P@3) of the recommendation candidates. We also calculated the Mean Reciprocal Rank (MRR), i.e. the inverse of the rank at which the removed paper was recommended. In addition, we calculated nDCG based on the ten most recently added papers and 50 recommendation candidates. Our evaluation method is similar to other offline evaluations in the field of research-paper recommender systems, where the citations made in research papers are used as ground-truth.

We intended to conduct two user studies – one lab study and one real-world study. For the lab study, we wanted to recruit participants through our blog[1]. In our blog we asked Docear's users to start Docear, request recommendations, click each of them, and read at least the abstract of each recommended paper. Users should then rate the relevance of the recommendations from 1 to 5 stars, and if they wish, request new recommendations and continue this process for as long as they like. The study was intended to run from April to July 2014. We promoted the study in our newsletter (8,676 recipients), on Facebook (828 followers), con Twitter (551 followers), and on Docear's homepage (around 10,000 visitors per month). Despite 248 people reading the blog post, only a single user participated in the study. He rated three recommendation sets, each with ten recommendations. Ratings of a single user are not suitable to receive meaningful results. Hence, we consider this user study as a failure, and focus on results of the real-world study. The real-world study was based on ratings that users provided during their normal work with Docear. The best possible rating was 5, the worst possible rating 1. Overall, 379 users rated 903 recommendation sets with 8,010 recommendations. The average rating was 2.82.

For the online evaluation, we analyzed data from Docear's recommender system, which displayed 45,208 recommendation sets with 430,893 recommendations to 4,700 users from March 2013 to August 2014. The acceptance rate was measured with the

---

[1] http://www.docear.org/2014/04/10/wanted-participants-for-a-user-study-about-docears-recommender-system/.

following metrics: Click-Through Rate (**CTR**) measured the ratio of clicked recommendations vs. delivered recommendations. Click-Through Rate over sets (**CTR$_{Set}$**) is the mean of the sets' individual CTRs. For instance, if eight out of ten recommendations had been clicked in *set I*, and two out of five recommendations in *set II*, then CTR would be $\frac{8+2}{10+5} = 66.67\%$ but CTR$_{Set}$ would be $\frac{8/10 + 2/5}{2} = 60\%$. We also calculated CTR over users (**CTR$_{User}$**). CTR$_{User}$ levels the effect that a few power users might have. For instance, if users A, B, and C saw 100, 200, and 1,000 recommendations, and user A clicked seven, user B 16, and user C 300 recommendations, CTR would be $\frac{7+16+300}{100+200+1000} = 24.85\%$, but CTR$_{User}$ would be $\frac{7}{100} + \frac{16}{200} + \frac{300}{1000} = 12.36\%$, i.e. the impact of user C would be weaker. However, CTR$_{User}$ was only calculated for two analyses (the reason is further discussed below). Link-Through Rate (**LTR**) describes the ratio of displayed recommendations against those recommendations that actually had been clicked, downloaded and linked in the user's mind-map. Annotate-Through Rate (**ATR**) describes the ratio of recommendations that were annotated, i.e. a user opened the linked PDF in a PDF viewer, created at least one annotation (bookmark, comment, or highlighted text), and imported that annotation in Docear. Cite-Through Rate (**CiTR**) describes the ratio of documents for which the user added some bibliographic data in the mind-map, which strongly indicates that the user plans to cite that document in a future research paper, assignment, or other piece of academic work.

If not otherwise stated, all reported differences are statistically significant ($p < 0.05$). Significance was calculated with a two-tailed t-test and χ^2 test where appropriate.

3 Results

We calculated the Pearson correlation coefficient for the different evaluation metrics. Both CTR and CTR$_{Set}$ show a strong positive correlation with ratings ($r = 0.78$). Correlation of all other metrics, both offline and online, with user ratings is between 0.52 (CiTR) and 0.67 (nDCG). This means that CTR and CTR$_{Set}$ are the most adequate metrics to approximate ratings. If the goal is to approximate CTR, then ratings, obviously, is the most adequate metric ($r = 0.78$), followed by LTR ($r = 0.73$). The other metrics have rather low correlation coefficients; the worst are nDCG ($r = 0.28$) and MRR ($r = 0.30$). Among the offline metrics, P@3 and P@10 correlate well with each other ($r = 0.92$), which is to expect. MRR and nDCG also show a reasonable strong correlation with each other ($r = 0.71$), while correlation of P@10 and MRR ($r = 0.56$) and P@10 and nDCG ($r = 0.55$) is rather weak.

The user study and online evaluation both led to the same ranking of the three recommendation approaches (Fig. 2): Term-based CBF performed best, i.e. CTR, CTR$_{Set}$, DTR, LTR, CiTR, and ratings were highest, citation-based CBF performed second best, and the stereotype approach performed worst, but still had a reasonable effectiveness. On average, LTR was around one third of CTR. For instance, LTR for the stereotype approach was 1.46 % while CTR was 4.11 %. This means that one third of the recommendations that had been clicked were actually downloaded and linked in the mind-maps. ATR was around half of LTR for the CBF approaches. This means that users annotated about half of the recommendations that they downloaded. However, for

	Approach			User Model Size						Number of Utilized Nodes						Node Selection Method				Stop Words		User Type		Labelling			Trigger Type	
	CBF (Terms)	CBF (Cit.)	Stereo-type type	[1;10]	[11;25]	[26;100]	[101;250]	[251;500]	[501;1,000]	[1;9]	[10;49]	[50;99]	[100;499]	[500;999]	[1,000+]	Modi-fied	Edit-ed	Creat-ed	Mov-ed	Rem-oved	Kept	Regis-tered	Anon.	Org.	Com.	No	Requ-ested	Auto
Ratings	2.91	2.48	2.10	2.62	2.94	3.26	2.92	3.00	2.94	2.56	3.17	3.46	3.16	3.07	2.58	2.85	2.82	3.04	3.31	3.16	2.88	2.90	--	2.82	2.92	2.88	2.83	2.93
CTR	6.53%	5.25%	4.11%	3.92%	6.27%	7.48%	5.40%	6.09%	4.84%	3.64%	6.62%	7.50%	7.08%	6.09%	6.38%	4.99%	5.38%	5.09%	7.40%	6.31%	5.94%	5.32%	3.86%	4.82%	4.92%	5.39%	9.14%	3.67%
CTR(Set)	5.33%	5.00%	4.04%	3.78%	6.27%	7.81%	5.60%	6.32%	4.85%	3.63%	6.62%	7.49%	7.18%	6.20%	6.60%	4.98%	5.57%	5.18%	7.46%	6.35%	6.01%	5.38%	3.83%	5.21%	5.33%	6.46%	9.23%	3.71%
CTR(Usr)	--	--	--	--	--	--	--	--	--	--	--	--	--	--	--	--	--	--	--	--	--	4.00%	3.77%	3.68%	3.18%	3.57%	--	--
LTR	2.32%	1.89%	1.46%	1.33%	2.18%	2.81%	1.86%	2.03%	1.60%	1.24%	1.23%	2.60%	2.42%	2.39%	1.87%	2.24%	1.96%	2.33%	2.57%	2.51%	2.33%	2.47%	2.30%	1.75%	1.76%	2.21%	3.14%	1.33%
ATR	1.20%	1.15%	0.18%	0.61%	0.59%	1.53%	0.98%	1.18%	1.26%	0.91%	0.88%	1.18%	1.28%	1.38%	1.11%	0.95%	1.38%	1.20%	1.18%	1.14%	1.21%	1.34%	1.34%	1.00%	1.23%	0.97%	1.32%	1.24%
CiTR	0.53%	0.52%	0.02%	0.49%	0.15%	0.58%	0.69%	0.55%	0.24%	0.25%	0.58%	0.72%	0.56%	0.45%	0.44%	0.52%	0.79%	0.48%	0.37%	0.58%	0.50%	0.52%	0.55%	0.59%	0.38%	0.48%	0.71%	0.38%
P@3	2.20%	0.19%	0.00%	2.01%	1.97%	2.25%	3.48%	2.07%	1.85%	1.57%	1.91%	2.15%	3.69%	2.97%	1.85%	3.68%	2.40%	1.52%	2.04%	2.71%	1.54%	2.71%	2.27%	2.03%	2.15%	2.15%	--	--
P@10	6.21%	0.41%	0.03%	4.63%	5.82%	6.49%	8.87%	4.18%	2.16%	4.44%	5.81%	6.19%	8.46%	7.14%	5.27%	7.08%	7.07%	5.41%	5.81%	7.98%	5.17%	4.24%	7.50%	6.31%	6.26%	6.30%	--	--
MRR	1.71%	0.31%	0.04%	0.58%	0.62%	2.16%	4.04%	3.17%	1.20%	1.61%	1.45%	1.86%	1.92%	1.22%	0.45%	1.61%	1.07%	1.94%	2.00%	1.73%	1.68%	1.69%	2.30%	1.83%	1.60%	1.63%	--	--
nDCG	1.37%	0.25%	0.03%	0.90%	1.20%	1.49%	1.63%	2.08%	1.92%	1.09%	1.16%	1.38%	1.54%	1.44%	0.61%	1.17%	1.49%	1.61%	1.00%	1.44%	1.29%	1.35%	1.31%	1.24%	1.38%	1.33%	--	--

Fig. 2. Results of the Evaluations

the stereotype approach, ATR was only 0.18 %, i.e. $\frac{1}{8}$ of LTR. Similarly, CiTR for the stereotype approach was only $\frac{1}{75}$ of LTR, while CiTR for term- and citation-based CBF was around $\frac{1}{4}$ of LTR. Apparently, stereotype recommendations were rarely annotated or cited, yet users cited every fourth content-based recommendation that they downloaded. The offline evaluation led to the same overall ranking than the online evaluation and user study. However, all four offline metrics attest that term-based CBF has significantly better effectiveness than citation based CBF (around four to ten times as effective), while user study and online evaluation only attest a slightly higher effectiveness. In addition, the effectiveness of the stereotype approach in the offline evaluation is close to zero, while user study and online evaluation show a reasonable effectiveness.

We researched not only the effectiveness of distinct recommendation approaches, but also variables such as the extent of a user's model (user model size). The user model size describes how many terms (or citations) are stored to represent the users' information needs. Whenever recommendations are requested, Docear randomly selected a user model size between 1 and 1000 terms. For term-based CBF, the highest ratings (3.26) were given for recommendations that were based on user models containing 26 to 100 terms (Fig. 2). The offline metrics led to slightly different results and showed the highest effectiveness for user models containing 101 to 250 terms.

Docear's mind-maps often contain thousands of nodes. We assumed that analyzing too many nodes might introduce noise into the user models. Therefore, Docear randomly selected how many of the x most recently modified nodes, should be utilized for extracting terms. Based on user ratings, analyzing between 50 and 99 nodes is most effective (Fig. 2). As more nodes were analyzed, the average ratings decreased. CTR, CTR_{Set}, LTR and CiTR also showed an optimal effectiveness for analyzing 50 to 99 nodes. Based on ATR, the optimal number of nodes is larger, but results were statistically not significant. The offline metrics indicate that analyzing a larger number of nodes might be sensible, namely 100 to 499 nodes.

Another variable we tested was the node modification type (Fig. 2). The recommender system chose randomly, whether to utilize only nodes that were newly *created*, nodes that were *moved*, nodes that were *edited*, or nodes with any type of *modification* (created, edited, or moved). Utilizing moved nodes only, resulted in the highest ratings on average (3.31). The online metrics CTR, CTR_{Set}, and LTR as well as the offline metric MRR also have the highest effectiveness when utilizing moved nodes. Results for ATR and CiTR differ, but are statistically not significant. Based on P@N, utilizing all modified nodes is most effective, based on nDCG utilizing newly created nodes is most effective.

When the recommender system removed stop-words, the average rating was 3.16 compared to 2.88 when no stop-words were removed (Fig. 2). All other metrics, except ATR, also showed higher effectiveness when stop-words were removed, but, again, results for ATR were statistically insignificant.

Docear's recommender system is open to both registered and unregistered/anonymous users[2], and we were interested whether there would be differences in the two users groups with respect to recommendation effectiveness. CTR and CTR_{Set} show a clear difference between the two user types (Fig. 2). Registered users had an average CTR of 5.32 % while unregistered users had an average CTR of 3.86 %. CTR_{User} is also higher for registered users (4.00 %) than for anonymous users (3.77 %), but the difference is not that strong. LTR and ATR also show a (slightly) higher effectiveness for registered users. The offline evaluation contradicts the findings of the online evaluation: P@3, P@10, and MRR indicate that recommendations for registered users were about half as effective as for anonymous users, and nDCG showed no statistically significant difference between the user groups.

For each user, Docear randomly determined whether to display an organic label (e.g. "Free Research Papers", a commercial label (e.g. "Research Papers [Sponsored]"), or to display no label at all. For each user a fix label was randomly selected once, i.e. a particular user always saw the same label. The label had no impact on how recommendations were generated. This means, if recommendation effectiveness would differ for a particular label, then only because users would value different labels differently. In the user study, there were no significant differences for the three types of labels in terms of effectiveness: the ratings were around 2.9 on average (Fig. 2). Based on CTR, CTR_{Set}, and LTR, displaying no label was most effective. In addition, commercial labels were slightly, but statistical significantly, more effective than organic labels. Based on CTR_{User}, commercial recommendations were least effective, organic labels were most effective, and 'no label' was second most effective. ATR and CiTR led to statistically not significant results, and offline metrics could not be calculated for this kind of analysis.

Two triggers in Docear lead to displaying recommendations. First, Docear displays recommendations automatically every five days when Docear starts. Second, users may explicitly request recommendations at any time. The user ratings for the two triggers are similar (Fig. 2). Interestingly, the online evaluation shows a significantly higher effectiveness for requested recommendations than for automatically displayed recommendations. For instance, CTR for requested recommendations is 2.5 times higher than for automatically displayed recommendations (9.14 % vs. 3.67 %). An offline evaluation was not conducted because it had not been able to calculate any differences based on trigger.

---

[2] Registered users have a user account assigned to their email address. For users who want to receive recommendations, but do not want to register, an anonymous user account is automatically created. These accounts have a unique random ID and are bound to a user's computer.

4 Discussion and Conclusions

4.1 Adequacy of Online Evaluation Metrics

We used six metrics in the online evaluation, namely CTR, CTR_{Set}, CTR_{User}, LTR, ATR, and CiTR. Overall, CTR and CTR_{Set} seem to be the most adequate metrics for our scenario. They had the highest correlation with ratings, are easiest to calculate, were more often statistically significant than the other metrics and are commonly used in other research fields such as e-commerce and search engines. CTR also provided the most plausible results for the stereotype recommendations: based on CTR, the stereotype approach was reasonably effective, while the approach was ineffective based on ATR and CiTR. The result based on CTR seems more plausible since the recommendations were about academic writing and most of Docear's users should be interested in improving their writing skills. However, there is little reason for someone who is doing research in a particular research field, to annotate or even cite an article about academic writing even if the article was useful. Hence, ATR and CiTR were low, and judging stereotype recommendations based on ATR or CiTR seems inadequate to us.

However, there are other scenarios in which ATR and CiTR might be more sensible measures than CTR. For instance, imagine two algorithms called "A" and "B". Both are content-based filtering approaches but B also boosts papers published in reputable journals.[3] Most people would probably agree that algorithm B would be preferable to algorithm A. In the online evaluation, users would probably see no difference between the titles of the recommendations created with the two approaches, assuming that authors publishing in reputable journals do not formulate titles that are significantly different from titles in other journals. Consequently, recommendations of the two algorithms would appear to be similarly relevant and received similar CTR. In this example, CTR would be an inadequate measure of effectiveness and ATR and CiTR might be more appropriate.

Measuring CTR, while displaying only the title of recommendations, was criticized by some reviewers of our previous publications. The reviewers argued that titles alone would not allow thorough assessment of recommendations and CTR could therefore be misleading. In some scenarios, such as the example above with the two algorithms, one being boosted by journal reputation, this criticism could indeed apply. However, in the scenario of Docear, the results do not indicate that displaying only the title led to any problems or bias in the results. At least for the content-based recommendations, CTR correlated well with metrics such as LTR, i.e. metrics indicating that the users thoroughly investigated the recommendations.

Compared to CTR, CTR_{User} smoothed the effect of variables that strongly affected a few users. For instance, CTR_{User} was highest for organic labels, lowest for commercial labels, and mediocre for no labels – a result that one would probably expect. In contrast, CTR was highest for no label, second highest for commercial recommendations, and lowest for organic recommendations – a result that one would probably no expect. After looking at the data in detail, we found that a few users who received many

---

[3] For this example we ignore the question how reputability is measured.

recommendations (with no label) "spoiled" the results. Hence, if the objective of an evaluation was to measure overall user satisfaction, CTR_{user} was probably preferable to CTR because a few power users will not spoil the results. However, applying CTR_{user} requires more users than applying CTR since CTR_{user} requires that each user receives recommendation based on the same parameters of the variables and not per recommendation set. For instance, to calculate CTR_{user}, each user must always see the same label, user models must always be the same size for a user, or recommendations must always be based on terms *or* citations. In contrast, to calculate CTR, users may receive recommendations based on terms *and* on citations, or user models could differ in size. Consequently, to receive statistically significant results, CTR_{user} requires more users than CTR. At least for Docear, calculating CTR_{user} for variables such as user model size, number of nodes to analyze, features to utilize (terms or citations), and weighting schemes is not feasible since we would need many more users than Docear currently has.

Considering the strong correlation of CTR and ratings, the more plausible result for stereotype recommendations, and the rather low number of users being required, we conclude that CTR is the most appropriate online metric for our scenario. This is not to mean that in other scenarios other metrics might not be more sensible. Given our results and examples, we suggest that a careful justification is needed in online evaluations about which metric was chosen and why.

4.2   Online Evaluations vs. User Studies

Ratings in the user study correlated strongly with CTR ($r = 0.78$). This indicates that explicit user satisfaction (ratings) is a good approximation of the acceptance rate of recommendations (CTR), and vice versa. Only in two cases CTR and ratings contradicted each other, namely for the impact of labels and the trigger. Based on these two analyses, it seems that ratings and CTR may contradict each other when it comes to evaluating human factors. For analyses relating to the recommendation algorithms (user model size, number of nodes to analyze, etc.), CTR and ratings always led to the same conclusions.

We argue that none of the metrics is generally more authoritative than another. Ultimately, the authority of user studies and online evaluations depends on the objective of the evaluator, and operator of the recommender system respectively. If, for instance, the operator receives a commission per click on a recommendation, CTR was to prefer over ratings. If the operator is interested in user satisfaction, ratings were to prefer over CTR. Ideally, both CTR and ratings, should be considered when making a decision about which algorithm to apply in practice or to choose as baseline, since they both have some inherent value. Even if the operator's objective was revenue, and CTR was high, low user satisfaction would not be in the interest of the operator. Otherwise users would probably ignore recommendations in the long run, and also stop clicking them. Similarly, if the objective was user satisfaction, and ratings were high, a low CTR would not be in the interest of the operator: a low CTR means that many irrelevant recommendations are given, and if these could be filtered, user satisfaction would probably further increase. Therefore, ideally, researchers should evaluate their approaches with both online evaluation and user study. However, if researchers do not have the resources to conduct both types of evaluation, or the analysis clearly focuses on recommendation algorithms with low impact of human factors, we suggest that conducting either a user study or an online evaluation should still be considered "good practice".

4.3 Adequacy and Authority of Offline Evaluations

Our research shows only a mediocre correlation of offline evaluations with user studies and online evaluations. Sometimes, the offline evaluation could predict the effectiveness of an algorithm in the user study or online evaluation quite precisely. For instance, the offline evaluation was capable of predicting whether removing stop-words would increase the effectiveness. Also the optimal user model size and number of nodes to analyze were predicted rather accurately (though not perfectly). However, the offline evaluation remarkably failed to predict the effectiveness of citation-based and stereotype recommendations. If one had trusted the offline evaluation, one had never considered stereotype and citation-based recommendations to be a worthwhile option. The uncertain predictive power of offline evaluations, questions the often proclaimed purpose of offline evaluations, namely to identify a set of promising recommendation approaches for further analysis.

We can only speculate about why offline evaluations sometimes can predict effectiveness in user studies and online evaluations, and sometimes offline evaluations have no predictive power. One possible reason is the impact of human factors. For instance, on first glance we expected that Docear's recommendation approaches create equally relevant recommendations for both anonymous and registered users. However, the offline evaluation showed higher effectiveness for anonymous users than for registered users while we saw the opposite in the online evaluation. Although we find these results surprising, the influence of human factors *might* explain the difference: We would assume that anonymous users are more concerned about privacy than registered users[4]. Users concerned about their privacy, might worry that when they click a recommendation, some unknown, and potentially malicious website, opens. This could be the reason that anonymous users, who tend to be concerned about their privacy, click recommendations not as often as registered users, and CTR is lower on average. Nevertheless, the higher accuracy for anonymous users in the offline evaluation might still be plausible. If anonymous users tended to use Docear more intensively than registered users, the mind-maps of the anonymous users would be more comprehensive and hence more suitable for user modeling and generating recommendations, which would lead to the higher accuracy in offline evaluations. This means that although mind-maps of anonymous users might be more suitable for user modeling, the human factor "privacy concerns" causes the low effectiveness in online evaluations.

If human factors have an impact on recommendation effectiveness, we must question whether one can determine scenarios in which human factors have *no* impact. Only in these scenarios, offline evaluations would be an appropriate tool to

---

[4] If users register, they have to reveal private information such as name and email address. If users are concerned about revealing this information, they probably tend to use Docear as anonymous user.

approximate the effectiveness of recommendation approaches in online evaluations or user studies. We doubt that researchers will ever be able to reliably predict whether human factors affect the predictive power of offline evaluations. In scenarios like our analysis of registered vs. anonymous users, it is apparent that human factors may play a role, and that offline evaluations might be not appropriate. For some of our other experiments, such as whether to utilize terms or citations, we could see no plausible influence of human factors, yet offline evaluations could not predict the performance in the user study and online evaluation. Therefore, and assuming that results of offline evaluations have no inherent value, we would propose abandoning offline evaluations, as they cannot reliably fulfil their purpose. However, offline evaluations, online evaluations, and user studies typically measure different types of effectiveness. One might therefore argue that comparing the results of the three methods is like comparing apples, peaches, and oranges, and that the results of each method have some inherent value. For online evaluations and user studies, such an inherent value doubtlessly exists (see previous section), but does it exist for offline evaluations?

An inherent value would exist if those who compiled the ground-truth, better knew which items were relevant than current users who decide to click, download, or rate an item. This situation is comparable with a teacher-student situation. Teachers know which books their students should read, and although students might not like the books, or had not chosen the books themselves, the books might be the best possible choice to learn about a certain subject. Such a teacher-student situation might apply to offline evaluations.

Ground-truths inferred, for instance, from citations, theoretically could be more authoritative than online evaluations or user studies. For instance, before a researcher decides to cite a document – which would add the document to the ground-truth – the document was ideally carefully inspected and its relevance was judged according to many factors such as the publication venue, the article's citation count, or the soundness of its methodology. These characteristics usually cannot be evaluated in an online evaluation or user study. Thus, one might argue that results based on personal-collection datasets might be more authoritative than results from online evaluations and user studies where users just had a few seconds or minutes at best, to decide whether to download a paper. Assuming that offline evaluations could be more authoritative than user studies and online evaluations, the following question arises: How useful are recommendations that might objectively be most relevant to users when users do not click, read, or buy the recommended item, or when they rate the item negatively? In contrast to teachers telling their students to read a particular book, a recommender system cannot force a user to accept a recommendation. We argue that an algorithm that is not liked by users, or that achieves low CTR, can never be considered useful. Only if two algorithms performed similarly or if both approaches had at least a mediocre performance in an online evaluation or user study, an additional offline evaluation might be used to decide which of the two algorithms is more effective. However, this means that offline evaluations had to be conducted *in addition* to user studies or online evaluations, and not beforehand or as only evaluation method. Consequently, a change in the current practice of recommender systems evaluation was required.

While offline evaluations with ground-truths inferred from e.g. citations look promising on first glance, there is a fundamental problem: ground-truths are supposed to contain *all* items that are relevant for recommendation. To compile such a ground-truth, comprehensive knowledge of the domain is required. It should be apparent that most users do not have comprehensive knowledge of their domain (which is why they need a recommender system). Consequently, ground-truths are incomplete and contain only a fraction of relevant items, and perhaps even irrelevant items. If the ground-truth is inferred from citations, the problem becomes even more apparent. Many conferences and journals have space restrictions that limit the number of citations in a paper. This means that even if authors were aware of all relevant literature – which they are not – they would only cite a limited amount of relevant articles.

Citation bias enforces the imperfection of citation-based ground-truths. Authors cite papers for various reasons and these do not always relate to the paper's relevance to that author [31]. Some researchers prefer citing the most recent papers to show they are "up-to-date" in their field. Other authors tend to cite authoritative papers because they believe this makes their own paper more authoritative or because it is the popular thing to do. In other situations, researchers already know what they wish to write but require a reference to back up their claim. In this case, they tend to cite the first appropriate paper they find that supports the claim, although there may have been more fitting papers to cite. Citations may also indicate a "negative" quality assessment. For instance, in a recent literature review we cited several papers that we considered of little significance and excluded from an in-depth review [16]. These papers certainly would not be good recommendations. This means that even if authors were aware of all relevant literature, they will not always select the most relevant literature to cite.

When incomplete or even biased datasets are used as ground-truth, recommender systems are evaluated based on how well they can calculate such an imperfect ground-truth. Recommender systems that recommend papers that are not contained in the imperfect dataset, but that might be equally relevant, would receive a poor rating. A recommender system might even recommend papers of higher relevance than those in the offline dataset, but the evaluation would give the algorithm a poor rating. In other words, if the incomplete status quo – that is, a document collection compiled by researchers who are not aware of all literature, who are restricted by space and time constraints, and who typically do biased citing – is used as ground-truth, a recommender system can never perform better than the imperfect status quo.

We consider the imperfection to be a fundamental problem. To us, it seems that the imperfection is also the most plausible reason why the offline metrics could not predict e.g. the effectiveness of citation-based and stereotype recommendations in the online evaluations and user study. As long as one cannot identify the situations in which the imperfection will affect the results, we propose that inferred ground-truths should not be used to evaluate research-paper recommender systems.

4.4   Limitations

We would like to note some limitations of our research. The offline dataset by Docear may not be considered an optimal dataset due to the large number of novice users.

A repetition of our analysis on other datasets, with more advanced users, may lead to more favorable results for offline evaluations. Nevertheless, as pointed out, inferred ground-truths may never be perfect and probably always suffer from some bias, space restrictions, etc. We also focused on research-paper recommender systems. Future research should analyze the extent to which the limitations of offline datasets for research-paper recommender systems apply to other domains. This is of particular importance since ground-truths in other domains may fundamentally differ from the ground truths being used for research-paper recommender systems. For instance, we could imagine that datasets with real ratings of movies are more appropriate than some ground-truth that was inferred from e.g. citations. We also believe that the adequacy of CTR, LTR, ATR, and CiTR need more research and discussion. Although we are quite certain that CTR is the most adequate online metric for our scenario, other scenarios might require different metrics.

Please note: We make most of the data that we used for our analysis publicly available [23]. The data should allow replicating our calculations, and performing new analyses beyond the results that we presented in this paper.

References

1. Ricci, F., Rokach, L., Shapira, B., Kantor, B.P. (eds.): Recommender systems handbook, pp. 1–35. Springer, Heidelberg (2011)
2. Torres, R., McNee, S.M., Abel, M., Konstan, J.A., Riedl, J.: Enhancing digital libraries with TechLens +. In: Proceedings of the 4th ACM/IEEE-CS Joint Conference on Digital Libraries, pp. 228–236 (2004)
3. Küçüktunç, O., Saule, E., Kaya, K., Çatalyürek, Ü.V.: Recommendation on Academic Networks using Direction Aware Citation Analysis, pp. 1–10 (2012). arXiv preprint arXiv:1205.1143
4. Gorrell, G., Ford, N., Madden, A., Holdridge, P., Eaglestone, B.: Countering method bias in questionnaire-based user studies. Journal of Documentation **67**(3), 507–524 (2011)
5. Leroy, G.: Designing User Studies in Informatics. Springer, Heidelberg (2011)
6. Ge, M., Delgado-Battenfeld, C., Jannach, D.: Beyond accuracy: evaluating recommender systems by coverage and serendipity. In: Proceedings of the Fourth ACM RecSys Conference, pp. 257–260 (2010)
7. McNee, S.M., Albert, I., Cosley, D., Gopalkrishnan, P., Lam, S.K., Rashid, A.M., Konstan, J.A., Riedl, J.: On the recommending of citations for research papers. In: Proceedings of the ACM Conference on Computer Supported Cooperative Work, pp. 116–125 (2002)
8. Turpin, A.H., Hersh, W.: Why batch and user evaluations do not give the same results. In: Proceedings of the 24th Annual International ACM SIGIR Conference on Research and Development in Information Retrieval, pp. 225–231 (2001)
9. McNee, S.M., Kapoor, N., Konstan, J.A.: Don't look stupid: avoiding pitfalls when recommending research papers. In: Proceedings of the 2006 20th anniversary conference on Computer supported cooperative work, pp. 171–180 (2006)
10. Jannach, D., Lerche, L., Gedikli, F., Bonnin, G.: What recommenders recommend – an analysis of accuracy, popularity, and sales diversity effects. In: Carberry, S., Weibelzahl, S., Micarelli, A., Semeraro, G. (eds.) UMAP 2013. LNCS, vol. 7899, pp. 25–37. Springer, Heidelberg (2013)

11. Knijnenburg, B.P., Willemsen, M.C., Gantner, Z., Soncu, H., Newell, C.: Explaining the user experience of recommender systems. User Model. User-Adap. Inter. **22**, 441–504 (2012)
12. Said, A., Tikk, D., Shi, Y., Larson, M., Stumpf, K., Cremonesi, P.: Recommender systems evaluation: a 3D benchmark. In: ACM RecSys 2012 Workshop on Recommendation Utility Evaluation: Beyond RMSE, Dublin, Ireland, pp. 21–23 (2012)
13. Herlocker, J.L., Konstan, J.A., Terveen, L.G., Riedl, J.T.: Evaluating collaborative filtering recommender systems. ACM Trans. Inf. Syst. (TOIS) **22**(1), 5–53 (2004)
14. Jannach, D., Zanker, M., Ge, M., Gröning, M.: Recommender systems in computer science and information systems – a landscape of research. In: Huemer, C., Lops, P. (eds.) EC-Web 2012. LNBIP, vol. 123, pp. 76–87. Springer, Heidelberg (2012)
15. Beel, J., Gipp, B., Breitinger, C.: Research paper recommender systems: a literature survey. Int. J. Digit. Libr., 2015, to appear
16. Beel, J., Langer, S., Genzmehr, M., Gipp, B., Breitinger, C., Nürnberger, A.: Research paper recommender system evaluation: a quantitative literature survey. In: Proceedings of the Workshop on Reproducibility and Replication in Recommender Systems Evaluation (RepSys) at the ACM RecSys Conference (RecSys), pp. 15–22 (2013)
17. Cremonesi, P., Garzotto, F., Turrin, R.: Investigating the persuasion potential of recommender systems from a quality perspective: An empirical study. ACM Trans. Interact. Intell. Syst. (TiiS) **2**(2), 11 (2012)
18. Zheng, H., Wang, D., Zhang, Q., Li, H., Yang, T.: Do clicks measure recommendation relevancy?: an empirical user study. In: Proceedings of the Fourth ACM RecSys Conference, pp. 249–252 (2010)
19. Cremonesi, P., Garzotto, F., Negro, S., Papadopoulos, A.V., Turrin, R.: Looking for "Good" recommendations: a comparative evaluation of recommender systems. In: Campos, P., Graham, N., Jorge, J., Nunes, N., Palanque, P., Winckler, M. (eds.) INTERACT 2011, Part III. LNCS, vol. 6948, pp. 152–168. Springer, Heidelberg (2011)
20. Hersh, W., Turpin, A., Price, S., Chan, B., Kramer, D., Sacherek, L., Olson, D.: Do batch and user evaluations give the same results? In: Proceedings of the 23rd annual international ACM SIGIR conference on Research and development in information retrieval, pp. 17–24 (2000)
21. Beel, J., Langer, S., Genzmehr, M., Gipp, B., Nürnberger, A.: A comparative analysis of offline and online evaluations and discussion of research paper recommender system evaluation. In: Proceedings of the Workshop on Reproducibility and Replication in Recommender Systems Evaluation (RepSys) at the ACM Recommender System Conference (RecSys), pp. 7–14 (2013)
22. Beel, J., Gipp, B., Langer, S., Genzmehr, M.: Docear: an academic literature suite for searching, organizing and creating academic literature. In: Proceedings of the 11th Annual International ACM/IEEE Joint Conference on Digital Libraries (JCDL), pp. 465–466 (2011)
23. Beel, J., Langer, S., Gipp, B., Nürnberger, A.: The architecture and datasets of docear's research paper recommender system. D-Lib Mag. **20**(11/12) (2014). doi:10.1045/november14-beel
24. Beel, J., Langer, S., Genzmehr, M., Müller, C.: Docears PDF inspector: title extraction from PDF files. In: Proceedings of the 13th Joint Conference on Digital Libraries (JCDL 2013), pp. 443–444 (2013)
25. Lipinski, M., Yao, K., Breitinger, C., Beel, J., Gipp, B.: Evaluation of header metadata extraction approaches and tools for scientific PDF documents. In: Proceedings of the 13th ACM/IEEE-CS Joint Conference on Digital Libraries (JCDL 2013), pp. 385–386 (2013)

26. Beel, J., Langer, S., Genzmehr, M., Nürnberger, A.: Introducing docear's research paper recommender system. In: Proceedings of the 13th Joint Conference on Digital Libraries (JCDL 2013), pp. 459–460 (2013)
27. Beel, J.: Towards effective research-paper recommender systems and user modeling based on mind maps. Ph.D. Thesis. Otto-von-Guericke Universität Magdeburg (2015)
28. Beel, J., Langer, S., Kapitsaki, G., Breitinger, C., Gipp, B.: Exploring the potential of user modeling based on mind maps. In: Ricci, F., Bontcheva, K., Conlan, O., Lawless, S. (eds.) UMAP 2015. LNCS, vol. 9146, pp. 3–17. Springer, Heidelberg (2015)
29. Beel, J., Langer, S., Genzmehr, M., Gipp, B.: Utilizing mind-maps for information retrieval and user modelling. In: Dimitrova, V., Kuflik, T., Chin, D., Ricci, F., Dolog, P., Houben, G.-J. (eds.) UMAP 2014. LNCS, vol. 8538, pp. 301–313. Springer, Heidelberg (2014)
30. Rich, E.: User modeling via stereotypes. Cognitive science 3(4), 329–354 (1979)
31. MacRoberts, M.H., MacRoberts, B.: Problems of citation analysis. Scientometrics 36, 435–444 (1996)

Connecting Emotionally: Effectiveness and Acceptance of an Affective Information Literacy Tutorial

Yan Ru Guo[✉] and Dion Hoe-Lian Goh

Wee Kim Wee School of Communication and Information,
Nanyang Technological University, Singapore, Singapore
{W120030,ashlgoh}@ntu.edu.sg

Abstract. Recent developments in affective computing have provided more options in the way online education can be delivered. However, research on how affective computing can be used in online education is lacking. The research objectives are twofold: to investigate the influence of affective EAs on students' motivation, enjoyment, learning efficacy and intention to use, and to uncover factors influencing their intention to use an online tutorial with affective EAs. To achieve this, 190 tertiary students participated in a between-subjects experiment (text-only vs. affective-EAs). Students benefited from the affective EAs in the tutorial as indicated by the increased learning motivation and enjoyment. Moreover, relevance, confidence, satisfaction, affective enjoyment, and behavioral enjoyment were found to be significant predictors for intention to use.

Keywords: Affective embodied agent · Enjoyment · Information literacy · Intention to use · Knowledge retention · Motivation · Online tutorial

1 Introduction

With the proliferation of and increased accessibility to information brought by the Internet, information literacy (IL) has become a critical skill to be acquired. IL education has thus become the shared responsibility of all educators and information providers [1]. While face-to-face instruction is the most widely adopted method in IL education, online education has been on the rise for the past few years. Compared to face-to-face instruction, online education is not constrained by location or time, can be tailored to fit the learning pace of each student, and is a suitable conduit for lifelong learning. However, the affective aspects of online education are still a largely unresolved area although it is widely known that learning is associated not only with cognitive ability but also with affect [2]. Thus to be effective, online education cannot neglect the affective aspects of learning.

Affective aspects can be infused into online learning systems through interface characters, also referred to as "agents" [3]. An embodied agent (EA) refers to a life-like agent, i.e., one with a physical face and body [3]. Accordingly, an affective EA is defined as one that is capable of eliciting certain affective experiences from users through multiple modalities such as speech, facial expressions and body gestures [4].

© Springer International Publishing Switzerland 2015
S. Kapidakis et al. (Eds.): TPDL 2015, LNCS 9316, pp. 169–181, 2015.
DOI: 10.1007/978-3-319-24592-8_13

Studies have found that the use of affective EAs in a pedagogical role can not only increase students' learning motivation and learning efficacy, but also help them overcome negative feelings such as boredom or frustration during the learning process [3].

Despite their potential, there is a dearth of research in using affective EAs in online IL education. Students' development of IL skills is often impeded by negative affective feelings such as fear of the library, worry that the needed information cannot be found [7]. There is much to be learnt in this area, and the present study is a step in this direction, investigating students' learning attitudes, enjoyment and knowledge retention towards an online tutorial with affective EAs. In particular, the online IL tutorial addresses the affective aspects of IL education, and differs from existing ones which primarily focus on imparting cognitive knowledge to students.

Equally important, users' motivation and enjoyment have been found to be strong predictors of intention to use new technologies [6]. However, their influence in IL education has not been established [7]. Although previous studies have suggested that the use of affective EAs could significantly increase students' learning motivation and enjoyment [7], a deeper understanding of students' attitudes towards online IL education is required to improve its usage. IL education differs from other educational domains as it involves higher-order thinking skills, skills activated when individuals encounter unfamiliar problems, uncertainties, confusions or discrepancies, such as before they start information seeking [8]. The concern over students' affective states should also be reflected in online IL instruction. Therefore, the objectives of the present paper are two-fold. The first is to evaluate the impact of affective EAs in IL tutorials on students' learning motivation, enjoyment, knowledge retention, and intention to use. The second is to investigate the influence of motivation and enjoyment on students' intention to use such a tutorial. Additionally, a widely used IL model, the Information Search Process (ISP) Model, was employed to structure the tutorial [9].

2 Literature Review

Online education has been gaining popularity by libraries to teach IL. For example, Western Michigan University created an online IL tutorial called *ResearchPath*. It incorporated elements of animation, video, audio, and interactivity to provide a range of stimuli for students. *ResearchPath* was evaluated against a previous version, and students expressed a stronger satisfaction and preference for it [10]. Such satisfaction may lead to more positive perceptions and attitudes towards the library.

The relative lack of consideration of users' affect is a main limitation of online IL education [11]. In the new movement to take into consideration of affective states in online IL education, the ISP Model is one of the seminal works [9]. Although called an ISP Model, its scope covers most of what is included in IL education models and standards. Therefore, it is treated as an encompassing IL education model here [12].

The ISP Model highlighted a user's affective states during the information seeking process, and regarded information seeking as a constructive process, which involves not only thoughts and actions, but also feelings. It is divided into six stages: task initiation, topic selection, prefocus exploration, focus formulation, information collection and search closure. It predicts that in the early stages of searching, negative feelings are common, especially when users have little knowledge of what is available or when the search problem is not clear. However, as the search progresses, and the awareness increases, the level of satisfaction and confidence increases correspondingly. At the end, the users will feel a sense of relief or satisfaction when the required information is found, or disappointment and anxiety when it was not.

Studies have shown that the use of affective EAs can foster students' engagement in online learning [7]. It can minimize the communicative gaps in interactions between human and computers, thus increasing learners' motivation [3, 7]. With EAs, learners are more motivated to make sense of what is being presented to them, and more likely to process the information deeply. Moreover, affective EAs can ease possible negative affect during learning. For example, [16] designed an affective EA to help users recover from negative affective states (such as frustration) that resulted from interactions with computer systems. They found that the use of affective EAs eased such negative affective states more effectively compared with a text-only condition. The ability to reduce negative affect is especially important in our present study, as negative affect is common in the information seeking process [17]. In prior work, [7] studied the use of affective EAs in IL education, and found that students in the affective-EAs group were more motivated and derived more enjoyment in acquiring IL knowledge compared to those in neutral-EAs and no-EAs groups. However, that study did not investigate the EAs' impact on students' intention to use IL tutorials, and the roles that motivation and enjoyment play. Studying these relationships would be beneficial to facilitate the integration of affective EAs into IL education.

3 Methodology

3.1 Hypotheses Development

It has been argued that motivation is probably one of the most important components of learning, and a catalyst to achieving one's goals [15]. In our study, the ARCS Model was used as a tool for measuring the motivational support provided by IL tutorial. The model suggests that learning motivation is dependent of four perceptual components: *attention* (learners' response to perceived instructional stimuli), *relevance* (the extent that instructions are associated with prior learning), *confidence* (learners' positive expectation towards their performance) and *satisfaction* (the extent that learners are allowed to practice newly acquired knowledge) respectively.

Enjoyment refers to a general positive disposition towards and liking of the media content [21]. In this study, measuring enjoyment will help understand how the use of affective EAs in an IL tutorial will impact students' reactions on learning. It is a multifaceted concept, encompassing *affective enjoyment* (the willingness to invest emotionally in the experience), *cognitive enjoyment* (the willingness to develop skills and solve problems) and *behavioral enjoyment* (the willingness to participate on a behavioral level) [21]. Further, although it has been pointed out that the instructional effectiveness of using affective EAs is sometimes due to its novelty rather than to the real benefit of increased knowledge retention by students, knowledge retention is still an important construct to measure, because it is the most direct and immediate result from the intervention [3]. Lastly, intention to use refers to the degree to which the participants have formulated plans to perform or not perform a specified behavior in the future [6]. Here, it refers to students' intention to use the IL tutorial. Based on the discussion above, we propose the following hypotheses:

H1: The use of affective EAs in an IL tutorial will provide more (a) attention, (b) relevance, (c) confidence, (d) satisfaction to students in IL learning than without.

H2: The use of affective EAs in an IL tutorial will provide more (a) affective enjoyment, (b) cognitive enjoyment, (c) behavioral enjoyment to students than without.

H3: The use of affective EAs in an IL tutorial will improve knowledge retention to students in IL learning than without.

H4: The use of affective EAs in an IL tutorial will increase students' intention to use the tutorial than without.

With regards to the second research objective, intention to use has frequently been used as an equivalent for the actual performance of the behavior, as the two have an immediate causal relationship [6]. Studies have shown that if students can be motivated in the early stages of learning, they are more likely to engage in sustained learning behaviors [15]. Moreover, enjoyment has been found to be a determinant of behavioral intention to use information systems [6]. Therefore, we expect that students' usage intention is positively associated with motivation and enjoyment derived from the IL tutorial with affective EAs. Thus two more hypotheses are proposed:

H5: (a) Attention, (b) relevance, (c) confidence, (d) satisfaction, is positively associated with intention to use the affective IL tutorial.

H6: (a) Affective enjoyment, (b) cognitive enjoyment, (c) behavioral enjoyment, is positively associated with intention to use the affective IL tutorial.

3.2   Design of the Information Literacy Tutorial

A 15 min online IL tutorial was created, featuring a novice female student learning how to search for academic information and a teaching helping her: Amy (see Fig. 1) and Ms. Tan (see Fig. 2), respectively. They maintained a visual presence throughout the tutorial, and conversations and instructions were presented in speech bubbles (see Fig. 3). This mirrors the approach taken by [18].

Fig. 1. Student EA (Amy)

Fig. 2. Teacher EA (Ms. Tan)

The structure of the tutorial was adopted from the ISP Model, and its storyline mirrored its stages closely. At the start, Amy is presented with a list of topics from which she could choose to write a literature review. She selects the topic of storytelling, and is immediately aware of the lack of knowledge, and of the need for information. This awareness is frequently accompanied by feelings of uncertainty and apprehension [9]. Ms. Tan guided Amy from the start, assuring her that she is not alone in feeling uncertain. Amy starts to look for information, and this stage is typically marked by feelings of confusion, uncertainty, and doubt. Ms. Tan shows empathic emotions and employed empathic words, such as *"This is normal"* to encourage Amy. Additionally, Ms. Tan introduces Amy to the skills required to retrieve more relevant information, such as the use of Boolean operators, and the techniques of backward and forward chaining. For Amy, the next stage involves the gathering of more information, which is accompanied by an increased level of confidence. As the assignment nears completion, Amy feels relieved and a sense of satisfaction. Finally, as Amy prepares to begin writing, her search process concludes [9].

As part of our study, a text-only version of the tutorial was also created (see Fig. 4). In this version, the two EAs were absent, with only textual instructions presented in rectangular dialog boxes. The content remained exactly the same.

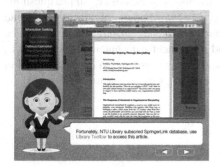 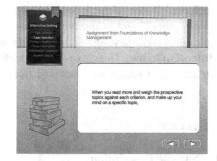

Fig. 3. Affective-EAs Tutorial   **Fig. 4.** Text-only Tutorial

3.3 Experimental Design and Research Sampling

A between-subjects design was used to address the first research objective and the participants were divided into the affective-EAs condition and text-only condition.

A matched-subjects design was employed, where individuals in one condition were roughly matched on their gender and educational background with those in the other condition. This approach is consistent with other studies such as [19, 20].

Recruitment was conducted in two major local universities. Participation was voluntary and anonymous, and no one withdrew during the study. The study was conducted with multiple batches of participants in a controlled laboratory setting. First, participants were briefed on the purpose of the study. After signing the online consent form, they were directed to the study's main website. From this site, they were then assigned to one of the two tutorials based on their gender and educational background. Upon completion of the tutorial, participants were asked to complete an online survey questionnaire. The whole study lasted approximately 25 min, and each participant was given $5 as a token of appreciation. In total, 190 students (both undergraduate and graduate) participated in the study.

3.4 Posttest Instrument

The questionnaire survey comprised six sections. The first collected demographic information such as age, gender, computer experience, and perceived computer knowledge. Next, participants were asked for their subjective opinions of the tutorial, and for suggestions on improvement. The other four sections are explicated below.

- **Motivation.** The ARCS Model was adapted to suit this study's purpose [15]. It consisted of 36 five-point Likert-scale items, in which 12 measure attention, nine measure relevance, nine measure confidence and six measure satisfaction.
- **Enjoyment.** Enjoyment was assessed with 12 five-point Likert-scale items from a survey questionnaire by [5] as it has been validated across various studies.
- **Knowledge retention.** Knowledge retention was evaluated via 13 questions, all of which were developed by the authors. It includes five multiple-answers questions, six true-or-false questions, and two fill in the blank questions. Example questions include: *"Which of the following are the typical information sources for academic information?"* and *"Citations are important in academic writing because ..."* The questions were developed after testing and refinement with potential participants.
- **Intention to use.** Three five-point Likert-scale items were used to measure intention to use, all of which were adopted from an existing survey instrument from [6].

4   Results

4.1   Sample Description

The sample consisted of 68 (35.79 %) males and 122 (64.21 %) females. There were 33 males and 62 females in the affective-EAs condition, and 35 male and 60 females in the text-only condition. The sample age ranged between 18 and 30, with an average of 21.21 years. With regards to their educational background, more than half (57.37 %) were from arts, humanities and social science, around a quarter (25.26 %) were from engineering, and the rest 18.42 % were from natural science.

4.2   Factor and Reliability Analyses

Prior to analyzing the data, principle component analyses were carried out to ensure that items under each construct were more related to its own construct. Only items that were highly loaded (0.5 and above) were retained in the following analyses.

- **Motivation.** Four factors emerged from the factor analysis, with Cronbach's alpha values of 0.805 ($M = 3.39$, $SD = 0.82$) for attention, 0.842 ($M = 3.20$, $SD = 0.73$) for relevance, 0.872 ($M = 3.22$, $SD = 0.90$) for confidence and 0.856 ($M = 3.53$, $SD = 0.76$) for satisfaction.
- **Enjoyment.** The factor analysis yielded three factors, with Cronbach's alpha value of 0.887 ($M = 2.62$, $SD = 0.68$) for affective, 0.903 ($M = 3.68$, $SD = 0.85$) for cognitive and 0.837 ($M = 2.79$, $SD = 0.30$) for behavioral enjoyment.
- **Intention to Use.** The factor analysis yielded a single category with Cronbach's alpha value of 0. 83 ($M = 3.30$, $SD = 0.94$) for intention to use.

4.3   Comparison of Tutorials

The results from the independent samples t-test are summarized in Table 1. There were significant differences between the affective-EAs and text-only conditions with respect to attention, t (188) = 2.80, $p < 0.01$; relevance t (188) = 2.01, $p = 0.04$; satisfaction, t (188) = 3.35, $p = 0.01$; affective enjoyment, t (188) = 3.27, $p = 0.01$; and intention to use t (188) = 2.97, $p < 0.01$. On the other hand, there were no statistically significant differences between confidence, cognitive enjoyment, behavioral enjoyment, and knowledge retention.

Table 1. Summary of independent t-test results (affective-EAs vs. text-only condition)

Variables		Affective-EAs ($n = 95$)		Text–only ($n = 95$)	
		Mean	SD	Mean	SD
Motivation					
H1a	Attention[**]	3.52	0.38	3.36	0.43
H1b	Relevance[*]	3.62	0.50	3.48	0.45
H1c	Confidence	3.23	0.44	3.24	0.56
H1d	Satisfaction[**]	3.40	0.45	3.16	0.52
Enjoyment					
H2a	Affective[**]	2.87	0.85	2.48	0.79
H2b	Cognitive	3.82	0.65	3.59	0.85
H2c	Behavioral	2.89	0.68	2.69	0.76
H3	**Knowledge Retention**	1.09	0.32	1.10	0.29
H4	**Intention to Use**[**]	3.24	0.74	2.88	0.61

Notes: [*]Statistically significant differences at $p < 0.05$; [**]Statistically significant differences at $p < 0.01$

4.4   Investigation of User Acceptance

A multiple linear regression analysis was performed, where the subconstructs of motivation and enjoyment were entered as the independent variables, and intention to use as the dependent variable. The results are summarized in Table 2.

Table 2. Summary of hypotheses test in affective-EAs condition (n = 95)

Hypotheses		Standardized β	t-values	R^2	Results
Motivation					
H5a	Attention	.03	.45		×
H5b	Relevance	.17	2.5^{**}		√
H5c	Confidence	.14	2.6^{*}		√
H5d	Satisfaction	.18	3.4^{**}		√
Enjoyment					
H6a	Affective Enjoyment	.20	3.8^{*}		√
H6b	Cognitive Enjoyment	.02	.31		×
H6c	Behavioral Enjoyment	.16	2.5^{*}		√
				$.65^{**}$	

Note: * Statistically significant differences at $p < .05$; ** Statistically significant differences at $p < .01$

The results revealed that 65 % of the variance was accounted for in the model. Among the motivation measures, hypotheses 5a, 5b, and 5d were supported. Only attention was not significantly associated with intention to use, while the rest were significant: relevance ($\beta = 0.17$, $p < 0.01$), confidence ($\beta = 0.14$, $p < 0.05$) and satisfaction ($\beta = 0.18$, $p < 0.01$). With regards to the measures on enjoyment, hypotheses 6a and 6c were supported. Affective enjoyment ($\beta = 0.20$, $p < 0.05$) and behavioral enjoyment ($\beta = 0.16$, $p < 0.05$) were significant, while cognitive enjoyment was not significantly positively associated with usage intention of the affective IL tutorial.

5   Discussion

5.1   Comparison of Tutorials

Motivation. Unsurprisingly, results showed that participants who used the affective EAs paid more attention to the tutorial than those who did not, consistent with [14] and [22]. The behaviors of the EAs in the tutorial contributed to the increased attention level, and prompted participants to engage in learning. For example, one participant from the affective-EAs condition remarked "*It is visually* appeasing *as it has a variety of graphics*". Interestingly, participants from both conditions rated high on relevance items, although the affective EAs condition seemed to have a slight, albeit statistically nonsignificant edge. Comments about the high relevance of the educational content in the two tutorials include "*It can be applicable to my academics*".

As in relevance, participants reported a low level of confidence across both experimental conditions. This result seems to differ from past work where the use of affective EAs was found to increase the confidence and perceived learning efficacy [20]. This might be because the overwhelming amount of educational content in the tutorial undermined the students' confidence, as many lamented that *"There is too much content in the tutorial"*. Additionally, participants in the affective EAs condition were more satisfied with the tutorial than those from the text-only condition. This is also confirmed by comments such as *"It is very informative and insightful at the same time in an engaging way I learnt something about what would have been really dry"*.

Enjoyment. Participants who interacted with the affective EAs derived more affective enjoyment from the tutorial than those who did not. This concurs with related research (e.g., [18]), demonstrating that affective EAs have a positive impact on affective enjoyment. The positive reinforcement of behaviors by EAs can also increase the level of user enjoyment in the tutorial. In this study, participants connected with the EAs in the tutorial, and this experience resulted in an increase in enjoyment level, which was confirmed by comments such as *"The avatar makes the presentation alive and more realistic"*, and *"It is fun to see avatars in an academic presentation"*.

In contrast, participants experienced high levels of cognitive enjoyment when going through the tutorials. This might be because the tutorial content was perceived to be highly relevant and educational to participants. For example, one participant remarked that *"It was very different from a plain PowerPoint slide, which was nice to see"*. Likewise, participants had similar levels of behavioral enjoyment. Given that this tutorial was deliberately simple by design, users only had to click to it. The behavioral enjoyment aspect will thus need to be further enhanced in the IL tutorial.

Knowledge Retention. There were no statistically significant differences between the two conditions. In fact, students from both conditions scored equally unsatisfactorily, suggesting that the increase in learning performance transcends affective EA mechanisms. A likely reason is that the novelty of incorporating EAs in IL instruction caused students to engage in superfluous cognitive activities. Students were more involved with how the tutorial was presented than what was presented in the tutorial [14]. Although many participants in the affective-EAs condition remarked that using affective EAs in IL tutorials was *"cute and attractive"*, and *"unusual and refreshing"*, their knowledge retention was no better than those from the text-only condition.

Another possible reason was that there was a lack of audio cues. It has been found that in order to understand the materials, learners had to hold auditory information in addition to the textual information on the screen [25]. When asked about how to improve the tutorial, some participants wrote that *"It would be nice to have sound effects"*. Notwithstanding the low level of knowledge retention, some participants wrote that they did benefit from the tutorial. For example, one participant mentioned: *"The tutorial answered some questions I have about finding references"*.

Intention to Use. Participants in the affective EAs condition were more likely to use the IL tutorial with affective EAs than those from text-only condition. Since intention to use has been found to be the single best predictor of actual system usage [6], the findings indicate that affective EAs can foster actual usage of the IL tutorial.

Participants in the affective-EAs condition were more motivated and derived more enjoyment, thus it is natural that they were more likely to use an IL tutorial. This finding is supported by participants' comments in affective-EAs condition such as *"Learning such a tutorial would be useful to us"*, contrasted with some negative feedback such as *"It could have been more engaging and interactive, rather than just lengthy words"* from text-only condition.

5.2   Investigation of User Acceptance

The Influence of Motivation. In terms of the influence of motivation on intention to use, three out of four subconstructs were found to be significant predictors of usage intention. The findings showed that relevance had a positive effect on intention to use an IL tutorial with affective EAs. Students were surprised that *"IL can be so relevant to our studies"*, and the knowledge *"will be useful for future"*. The result is consistent with previous studies [6]. In this study, the results could be explained by the deliberate tutorial design decisions. Compared to other IL tutorials focusing on teaching library-specific knowledge (e.g. location of collections, types of services), a general model was used in our tutorial as an overarching structure, and general information seeking strategies were taught. This enabled students to relate their own information seeking process to this tutorial, despite their diverse educational backgrounds.

Further, confidence was identified as a significant predictor of intention to use. The use of affective EAs positively impacted the perception of the difficulty level of the learning content, thus students felt more confident about the learning process [13]. Prior research has also found that confidence can impact perceived self-efficacy, and thus increases intention to use [6]. This point was reflected in the IL tutorial, where Ms. Tan provided encouragement to Amy, and emphasized that she could succeed if effort was put in. Another important finding is that satisfaction significantly predicted intention to use the affective EA tutorial, consistent with prior studies [23]. In this case, user satisfaction may come from the innovative use of affective EAs in teaching IL, as one participant commented that, *"There were encouraging comments after each slide to reinforce a positive attitude in the student"*.

Surprisingly, attention was found to have a non-significant influence on intention to use an IL tutorial with affective EAs. Although the tutorial with affective-EAs attracted more attention, participants did not view this as a signal in intention to use. We speculate that this demonstrates the pragmatism of students as and they rated relevance over attractiveness in terms of online learning systems. For instance, one participant stressed the importance of relevance in the tutorial *"The materials have to be very relevant, or no matter or interesting the tutorial is, it is not useful"*.

The Influence of Enjoyment. In terms of the impact of enjoyment on intention to use, affective and behavioral enjoyment were found to be significant predictors of students' intention to use the IL tutorial with affective EAs. This finding extends previous work which found that enjoyment was a significant predictor of intention to use educational systems [24], and provided a more nuanced perspective by closely examining the impact of the three different dimensions of enjoyment use.

The significant influence of affective enjoyment on intention to use confirms the importance of affect in online education [11]. Moreover, this finding affirms the importance of creating an affective learning environment where emotional support and help are provided to alleviate frustration and anxiety in the learning process [7, 15]. In the context of this study, such an environment was supported through the simulated conversation between the student and teacher, as well as the encouraging words from the teacher to reassure the student.

Moreover, behavioral enjoyment was found to be a significant predictor. It is reasonable to believe that the behaviors of the EAs in the tutorial prompted participants to engage in learning. Additionally, although the main action needed for going through the IL tutorial is to read and click on to get to the next page, the positive results indicates that simple interaction styles may be all that is necessary when creating effective online tutorials [16]. However, cognitive enjoyment was not significant in predicting usage intention. The reason might be similar to that of confidence. Students felt overloaded by the amount of IL knowledge provided in the tutorial. Thus an important lesson learnt in this study is that content should be appraised for quantity.

6 Conclusion

Informed by the recent developments in affective EAs, this study incorporated them into an IL tutorial to increase students' learning motivation and enjoyment, as well as to ease potential negative feelings during the information seeking process. The ISP Model was utilized as a theoretical framework in the IL tutorial.

From a research perspective, this study affirms the effectiveness of using affective EAs to increase students' motivation and enjoyment in online education. These empirical findings are important as they show how librarians can rethink library instruction programs to cater to the needs of students. Relatedly, this study highlights the role of affect in the context of online education, which has been neglected in previous educational systems. This study shows that when affect is incorporated, students perceive such systems to be more motivational and enjoyable. Therefore, it indicates that students indeed pay attention to the affective aspects in online learning.

In addition, the findings have implications for practitioners such as curriculum designers and library professionals. First, this study addresses the criticism of the lack of affect in online education by using affective EAs, thus aligning IL instruction with students' expectations. The learning environment created by affective EAs was found to be motivational and enjoyable, since it was an environment that students were accustomed to and could readily engage in. Second, the tutorial was constructed from rudimentary elements of a graphical user interface, such as dialog boxes, buttons, and text. It is possible to create affective EAs without employing complicated technologies but still be effective, illustrated by the improvement in user motivation and enjoyment. This has valuable implications to libraries or other interested institutions without resources for sophisticated software development.

Although the study has yielded valuable insights, there are several limitations that may reduce generalizability of the results. First, the topic in the IL tutorial was confined to storytelling, thus students from other educational backgrounds may not have been able to relate well to the content. Some participants may prefer topics from their own disciplines. Hence, future studies could incorporate other topics and domains, either general or library specific, to accommodate students with diverse backgrounds and interests. Further, the amount of content in the tutorials was probably heavy for students, as indicated by students' laments on the undermined confidence in the subjective feedback. Future work should balance the amount of knowledge in the tutorial and avoid overloading students. In spite of these limitations, this paper affirms the importance of affect in online education, and points to the promise of using affective EAs in increasing students' learning motivation and enjoyment in online education.

References

1. Thomas, N.P., Crow, S.R., Franklin, L.L.: Information Literacy and Information Skills Instruction: Applying Research to Practice in the 21st Century School Library. ABC-CLIO, Santa Barbara, CA (2011)
2. Lopatovska, I., Arapakis, I.: Theories, methods and current research on emotions in library and information science, information retrieval and human–computer interaction. Inf. Process. Manage. 47(4), 575–592 (2011)
3. Mumm, J., Mutlu, B.: Designing motivational agents: the role of praise, social comparison, and embodiment in computer feedback. Comput. Hum. Behav. 27(5), 1643–1650 (2011)
4. Beale, R., Creed, C.: Affective interaction: How emotional agents affect users. Int. J. Hum. Comput. Stud. 67(9), 755–776 (2009)
5. Pe-Than, E.P.P., Goh, D.H.-L., Lee, C.S.: Enjoyment of a Mobile Information Sharing Game: Perspectives from Needs Satisfaction and Information Quality. Springer-Verlag, Berlin Heidelberg (2012)
6. Venkatesh, V., Morris, M.G., Davis, G.B., Davis, F.D.: User acceptance of information technology: toward a unified view. MIS Q. 27(3), 425–478 (2003)
7. Guo, Y.R., Goh, D.H.-L., Luyt, B.: Using Affective Embodied Agents in Information Literacy Education. IEEE, London (2014)
8. King, F., Goodson, L., Rohani, F.: Higher Order Thinking Skills. Florida State University, Tallahassee (2009)
9. Kuhlthau, C.C.: Seeking Meaning: A Process Approach to Library and Information Services. Libraries Unlimited, Westport (2004)
10. Sachs, D.E., Langan, K.A., Leatherman, C.C., Walters, J.L.: Assessing the effectiveness of online information literacy tutorials for millennial undergraduates. Coll. Undergraduate Libr. 20(3–4), 327–351 (2013)
11. Zhang, P.: The affective response model: a theoretical framework of affective concepts and their relationships in ICT context. Manage. Inf. Syst. Q. 37(1), 247–274 (2013)
12. Wilson, T.D., Ford, N.J., Ellis, D., Foster, A.E., Spink, A.: Information seeking and mediated searching: Part 2. uncertainty and its correlates. J. Am. Soc. Inform. Sci. Technol. 53(9), 704–715 (2002)
13. Doumanis, I., Smith, S.: An empirical study on the effects of embodied conversational agents on user retention performance and perception in a simulated mobile environment (2013)
14. Dunsworth, Q., Atkinson, R.K.: Fostering multimedia learning of science: exploring the role of an animated agent's image. Comput. Educ. 49(3), 677–690 (2007)

15. Keller, J. M. Manual for instructional materials motivational survey (IMMS). Unpublished manuscript, Tallahassee, FL. (1993)
16. Klein, J., Moon, Y., Picard, R.W.: This computer responds to user frustration: theory, design, and results. Interac. Comput. **14**(2), 119–140 (2002)
17. Kracker, J.: Research anxiety and students' perceptions of research: an experiment. Part I. effect of teaching Kuhlthau's ISP model. J. Am. Soc. Inform. Sci. Technol. **53**(4), 282–294 (2002)
18. Maldonado, H., Lee, J.-E. R., Brave, S., Nass, C., Nakajima, H., Yamada, R., Iwamura, K., Morishima, Y.: We learn better together: enhancing eLearning with emotional characters. Int. Soc. Lear. Sci. 408–417 (2005)
19. Lin, L., Atkinson, R., Christopherson, R., Joseph, S., Harrison, C.: Animated agents and learning: does the type of verbal feedback they provide matter? Comput. Educ. **67**, 239–249 (2013)
20. Mayer, R.E., DaPra, C.S.: An embodiment effect in computer-based learning with animated pedagogical agents. J. Exp.l Psychol. Appl. **18**(3), 239–252 (2012)
21. Nabi, R.L., Krcmar, M.: Conceptualizing media enjoyment as attitude: implications for mass media effects research. Commun. Theory **14**(4), 288–310 (2004)
22. Lee, R.J.E., Nass, C., Brave, S.B., Morishima, Y., Nakajima, H., Yamada, R.: The case for caring colearners: the effects of a computer-mediated colearner agent on trust and learning. J. Commun. **57**(2), 183–204 (2007)
23. Liu, I.-F., Chen, M.C., Sun, Y.S., Wible, D., Kuo, C.-H.: Extending the tam model to explore the factors that affect intention to use an online learning community. Comput. Educ. **54**(2), 600–610 (2010)
24. Wu, J., Liu, D.: The effects of trust and enjoyment on intention to play online games. J. Electron. Commer. Res. **8**(2), 128–140 (2007)
25. Jeung, H.J., Chandler, P., Sweller, J.: The role of visual indicators in dual sensory mode instruction. Educ. Psychol. **17**(3), 329–345 (1997)

Applications of Digital Libraries

A Survey of FRBRization Techniques

Joffrey Decourselle[✉], Fabien Duchateau, and Nicolas Lumineau

LIRIS, UMR5205, Université Claude Bernard Lyon 1, Lyon, France
{joffrey.decourselle,fabien.duchateau,nicolas.lumineau}@liris.cnrs.fr

Abstract. The Functional Requirements for Bibliographic Records (FRBR), an emerging model in the bibliographic domain, provide interesting possibilities in terms of cataloguing, representation and semantic enrichment of bibliographic data. However, the automated transformation of existing catalogs to fit this model is a requirement towards a wide adoption of FRBR in libraries. The cultural heritage community proposed a notable amount of FRBRization tools and projects, thus making it difficult for practitioners to compare and evaluate them. In this paper, we propose a synthetic and relevant classification of the FRBRization techniques according to specific criteria of comparison such as model expressiveness or specific enhancements.

1 Introduction

A large majority of current IT systems used by librarians to manage their bibliographic records are based on old standards like the well-known MAchine Readable Cataloguing (MARC) format, designed in 1965 and used around the world in different versions like MARC21 or UNIMARC. Although these formats are references in libraries, they suffer from many fields with ambiguous semantic (e.g. 7xx fields in MARC21) and from a lack of mechanisms to represent complex relationships between records, i.e., bibliographic families composed by different executions and many related works. Furthermore, librarians nowadays have to deal with a large amount of storage formats (e.g., ebook, streams) and new data sources. This may result in bad cataloguing practices, numerous typing errors and finally increasing cataloguing costs [12]. However, best practices from Semantic Web encourage the reuse of data which implies, in the context of cultural heritage data, to improve interoperability of bibliographic catalogs [3].

In response to the discussions around major librarian's topics such as reducing cataloguing costs and enhancing the representation of bibliographic families, the IFLA Study Group published, in 1998, new recommendations for cataloguing and a set of concepts to represent bibliographic data in the context of the Functional Requirements for Bibliographic Records (FRBR) [20]. FRBR offers a flexible framework to represent any cultural content in library [18], new benefits

This work has been partially supported by the French Agency ANRT (www.anrt.asso.fr), the company PROGILONE (http://www.progilone.fr/), a PHC Aurora funding (#34047VH) and a CNRS PICS funding.

S. Kapidakis et al. (Eds.): TPDL 2015, LNCS 9316, pp. 185–196, 2015.
DOI: 10.1007/978-3-319-24592-8_14

for improving searching and visualization [13] and new possibilities for semantic enrichment of cultural entities [8]. The adoption of FRBR in cultural institutions requires different issues to be solved like the transformation of legacy data into the new entity-relationship model, a process mainly called **FRBRization** [1,31]. The last decade has seen the emergence of a significant amount of FRBRization tools and projects, thus making it difficult for practitioners to gather relevant information [34].

This paper presents a survey of FRBRization techniques, both research projects and full commercial tools. To the best of our knowledge, our work is the first classification of these techniques according to criteria such as the type of transformation, the model expressiveness and its enhancements (see Sect. 4). This classification is useful for practitioners who need the keys to quickly understand and identify suitable techniques w.r.t. their requirements. The rest of the paper is organized as follows: Sect. 2 introduces preliminaries about FRBR and FRBRization, then we present the related work in Sect. 3. Our classification is detailed in Sect. 4. Section 5 concludes the article.

2   Preliminaries About FRBR and FRBRization

FRBR provides a set of concepts organized in an entity-relationship structure to represent bibliographic families and subject authority data of any cultural material with no ambiguity [10]. The main contribution of FRBR is that it deeply redefines the notion of what an intellectual work is and its representation [5]. Entities are divided into three groups: the first group focuses on the record's representation, the second one is about related persons or corporate body and the third group represents what a work is about. Contrary to legacy models, several levels of metadata can be used to describe each bibliographic record [23].

As illustrated in Fig. 1, the first group is composed of entities **Work** (the most abstract object which represents an intellectual or artistic idea), **Expression** (the realization of a Work in various intellectual forms, such as a translation), **Manifestation** (the physical representation of an Expression in terms of material attributes) and **Item** (a single example of a Manifestation, for example a single book on a library's shelf). The components of the second group represent humans, groups or organizations and have been later gathered under the **Agent** name. In Fig. 1, each entity from the first group has a link with an Agent. The third and last group of the FRBR model concerns the subject relation of a work. It allows the model to answer the question of what a Work is about and to align several Works according to specific criteria (e.g., same topic, review of).

The Functional Requirements for Authority Data (FRAD) [24] and the Functional Requirements for Subject Authority Data (FRSAD) [32] have later been added to FRBR to cover authority data and complex relationships between entities respectively. For instance, Fig. 1 depicts examples of these improvements like the new entity Family from FRAD or the new concepts of Thema and Nomen from FRSAD. These three functional requirements, FRBR, FRAD and FRSAD are called the FRBR Family and must be seen as a single architecture for the

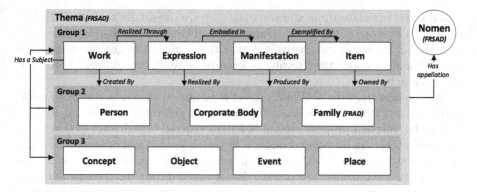

Fig. 1. Conceptual representation of the FRBR Family Model

new cataloguing concepts. Because the FRBR model have been created regardless of technological constraints a large amount of provided concepts covers a too large scope for being used in practical implementations. In other words, FRBR must not be used directly as a final object model. Thus, there is a branch of the FRBR researches which is focused on modelling practical ontologies based on FRBR but adapted to the user needs [11,27].

The adoption of a FRBR-based model in existing systems is known as the FRBRization process. Such a transformation relates to the well known challenges of data integration and data conversion [2]. This study focuses on a specific part of the process which aims at identifying, creating and connecting FRBR entities from existing records. In this context, FRBRization is widely regarded as a complete metadata migration framework rather than a simple algorithm [15]. The initial phases of a FRBRization process have been formalized as follows: pre-processing to harvest and normalize the input catalog, extraction to generate the FRBR entities and relationships and post-processing with deduplication and adaptation of data for visualization [1]. The next section presents the different attempts of surveys and classifications for FRBRization techniques.

3   Related Work

One of the earliest survey about FRBRization was conducted at the Online Computer Library Center (OCLC) [13]. Although focused on the benefits of FRBR cataloguing, it also reviews the major solutions which implemented this model. However, the projects are simply listed along with their properties and available technical details to mainly illustrate the contrast between a growing interest in FRBRization and the lack of successful tools.

Another notable work is the observation of the whole landscape of FRBR implementations in libraries by Zhang & Salaba [33]. The existing implementations have been classified according to the following categories: the full-scale systems (i.e., that are mainly FRBRized systems), the prototypes or experimental systems and the algorithms and software that are essentially tools to FRBRize catalogs. The scope of the study is large since all kinds of FRBR-related projects are covered. Although exhaustive (at that time), such a broad classification is not sufficient to understand technical features between two different FRBRization approaches.

Researchers from the TelPlus[1] project have also worked on a survey to introduce and list FRBR experiments [30]. They have presented, with a rich description, all the relevant projects related to FRBR ordered in high level sections. The conversion tools are introduced under a Research section and presented as a list. More recently, Aalberg et al. presented a list of FRBRization projects [2]. For each project, they studied the possible issues raised during the transformation of records, and more specifically in terms of common structures. The authors were able to propose solutions for interpreting these structures and to FRBRize them correctly. However, the list of projects is not exhaustive since it mainly aims at illustrating structural problems.

To summarize, existing surveys presented the state of the art of FRBRization techniques, but mainly as an unsorted list. In addition, the techniques can be described with a specific goal in mind (e.g., illustrating problems of conversion). The properties and features of each technique are provided textually, which does not facilitate a quick comparison. In the next section, we propose the first classification dedicated to FRBRization techniques.

4   A Novel Classification of FRBRization Techniques

This survey aims at exploring the various issues faced by existing FRBRization techniques, bringing a more precise view of this migration process and facilitating the comparison of existing solutions through a multi-criteria classification. Because many projects gravitate around FRBR, we do not consider the Integrated Library Systems (ILS) which do not include a transformation tool. This choice is justified with regards to current and future challenges on FRBRization such as scalability and semantic enrichment. Besides, our classification focuses on semi-automated techniques, i.e., any FRBRization solution in which most of the algorithm proceeds without human intervention. This excludes manual projects which have not detailed any automated process to transform their catalog, for instance the Austlit project [4] or Data.BNF [19].

Our classification makes use of three criteria of comparison:

- **Type of FRBRization** which relates to the methods for identifying FRBR entities from the legacy model. Two strategies are found in the literature. The former consists of grouping physical form of records by comparing descriptive keys (e.g., a concatenation of author names and title) to deduce more abstract levels (see Sect. 4.1). The latter aims at interpreting the original fields of the records (e.g., with mapping rules) to build the FRBR entities and relationships (see Sect. 4.2).

---

[1] http://www.theeuropeanlibrary.org/confluence/display/wiki/TELplus+project.

- **Model Expressiveness** that deals with the models designed to receive FRBRized data. Three variations are presented: a *limited model* means that the main entities from the first group of FRBR are not completely implemented, a *standard model* indicates that the FRBRization technique takes into account the entities and relations from the three groups of the basic FRBR model and an *enhanced model* corresponds to specific implementations where significant changes have been made to the initial FRBR model.
- **Specific Enhancements** used to improve the quality or the performance of the FRBRization process. It may include additional steps, improvements of algorithms or interoperability enhancements (see Sect. 4.3).

Figure 2 depicts the FRBRization techniques classified according to our criteria. For instance, the FRBRization project TelPlus is classified under *rule-based interpretation of fields, standard model* and it includes *clustered deduplication* and *evaluation metrics* as specific enhancements. The rest of this section provides details for each technique according to these criteria.

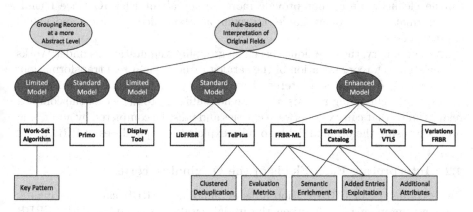

Fig. 2. Classification of the FRBRization Techniques, with the Type of FRBRization in light-grey circled boxes, the Model Expressiveness in dark-grey circled boxes, the Solutions in white squared boxes and the Specific Enhancements in light-grey squared boxes

4.1 Grouping Records at a More Abstract Level

When FRBRizing, it appears that a large part of the MARC records corresponds to the FRBR Manifestation entity. In such a case, the most complex task is to produce Work and Expression entities from the MARC records. Thus, the initial intuition is to group the derived Manifestations into relevant clusters which represent a more abstract level. A common solution deals with the detection of the Work entities based on generated description keys from Manifestation attributes, and to deduce the primal Expression that links the Works to the Manifestations.

Among the first projects about FRBRization, the studies from OCLC describe how to automatically group records at a more abstract level [5]. The well-known **Work-Set algorithm**, applied in FictionFinder and WorldCat's catalog [16], was designed to produce "sets of Works" for MARC records based on the generation and the comparison of normalized keys [17]. This process is based on three main steps to build these keys: constructing an author portion, constructing a title portion and grouping these portions according to four specific patterns. Each portion is built according to mappings with MARC fields and matching with authority files. These studies reveal two major issues from this technique: the precision about the semantic level of the deduced Works (e.g., a too large clustering may lead to groups at more higher level than the original FRBR Work) and the difficulty to identify the variations of a same Work at Expression level.

Another experience in the grouping records category is **PRIMO** from *Ex Libris* [25]. This commercial discovery tool provides FRBRization features by proposing transformation options like automatic grouping of records under a single Work. In addition, the FRBRization includes a pre-processing step which consists in harvesting catalog data into an intermediate model so that records can be cleaned. We cannot provide more details about PRIMO since technical information are not available, for example about deduplication and model expressiveness.

As a summary, the techniques based on grouping Manifestations under Works can be seen as a reorganization of the catalog rather than a real transformation. Furthermore, even if the patterns (based mainly on title and authors fields) are useful for clustering records, they are not sufficient to detect all the complex relationships that can exist between entities and must be completed by additional processes using the original fields to fully exploit the capabilities of FRBR.

4.2   Interpreting Each Fields of the Original Records

The second type of FRBRization aims at building a FRBR catalog by mainly applying mapping rules between the initial catalog metadata and the FRBR attributes. In the context of MARC catalogs, such a process is basically performed by reading each record's field and identifying whether it matches any rule. Depending on the quality and completeness of the input catalog data, the rules have to be more or less complex to create not only the FRBR entities but also the different relationships that link them. It is worth noting that a single MARC record usually generates more than one FRBR entity. Indeed, a MARC record contains information for different FRBR entities. Conversely, information about the same FRBR entity may appear in several MARC records, in particular when the second and the third FRBR groups are considered. In the rest of this section, we describe the major projects based on this kind of extraction technique.

The **LC Display Tool** is one of the first prototype for FRBRization provided by the researchers from the Library of Congress in 2003 [28]. This solution takes as input MARC21 records and it can produce FRBR XML and HTML. The process uses MARCXML as an intermediate format to benefit from the XML format and from the MARC leader fields. Rules stored in XSL files are applied to the MARCXML catalog to clean the data and to generate FRBR entities. The process returns a list of Works, each of them hierarchically linked to one

Expression and one Manifestation and containing title and author attributes. This representation focuses on the main entities from the first group of the FRBR model which is a very limited expression of the capabilities of FRBR. Such choices for displaying results have also raised early issues in terms of visualization of FRBR entities. Indeed the hierarchical organization of XML (e.g., if using HTML) combined with the large amount of relationships available in FRBR may produce very large files which makes it more difficult to read.

The **TelPlus** project [21] allowed the realization of a semi-automatic transformation process for MARC records. The FRBRization is based on a preprocessing stage for correcting and enriching records, a rule-based extraction step and a deduplication phase. Several issues are raised by this study such as the minimum information needed in each record to guarantee a good quality of the process, the large amount of duplicate FRBR entities generated and the high complexity of some mapping rules. They face the first issue by providing a list of requirements to filter the records in input. The second one (i.e., duplicates entities) is managed with a clustered deduplication process (see Sect. 4.3). They deal with the last issue (i.e., complex rules) by using an extension of the extraction tool from Aalberg [1] to enable the implementation of more complex rules. The experiments have been performed on a large number of heterogeneous records (from TelPlus project) and the output model used is a quite standard version of initial FRBR model designed in RDF[2]

LibFRBR is another rule-based implementation built in the context of FRBRizing the Chinese Koha system [9]. This project includes a FRBRization tool written in Perl which can harvest data in MARC21 or CMARC, extract FRBR entities from the three groups and store them in FRBRXML or in the Koha database. There are only few information about the model used by the tool which seems to be close of the standard initial FRBR model. A cataloguing interface for performing corrections on the transformed data and for clustering equivalent FRBR entities has been implemented and evaluated. Compared to previous tools, the advantages of LibFRFR reside both in the edition of the mapping rules (which are not hard-coded) and in the cataloguing interface which involves librarians in the process.

Virtua VTLS is also a solution which provide a rule-based interpretation of input fields to extract FRBR concepts [14]. A specificity of this commercial tool is that MARC records can coexist with FRBR entities, i.e., FRBR is implemented as an optional layer. Although only few technical informations are given about the FRBRization, many options and user interfaces have been presented to analyse MARC records and FRBRize the catalog. Furthermore, the FRBR model has been extended to allow complex relationships between different works of a same bibliographic family (e.g., *Super Works*).

The **Variations** or **V/FRBR** project, dedicated to musical content, brings a new FRBRization tool with an extended FRBR implementation [26]. A major proposal of this project deals with the model because three XML-based schemas have been released: a strict version of FRBR and FRAD entities, an extended

---

[2] FRBR in RDF, http://vocab.org/frbr/core.html.

model which globally allows to add more meta-information to the previous strict model and the V/FRBR model which is based on the extended model in which attributes have been adapted to the musical context. The other contribution is a Java-based tool that has been designed to FRBRize musical data to build a FRBRized OPAC called Scherzo [22]. The results of the project show that FRBRization may also succeed in real-world contexts by using sophisticated extraction rules. However, the mapping rules were apparently hardcoded even if a complete documentation have been freely provided with the project.

In **eXtensible Catalog** (XC) [6], FRBRization is handled by the Metadata Service Toolkit with the goal of improving the quality of the migration. This open-source tool is composed of a pre-processing phase with mechanisms to harvest and normalize records (from OAI-PMH repositories[3]), a transformation phase to migrate the normalized data (FRBRization) and an aggregation phase to detect and merge duplicate entities [7]. In this project, the Metadata Service Toolkit is a component inside a full ILS solution, and the implemented model (the XC schema) has been extended, mainly with the *Holding* entity, i.e., a specific implementation of the FRBR Item entity to handle MARC21 Holding records. Several interesting challenges of FRBRization are faced by XC in terms of input model management and normalization of extracted data (see Sect. 4.3).

FRBR-ML [29] is an extended version of Aalberg's FRBRization tool [1]. In addition to the transformation aspects, it aims at promoting interoperability between different models such as MARC and FRBR. Thus, this new system is built on an XML based intermediate model which was designed to ease exporting data in various semantic formats. The FRBRized records are represented with an hybrid semantic format which may use external attributes from the Linked Open Data.

To summarize, all these rule-based solutions cover various model needs. FRBRization techniques based on enhanced model may be more complex to design but offer a better completeness in terms of transformed data. Furthermore, a major part of the solutions has managed to improve the quality of their transformation at different steps. At the beginning of the process, some have chosen to involve the user feedback with an intuitive interface to create or refine the rules and to ease building the output model. At the end of the process, several solutions have provided an aggregation phase to find and merge duplicate records. It is a critical step especially when rules are applied on fields that may contain duplicate values.

4.3   Specific Enhancements

The last criterion of our classification is about specific enhancements, mainly designed for tackling quality or performance issues. These enhancements have been isolated from the standard features presented previously because they don't represent a fundamental step of the initial FRBRization process but are needed to face the metadata migration challenges (see Sect. 2).

---

[3] OAI-PMH, https://www.openarchives.org/pmh/.

The FRBR model and its extensions are suitable for the representation of many information in the cultural heritage domain. However, the original model only provides a limited set of attributes and relationships labels which are mainly concepts. Thus, to cover the practical requirements of librarians and to enhance related information, it may be useful to include **additional attributes** while modelling with FRBR. Several tools are able to integrate vocabularies from other models. For instance, Virtua VTLS uses RDA vocabulary and rules[4] to provide a more structured and interoperable catalog. Similarly, eXtensible Catalog extends its models with elements from RDA and Dublin Core [6].

The **exploitation of added entries** from the initial data has been widely studied to find solutions for extracting data stored in the MARC21 7xx fields[5]. Such fields may contain additional authors, geographic names, information about Agents at the Expression level (e.g., cover drawer, translator) and the challenge is to store this information in the correct FRBR entity. Variations [26] and eXtensible Catalog [6] roughly use the same method for exploiting these added entries: creation of new FRBR entities from specific fields (e.g., in MARC21 *700, 710* or *740* with a second indicator equal to *2*). Variations may create several levels of Works[6] while eXtensible Catalog can directly create a new Expression and its parent Work[7]. In FRBR-ML, the idea for disambiguating information contained in the MARC21 7xx fields is to analyse similar records that may not contain the ambiguity [29]. This search is performed either in the same collection, in external collections (z39.50 or SRU/SRW) or in knowledge bases such as the Linked Open Data cloud[8] (LOD).

An interesting enhancement for FRBR catalogs is the **semantic enrichment**, i.e., the addition of extra information such as new attributes or relationships. Using the potential of external data can both enhance the completeness of the process and also improve the validation of the results. However, this task involves many new issues because of the heterogeneity of the databases that can be used. For instance, using both information from structured sources (e.g., LOD) and non-structured sources (e.g., websites) involve effective and efficient matching tools at schema and instance levels. FRBR-ML performs matching to external knowledge bases (LOD cloud) for discovering the semantic type of an ambiguous field (e.g., person, location) and detecting missing relationships between FRBR entities [29]. As for eXtensible Catalog, it can perform a specific data fusion process to merge the attributes from equivalent entities [7].

Descriptive keys are used for comparing records and for detecting duplicates, thus improving quality [7,9,21,29]. The issue is to find the most relevant **patterns for descriptive keys**. In the Work-Set project, the OCLC has proposed four patterns to fulfil this goal [16]. For instance, one of the patterns is the com-

---

[4] Virtua does not provide any public specification about its use of RDA.

[5] MARC21 7xx fields, http://www.loc.gov/marc/bibliographic/bd70x75x.html.

[6] Converting MARC to FRBR, http://www.dlib.indiana.edu/projects/vfrbr/projectDoc.

[7] XC, https://code.google.com/p/xcmetadataservicestoolkit/wiki/TranserviceIntro.

[8] Linked Open Data cloud, http://linkeddata.org/.

bination of author name and title. These four OCLC patterns are still a reference for the generation of descriptive keys for bibliographic content, and they have been applied in the most recent FRBRization techniques.

The deduplication, which aims at detecting duplicate records, is one of the most time consuming step because of the Cartesian product applied between all records. The **clustered deduplication enhancement** impacts efficiency by reducing the execution time of the deduplication. In the TelPlus project, each FRBR entity produces a set of keys (according to the patterns for descriptive keys from OCLC [16]). The intuition is to group in the same cluster all entities that share at least one identical value for one of their keys. The last step identifies duplicates inside clusters by comparing their keys and computing similarity values with thresholds.

Evaluation of the FRBRization process is crucial, but it requires the **definition of evaluation metrics**. TelPlus proposes a metric for the evaluation of the aggregation level (i.e., the percentage of duplicate entities extracted). The FRBR-ML project [29] defines three metrics to measure the degree of completeness (i.e., the amount of information lost during FRBRization), the minimality rate (i.e., the amount of redundant information, at the property, record and collection levels) and the extension rate (i.e., the amount of enriched information).

Finally, our classification enables the identification of relevant techniques according to the type of FRBRization and the model expressiveness. Compared to the initial FRBRization process, most tools have proposed specific enhancements either for effectiveness (quality aspects) or efficiency (performance aspects).

5   Conclusion

This paper introduces a survey about FRBRization tools and projects to provide a more comprehensive view of this process. To the best of our knowledge, this is the first classification according to criteria such as the type of FRBRization and the model expressiveness. Our contribution is useful both for librarians who need to choose a FRBRization technique and for IT company working on cultural heritage data who plan to design mature FRBRization tools. According to the results of our study, we advocate that the FRBRization process must be refined to cover the most recent challenges of metadata migration. For instance, involving user feedback at the initialization phase (e.g., rule selection) can reduce human mistakes. Furthermore, the quality of the process can be enhanced by providing additional steps at the post-processing phase such as an automatic evaluation step or a semantic enrichment step based on external sources.

References

1. Aalberg, T.: A process and tool for the conversion of MARC records to a normalized FRBR implementation. In: Sugimoto, S., Hunter, J., Rauber, A., Morishima, A. (eds.) ICADL 2006. LNCS, vol. 4312, pp. 283–292. Springer, Heidelberg (2006)

2. Aalberg, T., Žumer, M.: The value of MARC data, or, challenges of frbrisation. J. Documentation **69**, 851–872 (2013)
3. Alemu, G., Stevens, B., Ross, P., Chandler, J.: Linked Data for libraries: benefits of a conceptual shift from library-specific record structures to RDF-based data models. New Libr. World **113**(11/12), 549–570 (2012)
4. Ayres, M.L., Kilner, K., Fitch, K., Scarvell, A.: Report on the successful austlit: Australian literature gateway implementation of the FRBR and INDECS event models, and implications for other FRBR implementations. Int. Cataloguing Bibliographic Control **32**, 8–13 (2003)
5. Bennett, R., Lavoie, B.F., O'neill, E.T.: The concept of a work in WorldCat: an application of FRBR. Libr. Collections Acquisitions Tech. Serv. **27**, 45–59 (2003)
6. Bowen, J.: Moving library metadata toward linked data: opportunities provided by the eXtensible catalog. In: International Conference on Dublin Core and Metadata Applications (2010)
7. Bowen, J., Investigat, C.P.: Supporting the eXtensible Catalog through Metadata Design and Services (2009)
8. Buchanan, G.: FRBR: enriching and integrating digital libraries. In: Proceedings of the 6th ACM/IEEE-CS Joint Conference on Digital Libraries, pp. 260–269 (2006)
9. Chang, N., Tsai, Y., Dunsire, G., Hopkinson, A.: Experimenting with implementing FRBR in a Chinese Koha system. Libr. Hi Tech News **30**, 10–20 (2013)
10. Committee, S., Group, I.S.: Functional Requirements for Bibliographic Records: Final Report, vol. 19. K.G. Saur Publisher, Munich (1998)
11. Coyle, K.: FRBR, twenty years on. Cataloging Classif. Q. **53**, 1–21 (2014)
12. Denton, W.: FRBR and the history of cataloging. In: Taylor, A.E. (ed.) Understanding FRBR: What It Is and How It Will Affect Our Retrieval Tools. Libraries Unlimited, Westport (2007)
13. Dickey, T.J.: FRBRization of a library catalog: better collocation of records, leading to enhanced search, retrieval, and display. Inf. Technol. Libr. **27**, 23–32 (2008)
14. Espley, J.L., Pillow, R.: The VTLS implementation of FRBR. Cataloging Classif. Q. **50**, 369–386 (2012)
15. Hegna, K., Murtomaa, E.: Data mining MARC to find: FRBR? Int. Cataloguing Bibliographic Control **32**, 52–55 (2003)
16. Hickey, T.B., O'Neill, E.T.: FRBRizing OCLC's WorldCat. Cataloging Classif. Q. **39**, 239–251 (2005)
17. Hickey, T.B., Toves, J.: FRBR Work-Set Algorithm (2.0). OCLC, Dublin (2009)
18. Le Bœuf, P.: FRBR: hype or cure-all? Introduction. Cataloging Classif. Q. **39**(3–4), 1–13 (2005)
19. Le Bœuf, P.: Customized OPACs on the semantic web: the OpenCat prototype. In: IFLA World Library and Information Congress, pp. 1–15 (2013)
20. Madison, O.M.: The origins of the IFLA study on functional requirements for bibliographic records. Cataloging Classif. Q. **39**, 15–37 (2005)
21. Manguinhas, H.M.A., Freire, N.M.A., Borbinha, J.L.B.: FRBRization of MARC records in multiple catalogs. In: Hunter, J., Lagoze, C., Giles, C.L., Li, Y.F. (eds.) JCDL, pp. 225–234. ACM (2010)
22. Notess, M., Dunn, J.W., Hardesty, J.L.: Scherzo: a FRBR-based music discovery system. In: International Conference on Dublin Core and Metadata Applications, pp. 182–183 (2011)
23. O'Neill, E.T.: FRBR: functional requirements for bibliographic records. Libr. Res. Tech. Serv. **46**, 150–159 (2002)

24. Patton, G.E.: An introduction to functional requirements for authority data (FRAD). In: Taylor, A.G. (ed.) Understanding FRBR: What It Is and How It Will Affect Our Retrieval Tools, pp. 21–27. Libraries Unlimited, Westport (2007)
25. Putz, M., Schaffner, V., Seidler, W.: FRBR the MAB2 perspective. Cataloging Classif. Q. **50**, 387–401 (2012)
26. Riley, J.: Enhancing interoperability of FRBR-based metadata. In: International Conference on Dublin Core and Metadata Applications (2010)
27. Riva, P., Doerr, M., Zumer, M.: FRBRoo: enabling a common view of information from memory institutions. In: World Library and Information Congress: 74th IFLA General Conference and Council (2008)
28. Schneider, J.: FRBRizing MARC records with the FRBR Display Tool (2008)
29. Takhirov, N., Aalberg, T., Duchateau, F., Žumer, M.: FRBR-ML: a FRBR-based framework for semantic interoperability. Semant. Web **3**, 23–43 (2012)
30. Teixeira, T., Lopes, M., Freire, N., José, B.: Report on FRBR experiments. Technical report, TELplus (2008)
31. Yee, M.M.: FRBRization: a method for turning online public finding lists into online public catalogs. Inf. Technol. Libr. **24**, 77–95 (2005)
32. Zeng, M.L., Žumer, M., Salaba, A.: Functional Requirements for Subject Authority Data (FRSAD): A Conceptual Model. Walter de Gruyter, Berlin (2011)
33. Zhang, Y., Salaba, A.: Implementing FRBR in Libraries: Key Issues and Future Directions. Neal-Schuman Publishers, New York (2009)
34. Zhang, Y., Salaba, A.: What is next for functional requirements for bibliographic records? A Delphi study1. Libr. Q. **79**, 233–255 (2009)

Are There Any Differences in Data Set Retrieval Compared to Well-Known Literature Retrieval?

Dagmar Kern[1(✉)] and Brigitte Mathiak[1,2]

[1] GESIS – Leibniz Institute for the Social Sciences, Cologne, Germany
dagmar.kern@gesis.org, b.mathiak@uni-koeln.de
[2] University of Cologne, Cologne, Germany

Abstract. Digital libraries are nowadays expected to contain more than books and articles. All relevant sources of information for a scholar should be available, including research data. However, does literature retrieval work for data sets as well? In the context of a requirement analysis of a data catalogue for quantitative Social Science research data, we tried to find answers to this question. We conducted two user studies with a total of 53 participants and found similarities and important differences in the users' needs when searching for data sets in comparison to those already known in literature search. In particular, quantity and quality of metadata are far more important in data set search than in literature search, where convenience is most important. In this paper, we present the methodology of these two user studies, their results and challenges for data set retrieval system that can be derived thereof. One of our key findings is that for empirical social scientists, the choice of research data is more relevant than the choice of literature; therefore they are willing to put more effort into the retrieval process. Due to our choice of use case, our initial findings are limited to the field of Social Sciences. However, because of the similar characteristics for data sets also in other research areas, such as Economics, we assume that our results are applicable for them as well.

1 Introduction

Inspired by the successes of Bioinformatics, much research infrastructure has been built in the last few years to allow all scientific disciplines to archive and share scientific data sets. This infrastructure has been extraordinarily successful, in the sense that there are now much more data sets available than ever before. The trend of journals to force authors to share their data sets before publishing their articles based on these data sets also contributes to that. DataCite[1], probably the biggest catalogue, lists over two million data sets. With the increase of available data sets, the finding of information becomes more difficult for the researcher and thus, an important issue for further research in the field of data set retrieval. Even with a large time budget simple reading through everything is no longer a viable option. Most of the current data catalogues

---

[1] https://www.datacite.org/.

© Springer International Publishing Switzerland 2015
S. Kapidakis et al. (Eds.): TPDL 2015, LNCS 9316, pp. 197–208, 2015.
DOI: 10.1007/978-3-319-24592-8_15

utilise systems which were originally designed for literature or web site searches, e.g. SOLR[2]. But do these data set retrieval systems really satisfy users' needs? We conducted two user studies and found answers to that question. In a lab study with 7 participants and in telephone interviews with 46 participants we focussed on data set retrieval for a specific domain, the Social Sciences.

In this paper, we describe the methodology of the user studies, discuss the results and derive challenges that developer of data set retrieval systems are facing. Although we assessed user behaviour in a particular system for social scientists, we believe that many of our observations are transferable to other similar systems in different domains. In particular, our key insight is that users are willing to put higher effort in the searching and selecting process of data sets than they do when searching for literature. They also expect high-quality metadata to support this process. We assume that this is true for other data retrieval scenarios as well.

2   Motivation

For the Social Sciences there are different relevant archives that provide research data sets. The most relevant international archives for Social Science research data are the ICPSR[3] in the USA, the UKDA[4] in Great Britain and the NESSTAR[5] software. One of the most known data catalogue in Germany is the DBK[6]. Other rather highly specialised data catalogues only provide one data set or a collection of data sets (SOEP[7]). Many archives are now linked to each other, exchange metadata and share infrastructure and sometimes even software. Almost all current archives now have some sort of online representation of their data. Typically, those systems are similar to nowadays digital literature retrieval systems. There are some systems that attempt to transcend this general trend and try to develop tailored services for data sets. E.g. OECD and NESSTAR focus on browsing and visualisation of the data. Others focus on rich metadata for the data sets, like the ICPSR and most specialised archives. The retrieval component is an integral part of all systems.

When treating metadata of data sets like metadata of publications the integration of such information in a library system can be made easily [15]. When taking a closer look at both types of metadata (cf. Fig. 1), it can be seen that there are some basic similarities. However, in order to use data for scientific analyses, researchers need more detailed information about the data set and its structure. It has to be understood which values belong to which question and in which context the data has been captured. Furthermore, for the reusability of data sets and the comparability of results gained

---

[2] http://lucene.apache.org/solr/.

[3] https://www.icpsr.umich.edu.

[4] http://www.data-archive.ac.uk/.

[5] http://www.nesstar.com/ NESSTAR is the meta-search portal of many European archives.

[6] https://dbk.gesis.org/dbksearch/index.asp?db=e.

[7] http://www.diw.de/en/diw_02.c.221180.en/research_data_center_soep.html.

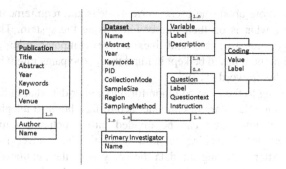

Fig. 1. Comparison of metadata between publications and data sets. In both cases, we restricted ourselves to what could arguably be the most relevant fields.

through analysis based on the data set it is essential to know which other analyses have been already performed with the data set.

3 Related Work

That there is a need for sharing data for reuse is an almost common agreement in the scientific community (e.g. see [9]) However, to the best of our knowledge no studies on the behaviour of users of data set repositories in the field of the Social Sciences have been published yet. However, studies on systems in related fields, such as History [7] or in the Humanities [6] have been made. The results of these studies show that the users asked first for more accessibility, then for more and richer metadata and then for more convenience, including domain-specific applications. This stands in strong contrast to literature retrieval, where convenience is by far the most commonly named requirement [4]. Established theory on information seeking [17] stresses the time constraints. Optimising task speed, usability and simplicity has ever since been a major driver in the development of information system user interfaces. Accordinglg, [5] points out that the importance of convenience is highly dependent on the context.

In the field of the Social Sciences, there are a few studies performed to understand the seeking behaviour of this target group in general (e.g. [14, 18]). However, data set retrieval is often not an explicit subject of discussion. Most of the presented results focus on literature retrieval. In our research, we try to fill this gap by applying standardised methods for collecting information about user requirements and user behaviour. While Diane Kelly [12] provides a good overview of different evaluation techniques in the field of interactive information retrieval in general, we present in the following the methods that are relevant for our conducted studies. One method is the "simulated work task situation" introduced by Borlund and Ingwersen [2]. A short cover story is used to describe a realistic information seeking situation that motivates the test person to use the system under investigation. Results gained through this method reflect real information needs [3]. In case of our data set retrieval system, this is a task such as "Find one or more relevant data sets that fit your actual research topic". In the context of the simulated work task situation the "Think-aloud" method [8] can be deployed to learn more about the users' current needs and requirements. Subjects are asked to articulate what is on their minds when using the system. This data can be captured by recording software and inexpensive computer microphones. While applying this technique it has to be kept in mind that participants might have difficulties to put their thoughts in words [12].

Interviews belong to the most common form to gather data that indicate user needs [16, S. 221]. The interviews are often made face-to-face but also telephone interviews serve this purpose. Interviews can be divided into structured, semi-structured or unstructured interviews depending on how strict the interviewer is following the interview guide. After gathering the data the analysis of the qualitative data follows. Usually, the responses are assigned to set of categories to be able to summarise them. The quality of the analysis depends on how the data are collected. A transcription of interviews followed by an analysis based on these notes is better than an analysis based on notes that are taken during a non-recorded interview [12].

4   Research Context

In an on-going project for integrating access to different sources of information (publications, data sets, web sites, etc.) we follow a user-centred design approach [10]. According to the first step "specify context of use" [10], we performed an analysis of the data catalogue of GESIS - Leibniz-Institute for the Social Sciences[8], called DBK. The DBK contains study descriptions and the associated data sets. Its main function is to serve as a central point for downloading and purchasing data sets, including the legal and commercial infrastructure for handling sensitive and confidential data. It also serves as a reference guide for additional documentation and metadata. The information about a data set in the database contains bibliographic citation, content and method-ology descriptions, lists of errata and versions, the list of primary publications about the data set and the link to the associated questionnaires and codebooks as well as the link to the data set download. So far, a comprehensive list about publications that are based on the data set is not available. This metadata for each data set is displayed in a detailed view (see Fig. 2). The search function of the DBK is a full text search based on a set of predefined fields of the relational database it is built upon. The underlying software is Open Source[9].

To gain some preliminary insights into the users' behaviours, we started our investigations with a query log and session analysis based on the web servers logging files. We were able to identify some differences to the typical query pattern that we know from our literature system sowiport[10]. Known-item queries constitute roughly two thirds of all queries, which is much more than in the case of our literature system. Additionally, session analysis shows that many users come in through direct links, but

---

[8] http://www.gesis.org/.

[9] https://dbk.gesis.org/DBKfree2.0/Defaulten.htm.

[10] http://sowiport.gesis.org/.

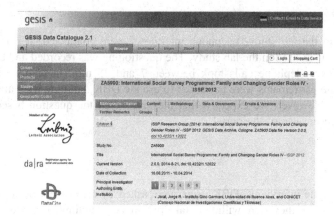

Fig. 2. Detailed view of a data set.

then spend a long time on the page with detailed information about the data set (see Fig. 2), by-passing the search engine completely.

When searching, author names, a kind of information users of our literature system often search for, are not used. Although these insights already exposed some information about the differences of literature and data set retrieval, we decided to perform more detailed user studies based on semi-structured interviews to get a clearer picture on the information seeking process and the users' intentions themselves.

5 Methodology of the User Studies

We conducted two studies, a lab study with seven participants and telephone interviews with 46 participants. Our goals were, to find out how Social Science researchers actually search for data sets, what their requirements are, how they achieve their goals and whether they are satisfied with the search path they used. The studies were conducted with German-speaking users of the DBK.

5.1 Lab Study Setup

The lab studies were performed in single sessions where each participant sat together with an experimenter in front of a computer. After a short introduction the participants were asked to use the DBK to search for one or more data sets fitting their current research questions. They were observed while using the DBK for ten minutes and were asked to think aloud, while working on their individual tasks. The audio as well as the screen were recorded. Afterwards, we conducted a semi-structured interview during which the participants were encouraged to use the data catalogue to recall some actions they had performed in the past or to examine functions they had not used before. These single sessions took about 45 min each.

5.2   Telephone Interview Setup

In addition to the lab study, telephone interviews were conducted. It was the same interview guide as used in the lab study. The questioning was recorded and transcribed afterwards. The telephone calls took between 7 and 25 min. In contrast to the lab interviews, the participants were not asked to use the DBK, but they were allowed to if they wanted. However, we expected that they answered the question based on their memory of recent use of the portal.

5.3   Subjects

For the telephone interviews, DBK users were asked by email (sent out to about 600 registered users), if they were willing to take part in a questioning. The interviews were conducted with 46 participants (11 female, 35 male). The youngest interview partner was 26 and the oldest 80 years old (mean age 41). Ten participants worked or used to work as a professor at a University, 18 were postdocs, 17 held a master or an equivalent degree and one held a bachelor degree. The main research interests lay in Social Sciences (20) or Political Sciences (12).

For the lab study, seven participants (3 female, 4 male) were recruited from the sample above, based on geographical closeness. They were between 26–43 years old. The group consisted of six PhD candidates and one postdoc. Their research interests were in Social Sciences (4) and Political Sciences (3). It must be mentioned that six of them had affiliations with GESIS, e.g. as research assistants, which may have introduced an institutional bias. However, since the lab participants were much more critical about the catalogue than the telephone interviewees, we believe that this bias at least did not result in underreporting of problems.

6   Results of the User Study

Based on the transcriptions of the interviews we summarised and analysed users' behaviour as well as their requirements on a data set retrieval system. In this paper, we focus mainly on the responses that help us to answer the question "Are there any Differences in Data Set Retrieval compared to well-known Literature Retrieval?" Detailed information about further results will be presented in another context. As the interview structure was the same in both studies the results are presented together. The interview guidelines were not followed strictly in all cases, e.g. when a preceding answer clearly made a follow-up question irrelevant. This occurred rarely, though. The numbers of responses to each question never dropped below 38.

6.1   Lab Study Observations

Participants of the lab study were asked to search for data sets relevant to their research by using the catalogue. The most often used strategy was to use the simple search as an entry point and then to browse through the often very long result list. In most cases the initial query was not refined. The subjects inspected each entry dutifully. They sometimes consulted additional documentation, such as questionnaires, to evaluate the relevance of entries. Two participants entered previously known survey programmes and then proceeded with browsing through the result list in a similar manner as the other participants did. All, except one participant, found a relevant data set within the allotted ten minutes, however, in many cases, it was implied that further investigations would be needed to properly assess its applicability. Most participants felt that there is a lack of additional filtering options. However, the filtering possibilities offered, were not used at all. It should be noted here that there were no complaints on the rather lengthy process of evaluating relevance. In digital library systems, the users are much quicker to judge positive document relevance, e.g. 26.49 s on average in the system described in [11]. From our experience of the lab study, we claim that for data sets this time is much higher. As our participants felt that it was part of the normal research process.

6.2   Frequency of Use

In our interview, the first questions aimed at assessing the participants' frequency of usage of the DBK. With the exception of those subjects, who use the DBK for teaching most participants used the system rarely (1-3 times per year, also cf. Fig. 3). Since all participants were active researchers, we assume that they use library systems much more often.

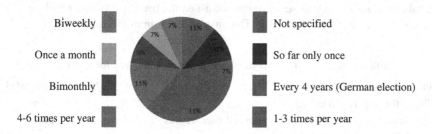

Fig. 3. Frequency of usage

6.3   Metadata of Data Sets Vs. Metadata of Literature

The results of our user study provide insight in the relevance of the different metadata for data sets. In the following we summarise the key findings in pointing out the differences of metadata for data sets and literature which seem to be quite similar at the first glance. For data sets, it can be said that the title is in most cases more a brand name than an accurate description of the contents. Furthermore, the publication year is irrelevant for data sets. Highly relevant, on the other hand, is the time frame to which the data refers to. What the abstract is for the publication the list of keywords and the table of content are for the data sets. Both metadata provide keywords but with a different granularity. For a publication, it is important in which journal or proceedings it is published and for the data set, this information can be seen as the context in which the data are collected for example under the umbrella of the comprehensive survey programme. While for publications the authors are very important, for data sets the primary investigators seem to play a rather minor role for the data set search process. Not one participant mentioned them in the interviews.

Furthermore, data set retrievers rely on the additional documentation like codebooks, method reports or questionnaires to be able to analyse the data. The quality of these documentations is essential. Other aspects that are very important for the users of data sets are the mode of questioning, the sampling size and the involved countries.

Similar to references for publications data set often based on specific item and scales. Documentation about these not only helps to understand the data collecting process, but also helps to estimate the quality of the data. Related literature that cited a data set is at least as important as it is in literature search systems. It might be even more important as the question if a research question has already been answered (a use case proposed by the participants) based on a certain data set could be answered by looking at the citing. The participants reported that searching for such publications is rather cumbersome, so far. Furthermore, in case the studies were performed in different countries and not all data sets are available in the presented catalogue, they would appreciate having a link to where to find the corresponding data sets.

6.4   Additional Functionalities

In addition to the questions about their search behaviour in the DBK and the more general questions about data set retrieval, we asked the participants about their opinion on a few new features for the DBK. The answers are summarised in Table 1.

Table 1. Answers to the questions about new feature ideas for the DBK.

Would you like to have...	Yes	No	Undecided
A list of the Top 10 Downloads?	27 %	62 %	11 %
The information of how often each individual data set was downloaded?	19 %	62 %	19 %
A comment function?	30 %	53 %	17 %
A user account?	55 %	45 %	–
A list that shows data sets downloaded by researches who also downloaded the currently observed data set?	82 %	18 %	–
An email-reminder asking for publications that you might have written based on the downloaded data set?	94 %	6 %	–

The features that we have prepared for judgement, were motivated by the experiences with our literature search engine. Most answers are unremarkable, except that both user accounts and email reminders have such a high acceptance rate. Most interviewees strongly demanded free access for as many things as possible and did not mind receiving emails.

6.5 Perceived Usability

It is important to mention that the comments regarding usability are quite different from participants of the lab study and the telephone interviews. In the lab study literally everyone complained about the inconvenience of the search functionality. In the telephone interviews, only one person seemed to be able to remember the inconvenient search. The general sentiment was that the DBK is an extremely useful tool. The users appreciate the access to the data sets as well as the high-quality metadata so much that any intricateness in the search functionality seems to be of marginal importance. Nevertheless, for starting the search process most of them expect a Google-like input field with Boolean search, autocomplete function, search term suggestion and an auto-correct on spelling mistakes as a bare minimum.

7 Open Challenges

From the general answers above, some challenges arise that were mentioned multiple times by the participants. In this section, we present a few selected issues, as they seem rather typical for data set retrieval.

7.1 Search Within the Metadata

Metadata of data sets can be quite extensive, so much that the line between metadata and digital objects in their own right is hard to draw. There are documentations like questionnaires and codebooks, but also method reports and even publications, which are linked because they describe the genesis of the data set. So far, these documents are often displayed together with a data set, but they are not included in the retrieval process of the DBK. However, at least 20 % of the participants of both studies mentioned that they would prefer a search opportunity for questionnaires, and 15 % would like to have this for codebooks.

This is a complicated issue. Variable labels and questions are both highly repetitive and much more common than the more relevant data sets. When providing data sets, questionnaires and codebooks are all together in one result list, the data sets thus recede into the background. This is a known problem for systems that offer such services. It works for catalogues with few studies, but the currently used visualisation approaches do not scale well. Probably, a complete new retrieval approach has to be chosen on another level.

7.2 Categorisation/Grouping of Data Sets

The most relevant data sets in the Social Sciences belong to comprehensive survey programmes in which the data collections are repeated periodically and often in several countries. For each individual collection, one or more new data sets are created and henceforth exist as separate records in the data catalogue. Although the DBK has a group structure and provides a group search function to address this, these functions were not used during the lab study. For example, the most popular study archived in the DBK is the Eurobarometer (it is also the most common query). Related to it are 621 data sets and 23 groups. Here most relationships are not clearly defined or sometimes even unknown. This complicates the creation of a suitable representation of the stored data. In addition, the way in which survey programmes publish their data is highly heterogeneous, making a standardised display even harder. The challenges of implementing such a system are outlined in [13]. Some of the participants suggested a hierarchical representation within the result list. So that they can easily skip results from survey programmes they are not interested in, or, on the other hand, so that they get an overview of results that belong together.

7.3   Interlinking Between Data Sets and Publications

The users are very interested to know how often and by whom a data set has been cited. Whenever possible, links to publications related to a data set should be included into the data set retrieval system. Both self-reported as well as automatically extracted citation [1] could significantly improve interlinking and, thus, improve the user experience of data set retrieval systems. Since these publication lists can become quite large, breaking the citation down to the specific variables was suggested, but this is again information that is not always available.

Another critical point is the timeliness of such publication lists. A few of the interviewees worry that the list in the data retrieval system is not as complete and current as a list that Google scholar could provide them. They suggested a link to Google scholar with predefined search terms including the DOI as an alternative and/or supplement to any list that might be compiled through self-reporting or text mining. All links should be clearly split to differentiate between secondary analysis and sources for the study. Many studies base some of their methodology on pre-existing research, e.g. they use scales and questions from databases or published research.

7.4   Registration and Personalisation

All forms of registration and login are explicitly frowned upon by 19 % of the participants. But this is a general issue and not limited to data set retrieval. Echoing similar principles in digital library systems, all metadata should be and actually is free, only the content is restricted, oftentimes for legal reasons. This implies that all documentations (including codebooks and questionnaires) should be freely accessible, as users tend to only register when they can be sure that the data set is relevant for them.

Inspecting the actual data set for quality and relevance is the last resort step, users avoid as long as possible. Not only is it time-consuming to go through registration and possibly legal requirements, without proper documentation, the data set is very likely unusable anyway. Yet, even with documentation, opening a data set for the first time often holds some surprises, ranging from software incompatibilities to mismatches between documentation and actual data. In this context, one participant even suggested providing an example data set with only a few fake data to get an impression what he/she has to expect and therefore to minimise the downloading effort.

Some sort of community function would allow users to help other users, but this is seen critical by almost half of the interviewees. They fear unqualified rumours to be given attention. A helpline towards qualified personal, or ideally the data set providers, would be much preferred. As the time user spend with analysing data sets is often quite long they are highly interested in notifications about changes in their downloaded data sets.

8   Conclusion

In this paper, we study the requirements users have for a data set retrieval system in the Social Sciences and compare those to the well-known requirements in Digital Libraries. Based on two user studies with altogether 53 participants, we could find similarities, but also differences. The need to know who cited a publication of interest respectively who published an article based on a data set of interest can be seen as the main common requirement users have on both retrieval systems. On the other hand, the underlying difference is: **Choosing a data set is a much more important decision for a researcher than choosing a piece of literature.** Therefore, they are willing to spend much more time on the selection process. As a consequence,

- Metadata quality and quantity is most important. Data without description is useless. Often, decision criteria for relevance are hidden inside lengthy documentation. Compared to the state-of-the-art, users wish for more metadata, in particular better structure and visualisation of the relationship between data sets, more links to additional resources, both on the context of the data set and its reception by the community.
- Researchers are dedicated to the research data they choose, even long after they have downloaded it. They appreciate updates and news, as well as suggestions for what might be new relevant data sets for them. Many are also willing to spend time on improving the metadata when so prompted. For example, they are willing to provide secondary literature.
- As for literature, users expect an optimal support while entering their search terms with search term suggestions, autocomplete function, etc. Furthermore, since users visit the catalogue so rarely it is important that a function is as self-explanatory as possible, or it will not be used.

We believe that the differences are mainly caused by the different importance papers and data sets have for the researchers. Research data sets are much more decisive for research activity than any literature is. Many empirical scientists spend their whole academic career analysing one essential study. Any time spent on choosing the correct data set is therefore time well-spent.

Typical information retrieval UIs, in contrast, are optimised to improve convenience, often understood as implementing time-saving mechanisms. The user of data set retrieval systems gives priority to the quality and completeness of the metadata as well as the data sets. While they prefer indeed a more comfortable search input field, they are less interested in getting fast results. This finding offers new possibilities for all the ideas in information retrieval that failed at the convenience barrier and spawn new research ideas in an environment where time does not matter (much).

References

1. Boland, K., Ritze, D., Eckert, K., Mathiak, B.: Identifying references to datasets in publications. In: Proceedings of the TDPL (2012)
2. Borlund, P., Ingwersen, P.: The development of a method for the evaluation of interactive information retrieval systems. J. Documentation **53**(3), 225–250 (1997)
3. Borlund, P.: Experimental components for the evaluation of interactive information retrieval systems. J. Documentation **56**(1), 71–90 (2000)
4. Connaway, L.S., Dickey, T.J., Radford, M.L.: If it is too inconvenient I'm not going after it: convenience as a critical factor in information-seeking behaviors. Libr. Inf. Sci. Res. **33**(3), 179–190 (2011)
5. Case, D.O. (ed.): Looking for Information: A Survey of Research on Information Seeking, Needs and Behavior. Emerald Group Publishing, Bingley (2012)
6. DeRidder, J.L., Matheny, K.G.: What Do Researchers Need? Feedback on use of Online Primary Source Materials. D-Lib Mag. **20**(7/8) (2014). http://doi.org/10.1045/july2014-deridder
7. Duff, W.M., Johnson, C.A.: Accidentally found on purpose: information-seeking behavior of historians in archives. Libr. Q. **72**, 472–496 (2002)
8. Ericsson, K.A., Simon, H.A.: Protocol Analysis. Verbal Reports as Data. The MIT Press, Cambridge (1984)
9. Faniel, I.M., Zimmerman, A.: Beyond the data deluge: a research agenda for large-scale data sharing and reuse. Int. J. Digit. Curation **6**(1), 58–69 (2011)
10. International Organization for Standardization. ISO 9241-210:2010 Ergonomics of Human System Interaction-Part 210: Human-Centred Design for Interactive Systems (Formerly Known as 13407) (2010)
11. Kelly, D., Belkin, N.J.: Reading time, scrolling and interaction: exploring implicit sources of user preferences for relevance feedback. In: Proceedings of the SIGIR. ACM (2001)
12. Kelly, D.: Methods for evaluating interactive information retrieval systems with users. Found. Trends Inf. Retrieval **3**(1–2), 1–224 (2009)
13. Mathiak, B., Boland, K.: Challenges in matching dataset citation strings to datasets in social science. D-Lib Mag. **21**(1/2) (2015). http://doi.org/10.1045/january2015-mathiak
14. Meho, L.I., Tibbo, H.R.: Modeling the information-seeking behavior of social scientists: Ellis's study revisited. J. Am. Soc. Inform. Sci. Technol. **54**(6), 570–587 (2003)
15. Ritze, D., Boland, K.: Integration of research data and research data links into library catalogues. In: DC 13 - The Lisbon Proceedings of the International Conference on Dublin Core and Metadata Applications: Papers, Project Reports and Posters for DC-2013, pp. 35–40 (2013)
16. Rogers, Y., Sharp, H., Preece, J.: Interaction Design: Beyond Human-Computer Interaction. Wiley, New York (2011)
17. Savolainen, R.: Time as a context of information seeking. Libr. Inf. Sci. Res. **28**(1), 110–127 (2006)
18. Shen, Y.: Information seeking in academic research: a study of the sociology faculty at the University of Wisconsin-Madison. Inf. Technol. Libr. **26**(1), 4–13 (2013)

tc-index: A New Research Productivity Index Based on Evolving Communities

Thiago H.P. Silva, Ana Paula Couto da Silva, and Mirella M. Moro[(✉)]

Computer Science Department, Universidade Federal de Minas Gerais,
Belo Horizonte, Brazil
{thps,ana.coutosilva,mirella}@dcc.ufmg.br

Abstract. Digital Libraries are used on contexts beyond organization, archival and search. Here, we use them to extract bibliography data for proposing a new productivity index that emphasizes the venue and the year of the publication. Also, it changes the evaluation perspective from a *researcher* alone (index based on one's own publications) to one's contribution to a whole *community*. Overall, our results show that the new index considers researchers' features that other well known indexes disregard, which allows a broader researchers' productivity analysis.

1 Introduction

Digital Libraries (DL) have been used on contexts beyond organization, archival and search. For instance, given the widespread research on social networks and the valuable data from DL, *academic social networks* may be built based on the coauthorship relations within DL: nodes represent scholars and edges their academic relationships. Then, it is possible to analyze those networks and uncover important information. Indeed, recent studies have considered the performance of each researcher individually as well as measured the degree of influence among research areas and contributions from researchers, providing interesting and meaningful visions of the academic environment [8,15,18,19].

Regarding a researcher performance, a very popular metric is the *h-index* [12]: the largest number h for which the researcher has h publications with at least h citations each. Despite its popularity, the h-index is inadequate for comparing the achievements of individual researchers, research groups, institutions, or countries [2,23]: (*i*) it disregards publications with less than h citations and more than one publication with h citations; (*ii*) it does not consider recent publications with potential impact but without enough citations; and (*iii*) it disregards venues' singularities, such as year of publication and citations per work ratio.

Here, we address such flaws by defining the *tc-index* (temporal community-based index), a novel approach for assessing researchers' productivity based on the evolution of communities. A *community* is the set of authors who published in the same venue (note that it can also be extended to other concepts of community). Overall, the tc-index explores two properties: (*i*) different communities have different patterns (e.g., an h-index of 10 in Networks is different from a 10 in

S. Kapidakis et al. (Eds.): TPDL 2015, LNCS 9316, pp. 209–221, 2015.
DOI: 10.1007/978-3-319-24592-8_16

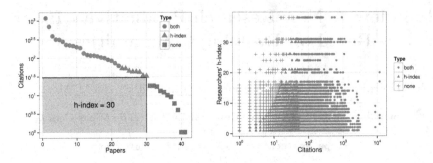

Fig. 1. Publications considered by *tc-index* and *h-index* (*both*) and *h-index* alone for **Christos Faloutsos** (left) and for everybody grouped by h-index (right) (Color figure online).

the Computing Theory community, as discussed in [16]); and (*ii*) the timing for citing a work depends on its publishing year (i.e., old publications have more time to be cited). Therefore, it is rather unfair to evaluate *research contribution* with only one bibliometric perspective [2,3,6,16,23].

The main contribution and novelty of *tc-index* is that each publication is compared to similar ones considering a *threshold* of community contribution. Such definition is illustrated by Fig. 1. In the left, we take 41 publications of one known researcher (*Christos Faloutsos*) and observe the following: (*i*) 30 of those 41 form his h-index (red circles and green triangles); (*ii*) 22 are considered important to their communities, and 8 that form his h-index (green triangles) do not outperform the similar ones from a community perspective. Then, at the right, for all works of all authors in our dataset distributed by h-index, it shows tc-index *filtering* the publications that are important due to their contribution to the communities. In addition, there is a large number of publications that form the researchers' h-index, but, in a community context, do not outperform the similar ones (green triangles). Hence, tc-index changes the evaluation *perspective* of one *researcher* (index based on one's own publications) to a whole *community*, i.e., it emphasizes the contributions of a publication to its community.

Next, Sect. 2 provides background on assessing academic research. Section 3 introduces *tc-index* for ranking researchers based on their communities impact over time. Sections 4 and 5 describe the experimental methodology and our analysis. Then, Sect. 6 addresses the experimental validation of *tc-index*. Finally, Sect. 7 discusses our conclusions and future work.

2   Related Work

Analyzing data extracted from Digital Libraries (e.g., citation analysis, ranking and bibliographic networks) is essential for understanding the evolution of science [1,4]. However, ranking researchers by publication impact is not easy. Based on the analysis of academic social networks, an earlier study proposed *weak ties* as links that join different groups by building bridges within the network [10]. Ranking is then possible by measuring the weak tie coefficient of each researcher. Likewise, Newman [19] measures the influence on the information flow among individuals through high scores of *brokering*; Burt [7] calls *brokers* the people who build social capital through structural holes in the networks; and Guimerà et al. [11] show that a newcomer on the network improves the research teams success. Similar to those, our approach captures the importance of researchers involved in different communities (parts of a network). As shown in Sect. 6, our metric aggregates a more diverse set of researchers from multiple communities in a global ranking. Hence, it is able to *capture* the researchers who are building bridges and connecting different network communities.

Broadly, studies of academic social networks emphasize the collaborations between researchers [19], their productivity [1,9,16,17] and the communities formed [5,20–22]. For instance, Glauber et al. [9] explore the scholar popularity within CS communities; Biryukov and Dong [5] show that short-time researchers (less than five year careers) dominate the DBLP communities; and Silva et al. [21] analyze the flows of research contribution based on collaborations around venues according the venues' quality. Here, we further explore the relation *researcher-community* in order to propose a fairer ranking strategy based on the analysis of publications that are grouped according to their venue and year.

Regarding productivity indices, they usually consider the number of citations received, volume, h-index [12], its variations and others [6]. There are also new approaches based on networks structure, to improve fairness provided by bibliometry only. For instance, Silva et al. [22] propose a community-based index for ranking venues by the external relationships of their members. Lima et al. [16] focus on ranking researchers across multiple areas and their different publication patterns. Here, we propose a new indicator for ranking researchers by exploring the community context of a publication as well as its visibility.

Overall, the aforementioned studies explored the Digital Libraries in their publications attributes. Our approach is one step closer to more robust production evaluation, since it considers the link *researcher-community* and the sets formed by similar publications regarding venue and time.

3 Temporal Community-Based Index

Base Concepts. Let $\mathcal{R}$ be a set of researchers, and $\mathcal{W}$ the set of works produced by $\mathcal{R}$. Each researcher $r \in \mathcal{R}$ has a set of published works $\mathcal{W}_r \subset \mathcal{W}$, for $1 \leq r \leq |\mathcal{R}|$. Also, let $\mathcal{V}$ be the set of all venues in a digital library. Then, a **community** C_v, for $1 \leq v \leq |\mathcal{V}|$, is a partition of $\mathcal{R}$ that contains the set of researchers who published in venue v. Hence, $\mathcal{W}$ can also be partitioned into sets $\mathcal{W}_v$, for $1 \leq v \leq |\mathcal{V}|$, with the works published in venue v for all researchers at C_v.

The main novelty of tc-index is to consider the temporal aspect of community evolution in a given time window t. The year is the time granularity so that $C_v(y)$ is the community formed at year y and $\mathcal{W}_v(y)$ the works published by each member of community $C_v(y)$[1]. The set of all years in which v was published is $\mathcal{Y}_v$.

---

[1] Note that other time granularities are possible (bi-annual, etc.).

The goal is then to consider the same venue at the same year for a fairer comparison of indicators from all researchers' works in $\mathcal{W}_v(y)$.

tc-index. We introduce the *temporal community-based index* (tc-index), a new metric for assessing a researcher r productivity, given by

$$\text{tc-index(r)} = \sum_{\forall v \in \mathcal{V}} \sum_{\forall y \in \mathcal{Y}_v} \sum_{\forall w \in \mathcal{W}_v(y)} \mathbb{1}(.), \tag{1}$$

where:

$$\mathbb{1}(.) = \begin{cases} 1, & \text{if } w \in \mathcal{W}_r \text{ and } ct_w(y) \geq \text{h-index}(\mathcal{W}_v(y)) \\ 0, & \text{otherwise} \end{cases}, \tag{2}$$

$ct_w(y)$ is the total number of citations of work w, and h-index($\mathcal{W}_v(y)$) is the h-index of venue v, both at year y. The *h-index*() function relates to the *venue* and is used as *threshold* to determine how w compares to others in the set $\mathcal{W}_v(y)$.

By definition, tc-index enables assessing the researcher productivity for comparing the *impact* of his/her works with a group formed by similar works (venue and year of publication). Thus, the productivity evaluation is fairer because it enables to compare recent works to other recent works (same temporal visibility) and, at same time, it avoids comparing works with different publication patterns. In other words, we assume that the better a researcher productivity, the more likely his/her works tend to outperform others *within their communities*.

4   Experimental Methodology

We now discuss the methodology underlying our experimental evaluation.

Publication Dataset. Without loss of generality, we use the communities from the ACM's Special Interest Groups[2], because most ACM's SIGs conferences are top-tier in Computer Science (CS) and top researchers on CS publish in them. Table 1 shows the set of SIGs considered, their related conferences and statistics. We collected their sets of publications within [2001,2010] from DBLP[3]. Citations were collected from Google Scholar by matching a 4-tuple (title, year, authors, venue). At the end, the dataset contains about 20 thousand authors, slightly more than 11 thousand papers with more than a half million citations.

As expected, each SIG has its own characteristics and publication profiles. For instance, SIGAPP has the expressive number of approximately 6,7 thousand authors, whereas SIGAda has only 169 different authors. Moreover, SIGBio has only 123 papers and the lowest rate of total of citation by authors, whereas SIGCOMM has highest rates of total of citations by paper and by authors.

Ranking Baseline. We also consider two well known bibliometric indices as baselines: citation count and h-index (widely used in online research-oriented search engines such as Google Scholar, Microsoft Academic Search and AMiner).

---

[2] ACM Special Interest Groups: http://www.acm.org/sigs.
[3] DBLP: http://www.informatik.uni-trier.de/~ley/db.

Table 1. DBLP statistics of ACM SIGs conferences within [2001,2010].

SIG/Conf.	Aut.	Pap.	$\frac{Aut.}{Pap.}$	Cit. (THS)	$\frac{Cit.}{Aut.}$	$\frac{Cit.}{Pap.}$
SIGAPP/SAC	6732	3078	2.19	45.19	6.71	14.68
SIGIR	2313	1528	1.51	85.25	36.86	55.79
SIGMOD	2196	1098	2	103.92	47.32	94.65
SIGKDD/KDD	2172	1075	2.02	109.31	50.33	101.69
SIGCSE	1957	1143	1.71	23.06	11.78	20.17
SIGMETRICS	1053	449	2.35	31.20	29.63	69.49
SIGSPATIAL/GIS	1011	423	2.39	9.03	8.93	21.35
SIGACCESS/ASSETS	849	418	2.03	7.86	9.25	18.79
SIGUCCS	838	655	1.28	1.68	2	2.56
SIGCOMM	816	338	2.41	104.73	128.35	309.86
SIGMOBILE/MOBICOM	748	276	2.71	66.88	89.41	242.31
SIGecom	581	347	1.67	17.83	30.69	51.39
SIGITE	468	317	1.48	2.01	4.29	6.34
SIGDOC	460	316	1.46	3.04	6.61	9.62
SIGBio/BCB	394	123	3.2	0.29	0.74	2.37
SIGAda	169	135	1.25	0.71	4.21	5.27
All	20470	11719	1.75	611.99	29.9	52.22

Evaluation Ground-Truth. Ranking the researchers is a challenging task and there is no world wide ranking for scientists in their research field. Nonetheless, alternative ranks appeared in: Hirsch [12] used the Nobel prize winners and Lima et al. [16] used grant receivers for Brazilian researchers. Likewise, we use a ranking based on the recognition of contributions and/or innovations from ACM (Association for Computing Machinery), such as the ACM Awards (given by specifics SIGs) and the distinct member grades to recognize the professional accomplishments of ACM's members (fellow, distinguished and senior)[4].

Evaluation Metrics. The validation process is to compare the ranking formed by the 137 winners of ACM Awards against our index and the baselines. We expect the winners appearing in the first positions of any ranking. For example, a perfect top-50 ranking would be formed by 50 winners of ACM Awards. The comparative metrics are defined as follows.

Ranking Comparison. To compare our metric to the baselines, we apply Spearman's rank correlation to analyze the rank positions [13], ρ, and Jaccard distance to measure dissimilarity between two sets (intersection over union). The main goal is to verify if our approach brings novelty, both in terms of new researchers and new works, to ranking task.

Ranking Precision. To further investigate effectiveness, we use precision and the discounted cumulative gain (DCG). The former is defined as the rate between researchers' relevance and number of rank positions n (precision@n). The latter measures the ranking quality and applies a log-based discount factor to penalize relevant items in lower positions. Formally, the DCG at position k is defined as $DCG@k = \sum_{i=1}^{k} \frac{2^{rel_i}-1}{log_2(i+1)}$, where rel_i is the relevance of the item at the i-th position [14].

The performance evaluation is presented from two perspectives. We analyze the dissimilarities for tc-index and the baselines, and investigate if the tc-index brings novelty to the rankings (Sect. 5). Then we validate it as to ranking relevant researchers against the baselines (Sect. 6).

---

[4] ACM memberships: http://awards.acm.org/grades-of-membership.cfm.

Table 2. Spearman correlation of top 50 researchers: tc-index and h-index (ρ_{tch}), tc-index and citation count (ρ_{tcc}), h-index and citation count (ρ_{hc}). Results in *bold* for strong and very strong, in *black* for moderate value, and *shaded* cells for weak correlation.

	ρ_{tch}	ρ_{tcc}	ρ_{hc}
SIGAda	0.94	0.98	0.90
SIGCOMM	0.82	0.76	0.83
SIGUCCS	0.70	0.87	0.85
SIGMOBILE	0.68	0.61	0.49
SIGBio	0.67	1.00	0.67
SIGecom	0.55	0.75	0.49
SIGMOD	0.54	0.74	0.39
SIGAPP	0.47	0.71	0.63
SIGCSE	0.44	0.79	0.69
SIGITE	0.43	0.82	0.66
SIGKDD	0.42	0.89	0.41
SIGDOC	0.38	0.77	0.67
SIGSPATIAL	0.33	0.89	0.44
SIGMETRICS	0.31	0.71	0.28
SIGIR	0.26	0.80	0.54
SIGACCESS	0.20	0.95	0.23
Global	0.61	0.39	0.57

Table 3. Jaccard distance between 50 top-ranked researchers of tc-index and h-index ($\mathcal{TH}$), tc-index and citation count ($\mathcal{TC}$), and h-index and citation count ($\mathcal{HC}$).

	$\mathcal{TH}$	$\mathcal{TC}$	$\mathcal{HC}$
SIGAPP	0.64	0.56	0.78
SIGCSE	0.54	0.38	0.6
SIGUCCS	0.46	0.02	0.46
SIGITE	0.42	0.02	0.4
SIGSPATIAL	0.4	0.5	0.54
SIGIR	0.32	0.32	0.5
SIGACCESS	0.26	0.3	0.4
SIGDOC	0.26	0.18	0.38
SIGMOD	0.26	0.38	0.32
SIGKDD	0.22	0.42	0.46
SIGMETRICS	0.18	0.5	0.48
SIGecom	0.16	0.34	0.44
SIGMOBILE	0.12	0.62	0.68
SIGCOMM	0.08	0.44	0.5
SIGAda	0.06	0.04	0.04
SIGBio	0.04	0	0.04
Global	0.26	0.58	0.62

5   Analysis

This section presents a broad analysis of *tc-index*. First, we analyze the correlation between its ranking and the baselines (Sect. 5.1) and discuss the dissimilarities among top-ranked researchers (Sect. 5.2). We also analyze the publications that contribute to each index (Sect. 5.3).

5.1   Global and Community Ranking Correlations

We quantify the correlations between the 50 top-ranked researchers according to tc-index, h-index and citation count using the Spearman's coefficient. We analyze both the global researcher ranking, which considers all publications in the dataset, as well as each SIG separately. Table 2 presents the results: correlations in *bold* refer to strong and very strong (≥ 0.7), in *black* refer to moderate ($0.4 \leq \rho < 0.7$), and *shaded* refer to weak correlation value ($\rho < 0.4$).

On the global analysis (table's last line), the correlation between tc-index and h-index (ρ_{tch}) is moderate (0.61). It means that tc-index has a high agreement in ranking positions with h-index, but brings new positioning to researchers in the ranking (Sect. 6 further analyzes effectiveness). On the other hand, the correlation between tc-index and citation (ρ_{tcc}) is just 0.39, whereas the correlation between h-index and citation (ρ_{hc}) is 0.57 (close to tc and h indices).

Most correlation values ρ_{tch} for SIGs are moderate and weak, suggesting a tendency of independence between tc and h indices. Exceptions are SIGAda and SIGCOMM, with very strong ρ_{tch} values. SIGAda correlation can be explained by its small number of researchers (169), which does not allows much space for changes in the top-50 ranking. SIGCOMM strong value is due to the community profile: the majority of works considered by h-index surpasses similar works in its own community. Considering tc-index and citation, most correlations are strong and very strong. Such result is very interesting because the global correlation is *not* valid for specific SIGs. Note that for global values, citation ranking is more likely to present outliers (many citations received by few works).

Overall, tc-index differs from the baselines by bringing different positioning and new researchers to the top-ranking. Hence, there is an independence between tc and h indices in most of cases, and an independence between tc-index and citation in the global context. Also, the results clearly show the differences and peculiarities of each SIG according to the rankings and, thus, corroborate to a complementary use of bibliometric indicators for a more complete analysis.

5.2 Ranking Similarities

In this section we verify if tc-index brings new researchers to its resulting ranking, i.e., if the sets of researchers at the top rankings are dissimilar. Specifically, let T be the set of 50 top-ranked researchers sorted by tc-index, H by h-index and C by citation count. Table 3 shows their dissimilarity according to Jaccard. As expected, most dissimilarities are low because the top-researchers tend to have theirs works outperforming the others in each community. Note also that there are 26 % different researchers between tc-index and h-index sets (TH) in the global context, as well as more than 20 % for most SIGs. The dissimilarities are bigger for comparisons against the citation count.

There are very low dissimilarities (less than 10 %) between tc-index and h-index for SIGAda, SIGBio and SIGCOMM. Except for SIGCOMM, the dissimilarities of these SIGs considering the citation count are lower, showing a common behavior for all. Moreover, SIGAda and SIGBio have few reaserchers (less than 400) and the Spearman correlation among all indices (as in the previous section) is very high for SIGAda and at least moderate for SIGBio. Considering SIGCOMM, in fact, there is a common behavior in the ranking by tc-index and h-index. SIGCOMM is further discussed in the upcoming sections.

A peculiar case is SIGAPP with more than 50 % of dissimilarity for all indices. SIGGAPP is the biggest community (6.7+ thousand researchers), but such behavior is not noticeable in other large SIGs. One explanation is the SIG structure, which has a large and diverse number of tracks and, hence, a non-concentraded *core* of researchers (as those in specific-area SIGs).

5.3 Dissimilarities Between Index Contributing Publications

We now analyze the level of dissimilarity between the set of works that contribute to tc-index and h-index scores, denoted by W_{tc} and W_h, respectively. Table 4 shows the Jaccard dissimilarities between sets W_{tc} and W_h for researchers who are in the top-50 (R_{50}), top-100 (R_{100}) and complete (R_{all}) rankings. Dissimilarity tends to increase when more researchers are considered and consequently more works contribute to indices values. Hence, in the global context, the dissimilarities are 41 % for R_{50} and 92 % for R_{100}. Such result is expected because the better the researcher is, the more likely his/her works are to outperform the others within the same community.

Table 4. Jaccard distance (dissimilarity) between sets formed by works considered in tc-index and h-index on: top 50, top 100, and for *All* researchers.

	R_{50}	R_{100}	R_{all}
SIGBio	0.41	0.69	0.92
SIGAPP	0.67	0.74	0.89
SIGUCCS	0.58	0.61	0.89
SIGCSE	0.61	0.64	0.76
SIGITE	0.63	0.64	0.76
SIGAda	0.18	0.42	0.68
SIGIR	0.47	0.51	0.68
SIGSPATIAL	0.5	0.54	0.67
SIGDOC	0.42	0.49	0.66
SIGACCESS	0.4	0.46	0.63
SIGMOD	0.38	0.41	0.55
SIGKDD	0.39	0.43	0.54
SIGMETRICS	0.26	0.3	0.42
SIGecom	0.25	0.28	0.41
SIGCOMM	0.06	0.08	0.17
SIGMOBILE	0.12	0.09	0.14
Global	0.23	0.27	0.68

Table 5. Global ranking by *tc-index* and relative positions on the baselines: bold for the winners of ACM Awards (by innovations and/or contributions), ♮ ACM fellow, ‡ ACM distinguished scientist, † ACM senior member, * winner of *test of time paper award*.

tc-index		h-index	Citation
Jiawei Han‡	1^O	1^O	12°
Scott Shenker*♮	2^O	4°	1^O
Philip S. Yu♮	3°	2^O	13°
ChengXiang Zhai*‡	4°	3°	18°
W. Bruce Croft*♮	5°	6°	19°
Christos Faloutsos*♮	6°	5°	17°
Wei-Ying Ma‡	7°	7°	44°
Ion Stoica*♮	8°	10°	2^O
Zheng Chen†	9°	8°	56°
Hari Balakrishnan*♮	10°	12°	4°
Surajit Chaudhuri♮	11°	9°	85°
Jennifer Rexford♮	12°	20°	55°
Dina Katabi*♮	13°	32°	14°
Qiang Yang‡	14°	23°	89°
Donald F. Towsley*♮	15°	41°	137°

Except for SIGCOMM and SIGMOBILE, dissimilarities are large, and low values respond to each community's features. For example, researchers constantly publish in the same SIGs and their works tend to have more impact when compared to the other works published by their peers at the same year. Overall, there are many cases of dissimilarity that reinforce the independence between *tc-index* and h-index/citation. This means that tc-index may be used as a complementary measure for broadly understanding researchers' productivity.

6 Experimental Validation

After comparing tc-index to h-index and citation count, we now validate tc-index against a ground truth composed by award winning researchers. We consider the accuracy of the global ranking (all 16 SIGs) and the representativeness of SIGs.

6.1 Global Ranking

This set of experiments is divided in two parts. First, we consider the 20,470 researchers from all 16 SIGs and their global rankings. Then, we consider the 137 winners of the ACM Awards given by the SIGs.

Evaluating top-rank researchers. We rank 20,470 researchers (16 SIGs) with tc-index and baselines. We then contrast the top-positions to the winners of ACM SIG Awards and ACM member distinctions in Table 5. Note that tc-index ranks *Jiawei Han* at the first position. Likewise, h-index also ranks him at the first position, whereas citation count ranks him at 10th position. A more subtle finding is that tc-index ranks 6 (40 %) winners of ACM SIG awards, 11 (73 %) ACM fellow members, 3 (20 %) ACM distinguished scientists, 1 (6.7 %) ACM senior member, and 8 (53 %) winners of *test of time paper award*.

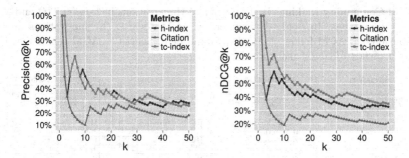

Fig. 2. Comparison among h-index, citation count and tc-index rankings by precision@k (left) and nDCG@k (right).

Another interesting difference is that the top-15 researchers according to tc-index are placed between the $[1^{st}, 41^{st}]$ positions according to h-index, and between $[1^{st}, 137^{th}]$ positions according to the citation count ranking. Likewise, there is a significant overlap between the top-10 researchers by tc-index and h-index (although in a slightly different order). Another large difference is given to *Donald F. Towsley* ranking positions: 15th by tc-index, 41st by h-index and 137th by citation count. This case suggests that tc-index is able to capture relevant information apparently ignored by both h-index and citation count.

Ranking winners of ACM SIG Awards. We now analyze ranking the 137 ACM award winners. We expect the winners to appear in the first positions of any ranking (i.e., a perfect top-50 rank would be formed by 50 award winners). Figure 2 compares the ranking produced by tc-index to those by h-index and citation in terms of attained precision@k (Fig. 2 left) and nDCG@k (Fig. 2 right). For precision, tc-index outperforms (slightly) or ties with h-index up to $k = 7$ and there are variations within the last positions. However, precision gives the same weight to everybody in the rank. For example, in precision@25, a relevant researcher at 1st and in 25th positions have the same importance. Nonetheless, our goal is to get the relevant researchers in the first positions. Then, nDCG complements such analysis by giving logarithmic penalties to researchers according to their position. Now, tc-index consistently outperforms both h-index and citation; with the exception of nDCG@1 in which all metrics are equal. Both h-index and citation have worse results because they rank relevant researchers in lower positions. Overall, tc-index successfully measures the impact of researchers' contributions and identifies outstanding researchers in their communities.

6.2 Community-Based Ranking

We now divide the rankings according to theirs SIGs (i.e., we rank the researchers within each SIG individually). Table 6 shows the top-10 rankings provided by tc-index and their related positions for h-index and citation. Overall, tc-index produces top-10 rankings that include 40 % (SIGKDD) and 20 % (SIGCOMM, SIGIR and SIGMOD) of ACM Award winners, and 70 % (SIGCOMM) and 20 %

Table 6. Community rankings by *tc-index* and the positions in *h-index* ($\mathcal{H}$) and *citation count* rankings. In bold, the winners of ACM Awards (by innovations and/or contributions). ♮ ACM fellow. ‡ ACM distinguished scientist. † ACM senior member. ✳ Winner of *test of time paper award*.

n°	SIGKDD tc-index	$\mathcal{H}$	Cit.	SIGCOMM tc-index	$\mathcal{H}$	Cit.	SIGIR tc-index	$\mathcal{H}$	Cit.	SIGMOD tc-index	$\mathcal{H}$	Cit.
1	**Jiawei Han**♮	1°	3°	**Scott Shenker**✳♮	1°	1°	**W. B. Croft**✳♮	1°	1°	S. Chaudhuri♮	1°	16°
2	C. Faloutsos♮	2°	2°	Ion Stoica*♮	2°	2°	C. Zhai*‡	2°	2°	Alon Y. Halevy♮	4°	4°
3	Philip S. Yu♮	3°	9°	Dina Katabi*♮	3°	11°	Wei-Ying Ma‡	3°	6°	M. Garofalakis‡†	5°	12°
4	E. J. Keogh	7°	10°	H. Balakrishnan*♮	4°	4°	**S. T. Dumais**♮	4°	3°	**H. V. Jagadish**♮	12°	36°
5	**J. Kleinberg**✳♮	19°	1°	Ratul Mahajan*	5°	14°	Qiang Yang‡	10°	10°	Samuel Madden*	8°	1°
6	Bing Liu*	10°	5°	G. Varghese♮	6°	15°	Hang Li	11°	13°	Philip S. Yu♮	17°	3°
7	I. S. Dhillon♮	15°	8°	Sylvia Ratnasamy*	7°	6°	Tie-Yan Liu†	17°	11°	J. F. Naughton*♮	13°	6°
8	**P. Smyth**♮	11°	26°	Srinivasan Seshan	8°	17°	Hua-Jun Zeng	37°	9°	Kian-Lee Tan	11°	26°
9	Jian Pei†	4°	27°	Jennifer Rexford♮	9°	21°	James Allan	5°	19°	Divesh Srivastava♮	6°	13°
10	C. Zhai‡	16°	29°	Paul Francis*	13°	9°	C. L. A. Clarke	6°	20°	Dimitris Papadias	9°	15°

(SIGKDD, SIGIR and SIGMOD) of *test of time award*. For instance, *Jon M. Kleinberg* (SIGKDD) is 15th by tc-index and 19th by h-index, whereas *H.V. Jagadish* (SIGMOD) is 4th by tc-index and 12nd by h-index. Regarding the number of citations, *Jon M. Kleinberg* is ranked at first, but the advantage of citation count is not noticed for the remaining winners.

Moreover, both tc-index and h-index have a perfect match for the first nine positions of SIGCOMM. For other SIGs, the top-10 ranked by tc-index are placed between the $[1^{st}, 19^{th}]$ (SIGKDD), $[1^{st}, 13^{rd}]$ (SIGCOMM), $[1^{st}, 37^{th}]$ (SIGIR), and $[1^{st}, 17^{th}]$ (SIGMOD) by h-index rankings. Ranges for citation rankings are even broader for all cases, mostly because its sensitivity to few works with many citations. Also, one of the most discrepant case is *H. V. Jagadish*, who is placed in 4th by tc-index, 12nd by h-index and 36th by citation. Finally, except for SIGCOMM, the results show diversity in the top rankings, which reinforces the complementary use of tc-index for a more complete analysis.

6.3   Representativeness of SIGs

Finally, we investigate the presence of researchers from specific SIGs in the global rank context. Hence, we count the number of researchers in a 202 (of a total of 20,470 researchers) top-rank according with each SIG[5]. The link researcher-SIG exists if the author has at least one work published there. For instance, if a researcher has twenty publications in SIGMOD and one in SIGKDD, then he/she contributes to both SIGs (independent of the number of publications).

Table 7 shows the number of authors in the global ranking according with each SIG. For instance, SIGACCESS has 5 researchers at global ranking for tc-index, 4 for h-index and no one for citation. In general, tc-index has a more *democratic* presence in the global ranking with 460 counts, whereas h-index has 449 and citation has 407. Also, there are at least 29 counts in all metrics for researchers from SIGCOMM, SIGIR, SIGKDD, SIGMETRICS, SIGMOBILE and SIGMOD. Nonetheless, tc-index is the only that includes at least one researcher from all

---

[5] We consider the first 202 researchers because at position 202nd, there is at least one representative from each SIG according to at least one metric.

Table 7. Number of researchers from each SIG in the 202 top-ranked for a global context for tc-index ($\mathcal{T}$), h-index ($\mathcal{H}$) and citation ($\mathcal{C}$).

	$\mathcal{T}$	$\mathcal{H}$	$\mathcal{C}$
SIGKDD	90	91	65
SIGMOD	87	93	75
SIGIR	64	84	44
SIGCOMM	55	39	76
SIGMETRICS	50	38	46
SIGMOBILE	36	29	68
SIGAPP	22	30	9
SIGecom	19	14	13
SIGSPATIAL	15	15	5
SIGCSE	8	6	4
SIGACCESS	5	4	0
SIGBio	3	4	2
SIGDOC	3	2	0
SIGAda	1	0	0
SIGITE	1	0	0
SIGUCCS	1	0	0
Sum	**460**	**449**	**407**

SIGs; whereas h-index does not include researchers from SIGAda, SIGITE and SIGUCCS, and citation the same ones plus SIGACCESS and SIGDOC. These results imply that tc-index, by considering the same *year-base community*, can capture the impact of researchers from different SIGs better than h-index and citation. Also, tc-index has the largest representativeness and, hence, the community concept brings a more diverse set of researchers in the top ranks.

7 Concluding Remarks

We have proposed a new productivity index that emphasizes the venue and the year of the publication, exploiting community aspects of research work. Our main findings are summarized as: tc-index differs from the baselines by bringing new positioning and researchers to the top-ranking; there is an independence between tc and h indices in most cases for SIGs analysis, and an independence between tc-index and citation in the global context; tc-index captures differences and peculiarities of each SIG according to the rankings, corroborating a complementary use of bibliometric indicators for a more in depth analysis; and tc-index has the largest SIGs representativeness, bringing a more diverse set of researchers to the top rankings. Furthermore, extending our approach to considering other digital libraries and analyzing other research communities should be straightforward.

For digital libraries, our work contributes in showing the potential for exploring their contents to acquire new knowledge from community-based analysis. In the future, we expect to add features as tc-index to existing academic-oriented systems that work over scientific digital libraries.

Acknowledgements. Work partially funded by CNPq and FAPEMIG, Brazil.

References

1. Alhoori, H., Furuta, R.: Can social reference management systems predict a ranking of scholarly venues? In: Aalberg, T., Papatheodorou, C., Dobreva, M., Tsakonas, G., Farrugia, C.J. (eds.) TPDL 2013. LNCS, vol. 8092, pp. 138–143. Springer, Heidelberg (2013)
2. Alonso, S., et al.: h-Index: a review focused in its variants, computation and standardization for different scientific fields. J. Informetrics **3**(4), 273–289 (2009)
3. Althouse, B.M., West, J.D., Bergstrom, C.T., Bergstrom, T.: Differences in impact factor across fields and over time. JASIST **60**(1), 27–34 (2009)
4. Barabási, A.L., Jeong, H., Néda, Z., Ravasz, E., Schubert, A., Vicsek, T.: Evolution of the social network of scientific collaborations. Physica A Stat. Mech. Appl. **311**(3–4), 590–614 (2002)
5. Biryukov, M., Dong, C.: Analysis of computer science communities based on DBLP. In: Lalmas, M., Jose, J., Rauber, A., Sebastiani, F., Frommholz, I. (eds.) ECDL 2010. LNCS, vol. 6273, pp. 228–235. Springer, Heidelberg (2010)
6. Bollen, J., de Sompel, H.V., Hagberg, A.A., Chute, R.: A principal component analysis of 39 scientific impact measures. PloS One **4**(6), e6022 (2009)
7. Burt, R.S.: Structural holes and good ideas. Am. J. Sociol. **110**(2), 349–399 (2004)
8. Deng et al, H.: Modeling and exploiting heterogeneous bibliographic networks for expertise ranking. In: JCDL, pp. 71–80 (2012)
9. Gonçalves, G.D., Figueiredo, F., Almeida, J.M., Gonçalves, M.A.: Characterizing scholar popularity: a case study in the computer science research community. In: JCDL, pp. 57–66 (2014)
10. Granovetter, M.S.: The strength of weak ties. Am. J. Sociol. **78**(6), 1360–1380 (1973)
11. Guimerá, R., et al.: Team assembly mechanisms determine collaboration network structure and team performanced. Science **308**, 697–702 (2005)
12. Hirsch, J.E.: An index to quantify an individual's scientific research output. PNAS **102**(46), 16569–16572 (2005)
13. Jain, R.: The Art of Computer Systems Performance Analysis - Techniques for Experimental Design, Measurement, Simulation, and Modeling. Wiley, New York (1991)
14. Järvelin, K., Kekäläinen, J.: Cumulated gain-based evaluation of IR techniques. ACM Trans. Inf. Syst. **20**(4), 422–446 (2002)
15. Laender, A.H.F., et al.: Building a research social network from an individual perspective. In: JCDL, pp. 427–428 (2011)
16. Lima, H., Silva, T.H.P., Moro, M.M., Santos Jr., R.L.T., W.M., Laender, A.H.F.: Aggregating productivity indices for ranking researchers across multiple areas. In: JCDL, pp. 97–106 (2013)
17. Lima, H., Silva, T.H.P., Moro, M.M., Santos, R.L.T., Wagner, J.M., Laender, A.H.F.: Assessing the profile of top Brazilian computer science researchers. Scientometrics **103**(3), 879–896 (2015)
18. Lopes, G.R., et al.: Ranking Strategy for graduate programs evaluation. In: ICITA, pp. 253–260 (2011)
19. Newman, M.E.: Who is the best connected scientist? A study of scientific coauthorship networks. In: Ben-Naim, E., Frauenfelder, H., Toroczkai, Z. (eds.) Complex Networks. Lecture Notes in Physics, vol. 650, pp. 337–370. Springer, Berlin (2004)

20. Reitz, F., Hoffmann, O.: An analysis of the evolving coverage of computer science sub-fields in the DBLP digital library. In: Lalmas, M., Jose, J., Rauber, A., Sebastiani, F., Frommholz, I. (eds.) ECDL 2010. LNCS, vol. 6273, pp. 216–227. Springer, Heidelberg (2010)
21. Silva, T.H.P., Moro, M.M., Silva, A.P.C.: Authorship contribution dynamics on publication venues in computer science: an aggregated quality analysis. In: SAC (2015)
22. Silva, T.H.P., Moro, M.M., Silva, A.P.C., Meira Jr., W., Laender, A.H.F.: Community-based endogamy as an influence indicator. In: JCDL (2014)
23. Zhang, C.T.: The e-index, complementing the h-index for excess citations. PLoS ONE 4(5), e5429 (2009)

Digital Humanities

Detecting Off-Topic Pages in Web Archives

Yasmin AlNoamany[(✉)], Michele C. Weigle, and Michael L. Nelson

Department of Computer Science, Old Dominion University,
Norfolk, VA 23529, USA
{yasmin,mweigle,mln}@cs.odu.edu

Abstract. Web archives have become a significant repository of our
recent history and cultural heritage. Archival integrity and accuracy is a
precondition for future cultural research. Currently, there are no quantitative or content-based tools that allow archivists to judge the quality
of the Web archive captures. In this paper, we address the problems of
detecting off-topic pages in Web archive collections. We evaluate six different methods to detect when the page has gone off-topic through subsequent captures. Those predicted off-topic pages will be presented to the
collection's curator for possible elimination from the collection or cessa-
tion of crawling. We created a gold standard data set from three Archive-
It collections to evaluate the proposed methods at different thresholds.
We found that combining cosine similarity at threshold 0.10 and change
in size using word count at threshold -0.85 performs the best with accuracy $= 0.987$, F_1 score $= 0.906$, and AUC $= 0.968$. We evaluated the
performance of the proposed method on several Archive-It collections.
The average precision of detecting the off-topic pages is 0.92.

Keywords: Archived collections · Experiments · Analysis · Document
filtering

1 Introduction

The Internet Archive [1] (IA) is the largest and oldest of the various Web
archives, holding over 400 billion Web pages with archives as far back as 1996 [2].
Archive-It[1] is a collection development service operated by the Internet Archive
since 2006. Archive-It is currently used by over 340 institutions in 48 states, and
features over 9B archived Web pages in nearly 2800 separate collections.

Archive-It provides their partners with tools that allow them to build themed
collections of archived Web pages hosted at Archive-It. This is done by the user
manually specifying a set of *seeds*, Uniform Resource Identifiers (URIs) that
should be crawled periodically (the frequency is tunable by the user), and to what
depth (e.g., follow the pages linked to from the seeds two levels out). Archive-It
also creates collections of global events under the name Internet Archive Global
Events. The seed URIs are manually collected by asking people to nominate URIs
that are related to these events, or are selected by the collection's curator(s).

---

[1] https://archive-it.org/.

© Springer International Publishing Switzerland 2015
S. Kapidakis et al. (Eds.): TPDL 2015, LNCS 9316, pp. 225–237, 2015.
DOI: 10.1007/978-3-319-24592-8_17

The Heritrix [3] crawler at Archive-It crawls/recrawls these seeds based on the predefined frequency and depth to build a collection of archived Web pages that the curator believes best exemplifies the topic of the collection. Archive-It has deployed tools that allow a collection's curators to perform quality control on their crawls. However, the tools are currently focused on issues such as the mechanics of HTTP (e.g., how many HTML files vs. PDFs and how many 404 missing URIs) and domain information (e.g., how many .uk sites vs. .com sites). Currently, there are no content-based tools that allow curators to detect when seed URIs go off-topic.

In this paper, we evaluate different approaches for detecting off-topic pages in the archives. The approaches depend on comparing versions of the pages through time. Three methods depend on the textual content (cosine similarity, intersection of the most frequent terms, and Jaccard coefficient), one method uses the semantics of the text (Web-based kernel function using a search engine), and two methods use the size of pages (the change in number of words and the content length). For evaluation purposes, we built our gold standard data set from three Archive-It collections, then we employed the following performance measurements: accuracy, F_1 score, and area under the ROC curve (AUC). Experimental results show that cosine similarity at the 0.15 threshold is the most effective single method in detecting the off-topic pages with 0.983 accuracy. We paired several of the methods and found that the best performing combined method across the three collections is cosine at threshold 0.10 with word count at threshold -0.85. Cosine and word count combined improved the performance over cosine alone with a 3 % increase in the F_1 score, 0.7 % increase in AUC, and 0.4 % increase in accuracy. We then used this combined method and evaluated the performance on a different set of Archive-It collections. Based on manual assessment of the detected off-topic pages, the average precision of the proposed technique for the tested collections is 0.92.

2   Motivating Example

We can define off-topic pages as the web pages that have changed through time to move away from the initial scope of the page. There are multiple reasons for pages to go off-topic, such as hacking, loss of account, domain expiration, owner deletion, or server/service discontinued [4]. Expired domains should return a 404 HTTP status that will be caught by Archive-It quality control methods. However, some expired domains may be purchased by spammers who desire all the incoming traffic that the site accrued while it was "legitimate". In this case, the Web page returns a 200 HTTP response but with unwanted content [5].

Figure 1 shows a scenario of a page going off-topic for several reasons. In May 2012, http://hamdeensabahy.com, which belonged to a candidate in Egypt's 2012 presidential election, is originally relevant to the "Egypt Revolution and Politics" collection (Fig. 1(a)). Then, the page went back and forth between on-topic and off-topic many times for different reasons. Note that there are on-topic pages between the off-topic ones in Fig. 1. In the example, the site went off-topic

(a) May 13, 2012: The page (b) May 24, 2012: Off-topic (c) June 5, 2014: The site has
started as on-topic.   due to a database error.   been hacked.

Fig. 1. A site for one of the candidates from Egypt's 2012 presidential election.

because of a database error on May 24, 2012 (Fig. 1(b)), then it returned on-topic again. After that, the page went off-topic between late March 2013 and early July 2013. The site went on-topic again for a period of time, then it was hacked (Fig. 1(c)), and then the domain was lost by late 2014. Today, http://hamdeensabahy.com is not available on the live Web.

The web page http://hamdeensabahy.com has 266 archived versions, or mementos. Of these, over 60 % are off-topic. While it might be useful for historians to track the change of the page in Web archives (possibly the hacked version is a good candidate for historians), the 60 % off-topic mementos such as the ones in Fig. 1(b)-(c) do not contribute to the Egypt Revolution collection in the same way that the on-topic archived Web site in Fig. 1(a) does. Although the former can be kept in the IA's general Web archive, it is a candidate to be purged from the Egyptian Revolution collection, or at the very least it should not be considered when summarizing the collection. Sites like http://hamdeensabahy.com that currently are not available on the live Web do not contribute to the collection, and assisting curators to identify and remove such pages is the focus of this paper.

3   Background

Despite the fact that Web archives present a great potential for knowledge discovery, there has been relatively little research that is explicitly aimed at mining content stored in Web archives [6]. In this section, we highlight the research that has been conducted on mining the past Web. First we define the terminology that will be adopted throughout the rest of the paper.

3.1   Memento Terminology

Memento [7] is an HTTP protocol extension which enables time travel on the Web by linking the current resources with their prior state. Memento defines several terms that we will use throughout. A URI-R identifies the original resource. It is the resource as it used to appear on the live Web. A URI-R may have 0

Table 1. Description of the Archive-It collections, including manual labeling of on and off-topic URI-Ms.

Collection Name	Occupy Movement 2011/2012	Egypt Revolution and Politics	Human Rights
Collection ID	2950	2358	1068
Curator	Internet Archive Global Events	American University in Cairo	Columbia University Libraries
Time span	12/03/2011- 10/09/2012	02/01/2011- 04/18/2013	05/15/2008- 03/21/2013
Total URI-Rs	728	182	560
Total URI-Ms	21,268	18,434	6,341
Sampled URI-Rs	255 (35 %)	136 (75 %)	198 (35 %)
Sampled URI-Ms	6,570	6,886	2,304
Off-topic URI-Ms	458 (7 %)	384 (9 %)	94 (4 %)
URI-Rs with off-topic URI-Ms	67 (26 %)	34 (25 %)	33 (17 %)

or more mementos (URI-Ms). A URI-M identifies an archived snapshot of the URI-R at a specific datetime, which is called Memento-Datetime, e.g., $URI\text{-}M_i = URI\text{-}R@t_i$. A URI-T identifies a TimeMap, a resource that provides a list of mementos (URI-Ms) for a URI-R with their Memento-Datetimes

3.2   Related Work

Mining the past Web is different from Web content mining because of the temporal dimension of the archived content [6,8]. Despite nearly two decades of Web history, there has not been much research conducted for mining Web archive data. The benefit of utilizing the Web archives for knowledge discovery has been discussed many times [6,9,10]. Below, we outline some of the approaches that have been used for mining the past Web using data in Web archives.

Jatowt and Tanaka [6] discussed the benefits of utilizing the content of the past Web for knowledge discovery. They discussed two mining tasks on Web archive data: temporal summarization and object history detection. They also presented different measures for analyzing the historical content of pages over a long time frame for choosing the important versions to be mined. They used a vector representation for the textual content of page versions using a weighting method, e.g., term frequency. They presented a change-detection algorithm for detecting the change in the past versions of a page through time. In a later study, Jatowt et al. [11] proposed an interactive visualization system called Page History Explorer (PHE), an application for providing overviews of historical content of pages and also exploring their histories. They used change detection algorithms based on the content of archived pages for summarizing the historical content of the page to present only the active content to users. They also extended the usage of term clouds for representing the content of the archived pages.

Francisco-Revilla et al. [12] described the Walden's Paths Path Manager system, which checks a list of Web pages for relevant changes using document signatures of paragraphs, headings, links, and keywords. Ben Saad et al. [13] claimed that using patterns is an effective way to predict changes, and then used this prediction to optimize the archiving process by crawling only important pages.

Spaniol and Weikum used Web archive data to track the evolution of entities (e.g., people, places, things) through time and visualize them [14]. This work is a part of the LAWA project (Longitudinal Analytics of Web Archive data), a focused research project for managing Web archive data and performing large-scale data analytics on Web archive collections. Jatowt et al. [10] also utilized the public archival repositories for automatically detecting the age of Web content through the past snapshots of pages.

Most of the previous papers used the change of web pages for optimizing the crawl or visualization. Despite the existence of crawl quality tools that focus on directly measurable things like MIME types, response codes, etc., there are no tools to assess if a page has stayed on-topic through time. The focus of this paper is assisting curators in identifying the pages that are off-topic in a TimeMap, so these pages can be excluded from the collections.

4 Data Set

In this section we describe our gold standard dataset. We evaluate our techniques using the ODU mirror of Archive-It's collections. ODU has received a copy of the Archive-It collections through April 2013 in Web ARchive file format (WARC) [15]. The three collections in our dataset differ in terms of the number of URI-Rs, number of URI-Ms, and time span over which the Web pages have been archived (ending in April 2013). The three collections, which include pages in English, Arabic, French, Russian, and Spanish, are described below. The details are provided in Table 1.

Occupy Movement 2011/2012 covers the Occupy Movement protests and the international branches of the Occupy Wall Street movement.

Egypt Revolution and Politics covers the January 25th Egyptian Revolution and Egyptian politics, contains different kinds of Web sites (e.g., social media, blogs, news, etc.).

Human Rights covers documentation and research about human rights that has been created by non-governmental organizations, national human rights institutions, and individuals.

We randomly sampled 588 URI-Rs from the three collections (excluding URI-Rs with only one memento). Together, the sampled URI-Rs had over 18,000 URI-Ms, so for each of the sampled URI-Rs, we randomly sampled from their URI-Ms. We manually labeled these 15,760 mementos as on-topic or off-topic. The bottom portion of Table 1 contains the results of this labeling.

5   Research Approach

In this section, we explain the methodology for preparing the data set and then the methodology for applying different measures to detect the off-topic pages.

5.1   Data Set Preprocessing

We applied the following steps to prepare the gold standard data set: (1) obtain the seed list of URIs from the front-end interface of Archive-It, (2) obtain the TimeMap of the seed URIs from the CDX file[2], (3) extract the HTML of the mementos from the WARC files (hosted at ODU), (4) extract the text of the page using the Boilerpipe library [16], and (5) extract terms from the page, using scikit-learn [17] to tokenize, remove stop words, and apply stemming.

5.2   Methods for Detecting Off-Topic Pages

In this section, we use different similarity measures between pages to detect when the *aboutness(URI-R)* over time changes and to define a threshold that separates the on-topic and the off-topic pages.

Cosine similarity. Cosine similarity is one of the most commonly used similarity measures in information retrieval. After text preprocessing, we calculated the TF-IDF, then we applied cosine similarity to compare the *aboutness(URI-R@t_0)* with *aboutness(URI-R@t)*.

Jaccard similarity coefficient. The Jaccard similarity coefficient is the size of the intersection of two sets divided by the size of their union. After preprocessing the text (step 5), we apply the Jaccard coefficient on the resulting terms to specify the similarity between the *URI-R@t* and *URI-R@t_0*.

Intersection of the most frequent terms. Term frequency (TF) refers to how often a term appears in a document. The aboutness of a document can be represented using the top-k most frequent terms. After text extraction, we calculated the TF of the text *URI-R@t*, and then compared the top 20 most frequent terms of the *URI-R@t* with the top 20 most frequent terms of the *URI-R@t_0*. The size of the intersection between the top 20 terms of *URI-R@t* and *URI-R@t_0* represents the similarity between the mementos. We name this method TF-Intersection.

Web-based kernel function. The previous methods are term-wise similarity measures, i.e., they use lexicographic term matching. But these methods may not suitable for archived collections with a large time span or pages that contain a small amount of text. For example, the Egyptian Revolution collection is from February 2011 until April 2013. Suppose a page in February 2011 has terms like "Mubarak, Tahrir, Square" and a page in April 2013 has terms like "Morsi, Egypt". The two pages are semantically relevant to each other, but term-wise

---

[2] http://archive.org/web/researcher/cdx_file_format.php.

the previous methods might not detect them as relevant. With a large evolution of pages through a long period of time, we need a method that focuses on the semantic context of the documents. The work by Sahami and Heilman [18] inspired us to augment the text of $URI\text{-}R@t_0$ with additional terms from the web using a search engine to increase its semantic context. This approach is based on query expansion techniques, which have been well-studied in information retrieval [19]. We used the contextually descriptive snippet text returned with results from the Bing Search API. We call this method "SEKernel".

We augment the terms of $URI\text{-}R@t_0$ with semantic context from the search engine as follows:

1. Format a query q from the top 5 words of the first memento ($URI\text{-}R@t_0$).
2. Issue q to the search engine SE.
3. Extract the terms p from the top 10 snippets returned for q.
4. Add the terms of the snippets p to the terms of the original text of the first memento to have a new list of terms, $ST = p \cup URI\text{-}R@t_0$.
5. $\forall t \geq 1$, calculate the Jaccard coefficient between the expanded document ST and the terms of $URI\text{-}R@t$.

If we apply this method on the previous example, we use terms "Mubarak, Tahrir, Square" as search keywords to generate semantic context. The resulting snippet will have new terms like "Egypt, President", which term-wise overlaps with the page that contains "Morsi, Egypt".

Change in size. We noticed that the size of off-topic mementos are often much smaller in size than the on-topic mementos. We used the relative change in size to detect when the page goes off-topic. The relative change of the page size can be represented by the content length or the total number of words (e.g., egypt, tahrir, the, square) in the page. For example, assume $URI\text{-}R@t_0$ contains 100 words and $URI\text{-}R@t$ contains 5 words. This represents a 95 % decrease in the number of words between $URI\text{-}R@t_0$ and $URI\text{-}R@t$.

We tried two methods for measuring the change in size: the content length (bytes) and the number of words (WordCount). Although using the content length, which can be extracted directly from the headers of the WARC files, saves the steps of extracting the text and tokenization, it fails to detect when the page goes off-topic in the case when the page has little to no textual content but the HTML forming the page template is still large. There are many cases where the page goes off-topic and the size of the page decreases or in some cases reaches 0 bytes, e.g., the account is suspended, transient errors, or no content in the page. One of the advantages of using the structural-based methods over the textual-based methods is that structural-based methods are language independent. Many of the collections are multi-lingual, and each language needs special processing. The structural methods are suitable for those collections.

Table 2. Evaluating the similarity approaches, averaged over the three collections.

Similarity Measure	Threshold	FP	FN	FP+FN	ACC	F_1	AUC
Cosine	0.15	31	22	53	**0.983**	**0.881**	**0.961**
WordCount	−0.85	6	44	50	0.982	0.806	0.870
SEKernel	0.05	64	83	147	0.965	0.683	0.865
Bytes	−0.65	28	133	161	0.962	0.584	0.746
Jaccard	0.05	74	86	159	0.962	0.538	0.809
TF-Intersection	0.00	49	104	153	0.967	0.537	0.740
Cosine|WordCount	0.10|−0.85	24	10	34	**0.987**	**0.906**	**0.968**
Cosine|SEKernel	0.10|0.00	6	35	40	0.990	0.901	0.934
WordCount|SEKernel	−0.80|0.00	14	27	42	0.985	0.818	0.885

6   Evaluation

In this section, we define how we evaluate the methods presented in Sect. 5.2 on our gold standard data set. Based on these results, we define a threshold th for each method for when a memento becomes off-topic.

6.1   Evaluation Metrics

We used multiple metrics to evaluate the performance of the similarity measures. We considered false positives (FP), the number of on-topic pages predicted as off-topic; false negatives (FN), the number of off-topic pages predicted as on-topic; accuracy (ACC), the fraction of the classifications that are correct; F_1 score (also known as F-measure), the weighted average of precision and recall; and the ROC AUC score (AUC), a single number that computes the area under the receiver operating characteristic curve [20].

6.2   Results

We tested each method with 21 thresholds (378 tests for three collections) on our gold standard data set to estimate which threshold for each method is able to separate the off-topic from the on-topic pages. In order to determine the best threshold, we used the evaluation metrics described in the previous section. To say that $URI\text{-}R@t$ is off-topic at $th = 0.15$ means that the similarity between $URI\text{-}R@t$ and $URI\text{-}R@t_0$ is < 0.15. On-topic means the similarity between $URI\text{-}R@t$ and $URI\text{-}R@t_0$ is ≥ 0.15.

For each similarity measure, there is an upper bound and lower bound for the value of similarity. For Cosine, TF-Intersection, Jaccard, and SEKernel, the highest value is at 1 and the lowest value is at 0. A similarity of 1 represents a perfect similarity, and 0 similarity represents that there is no similarity between the pages. The word count and content length measures can be from −1 to +1.

The negative values in change of size measures represent the decrease in size, so -1 means the page has a 100 % decrease from $URI\text{-}R@t_0$. When the change in size is 0 that means there is no change in the size of the page. We assume that a large decrease in size between $URI\text{-}R@t$ and $URI\text{-}R@t_0$ indicates that the page might be off-topic. Therefore, if the change in size between $URI\text{-}R@t$ and $URI\text{-}R@t_0$ is less than -95%, that means $URI\text{-}R@t$ is off-topic at $th = -95\%$.

Table 2 contains the summary of running the similarity approaches on the three collections. The table shows the best result based on the F_1 score at the underlying threshold measures averaged on all three collections. From the table, the best performing measure is Cosine with average $ACC = 0.983$, $F_1 = 0.881$, and $AUC = 0.961$, followed by WordCount. Using SEKernel performs better than TF-Intersection and Jaccard. Based on the F_1 score, we notice that TF-Intersection and Jaccard similarity are the least effective methods.

There was consistency among the values of th with the best performance of TF-Intersection, Jaccard, and SEKernel methods for the three collections, e.g., for all the collections, the best performance of the SEKernel method is at $th = 0.05$. However, there was inconsistency among the values of th with the best performance for each collection for Cosine, WordCount, and Bytes measures. For the methods with inconsistent threshold values, we averaged the best thresholds of each collection. For example, the best th values of Cosine for Occupy Movement collection, Egypt Revolution collection, and Human Rights collection are 0.2, 0.15, 0.1 respectively. We took the average of the three collections at $th = 0.2$, $th = 0.15$, and $th = 0.1$, then based on the best F_1 score, we specified the threshold that has the best average performance.

Specifying a threshold for detecting the off-topic pages is not easy due to differences in the nature of the collections. For example, long running collections such as the Human Rights collection (2009–present) have more opportunities for pages to change dramatically, while staying relevant to the collection. There is more research to be done in exploring the thresholds and methods. We plan to investigate different methods on larger sets of labeled collections, so that we can specify the features that affect choosing the value of the threshold.

6.3   Combining the Similarity Measures

We tested 6,615 pairwise combinations (15 method combinations $\times 21 \times 21$ threshold values). A page was considered off-topic if either of the two methods declared it off-topic. Performance results of combining the similarity approaches are presented in the bottom portion of Table 2. We present the three best average combinations of the similarity measures based on F_1 score and AUC. Performance increases with combining Cosine and WordCount (Cosine|WordCount) at $th = 0.1| - 0.85$. There is a 36 % decrease in errors (FP+FN) than the best performing single measure, Cosine. Furthermore, Cosine|WordCount has a 3 % increase in the F_1 score over Cosine. Cosine|SEKernel at $th = 0.1|0.0$ has a 2 % increase in F_1 over Cosine, while WordCount|SEKernel at $th = -0.80|0.00$ has lower performance than Cosine.

In summary Cosine|WordCount gives the best performance at $th = 0.1|-0.85$ across all the single and combined methods. Moreover, combining WordCount with Cosine does not cause much overhead in processing, because WordCount uses tokenized words and needs no extra text processing.

Table 3. The results of evaluating Archive-It collections through the assessment of the detected off-topic pages using Cosine|WordCount methods at $th = 0.10| - 0.85$.

Collection	ID	Time Span	URI-Rs	URI-Ms	Affected URI-Rs	TP	FP	P
Global Food Crisis	2893	10/19/2011-10/24/2012	65	3,063	7	22	0	1.00
Government in Alaska	1084	12/01/2006-04/13/2013	68	506	4	16	0	1.00
Virginia Tech Shootings	2966	12/08/2011-01/03/2012	239	1,670	2	24	0	1.00
Wikileaks 2010 Document Release Collection	2017	07/27/2010-08/27/2012	35	2,360	8	107	0	1.00
Jasmine Revolution - Tunisia 2011	2323	01/19/2011-12/24/2012	231	4,076	31	107	7	0.94
IT Historical Resource Sites	1827	2/23/2010-10/04/2012	1,459	10,283	34	45	14	0.76
Human Rights Documentation Initiative	1475	04/29/2009-10/31/2011	147	1,530	20	39	15	0.72
Maryland State Document Collection	1826	03/04/2010-12/03/2012	69	184	0	-	-	-
April 16 Archive	694	05/23/2007-04/28/2008	35	118	0	-	-	-
Brazilian School Shooting	2535	04/09/2011-04/14/2011	476	1,092	0	-	-	-
Russia Plane Crash Sept 7,2011	2823	09/08/2011-09/15/2011	65	447	0	-	-	-

7   Evaluating Archive-It Collections

We applied the best performing method (Cosine|WordCount) with the suggested thresholds $(0.1| - 0.85)$ on unlabeled Archive-It collections. We chose different types of collections, e.g., governmental collections (Maryland State Document Collection, Government in Alaska), event-based collections (Jasmine Revolution-Tunisia 2011, Virginia Tech Shootings), and theme-based collections (Wikileaks 2010 Document Release Collection, Human Rights Documentation Initiative). Table 3 contains the details of the 11 tested collections. We extracted the tested collections from the ODU mirror of Archive-It's collections. The number of URI-Rs in the table represents those URI-Rs with more than one memento.

The results of evaluating Cosine|WordCount are shown in Table 3. For the reported results for each TimeMap for each method, we manually assessed the FP and TP and then calculated the precision $P = TP/(TP + FP)$. We cannot compute recall since we cannot know how many off-topic mementos were not detected (FN). Precision is near 1.0 for five collections. Precision = 0.72 for the "Human Rights Documentation" collection, with 15 FP. Those 15 URI-Ms affected three TimeMaps. An example of one of the affected TimeMaps (https://wayback.archive-it.org/1475/*/http://www.fafg.org/) contains 12 FPs. The reason is that the home page of the site changed and newer versions use Adobe Flash. The 14

FPs from the "IT Historical Resource Sites" collection affected 5 URIs because the content of the pages changed dramatically through time. There are four collections that have no reported off-topic pages. Two of these collections, "Brazilian School Shooting" and "Russia Plane Crash", span only one week, which is typically not enough time for pages to go off-topic. The third collection with no detected off-topic mementos is the "Maryland State Document" collection. Perhaps this collection simply had well-chosen seed URIs.

In summary, Cosine|WordCount at $th = 0.1| - 0.85$ performed well on Archive-It collections with average $P = 0.92$. Studying the FP cases has given us an idea about the limitations of the current approach, such as detecting URIs with little to no textual content or URIs whose topics change significantly through time.

8   Conclusions and Future Work

In this paper, we present approaches for assisting the curator in identifying off-topic mementos in the archive. We presented six methods for measuring similarity between pages: cosine similarity, Jaccard similarity, intersection of the most 20 frequent terms, Web-based kernel function, change in number of words, and change in content length. We tested the approaches on three different labeled subsets of collections from Archive-It. We found that of the single methods, the cosine similarity measure is the most effective method for detecting the off-topic pages at $th = 0.15$. The change in size based on the word count comes next at $th = -0.85$. We also found that adding semantics to text using SEKernel enhanced the performance over Jaccard. We combined the suggested methods and found that, based on the F_1 score and the AUC, Cosine|WordCount at $th = 0.10| - 0.85$ enhances the performance to have the highest F_1 score at 0.906 and the highest AUC at 0.968. We tested the selected thresholds and methods on different Archive-It collections. We tested the performance of Cosine|WordCount at $th = 0.10| - 0.85$ by applying them on 11 Archive-It collections. We manually assessed the relevancy of the detected off-topic pages and found that the average precision $= 0.92$. In summary, evaluating the suggested approach, Cosine|WordCount at $th = 0.10| - 0.85$, for detecting the off-topic pages in the archives has shown good results with 0.92 precision. The presented approach will help curators to judge their crawls and also will obviate users from getting unexpected content when they access archived pages.

This is a preliminary investigation of automatically detecting the off-topic pages from web archives. There is more research to be done in exploring the thresholds and methods. For example, the nature of collection, such as the time span, might affect choosing the threshold. Users will be able to adjust the methods and thresholds as command-line parameters. The code and gold standard data set are available at https://github.com/yasmina85/OffTopic-Detection.

Our future work will continue to improve detection by using larger data sets and more collections with different features. The methods presented here detect off-topic pages within the context of a single TimeMap. The next step is to compute a model of the topic of the collection, in part to more easily detect seeds that begin off-topic.

Acknowledgments. This work supported in part by the Andrew Mellon Foundation. We thank Kristine Hanna from Internet Archive for facilitating obtaining the data set.

References

1. Negulescu, K.C.: Web Archiving @ the Internet Archive. Presentation at the 2010 Digital Preservation Partners Meeting (2010). http://www.digitalpreservation. gov/meetings/documents/ndiipp10/NDIIPP072110FinalIA.ppt
2. Kahle, B.: Wayback Machine Hits 400,000,000,000! http://blog.archive.org/2014/ 05/09/wayback-machine-hits-400000000000/ (2014)
3. Mohr, G., Stack, M., Ranitovic, I., Avery, D., Kimpton, M.: An introduction to Heritrix an open source archival quality web crawler. In: Proceedings of IWAW, pp. 43–49 (2004)
4. Marshall, C., McCown, F., Nelson, M.: Evaluating Personal archiving strategies for internet-based information. In: Proceedings of Archiving, pp. 151–156 (2007)
5. Bar-Yossef, Z., Broder, A.Z., Kumar, R., Tomkins, A.: Sic transit Gloria Telae: Towards an understanding of the web's decay. In: Proceedings of WWW, pp. 328–337 (2004)
6. Jatowt, A., Tanaka, K.: Towards mining past content of Web pages. New Rev. Hypermedia Multimed. **13**(1), 77–86 (2007)
7. Van de Sompel, H., Nelson, M.L., Sanderson, R.: RFC 7089 - HTTP framework for time-based access to resource states - Memento (2013)
8. Kosala, R., Blockeel, H.: Web mining research: a survey. SIGKDD Explor. Newsl. **2**(1), 1–15 (2000)
9. Arms, W.Y., Aya, S., Dmitriev, P., Kot, B.J., Mitchell, R., Walle, L.: Building a research library for the history of the web. In: Proceedings of ACM/IEEE JCDL, pp. 95–102 (2006)
10. Jatowt, A., Kawai, Y., Tanaka, K.: Detecting age of page content. In: Proceedings of ACM WIDM, pp. 137–144 (2007)
11. Jatowt, A., Kawai, Y., Tanaka, K.: Page history explorer: visualizing and comparing page histories. IEICE Trans. Inf. Syst. **94**(3), 564–577 (2011)
12. Francisco-Revilla, L., Shipman, F., Furuta, R., Karadkar, U., Arora, A.: Managing change on the web. In: Proceedings of ACM/IEEE JCDL, pp. 67–76 (2001)
13. Ben Saad, M., Gançarski, S.: Archiving the web using page changes patterns: a case study. In: Proceedings of ACM/IEEE JCDL, pp. 113–122 (2012)
14. Spaniol, M., Weikum, G.: Tracking entities in web archives: the LAWA project. In: Proceedings of WWW, pp. 287–290 (2012)
15. ISO: ISO 28500:2009 - Information and documentation - WARC file format (2009). http://www.iso.org/iso/iso_catalogue/catalogue_tc/catalogue_detail.htm? csnumber=44717
16. Kohlschütter, C., Fankhauser, P., Nejdl, W.: Boilerplate detection using shallow text features. In: Proceedings of ACM WSDM, pp. 441–450 (2010)
17. Pedregosa, F., Varoquaux, G., Gramfort, A., Michel, V., Thirion, B., Grisel, O., Blondel, M., Prettenhofer, P., Weiss, R., Dubourg, V., Vanderplas, J., Passos, A., Cournapeau, D., Brucher, M., Perrot, M., Duchesnay, E.: Scikit-learn: machine learning in Python. J. Mach. Learn. Res. **12**, 2825–2830 (2011)

18. Sahami, M., Heilman, T.D.: A web-based kernel function for measuring the similarity of short text snippets. In: Proceedings of WWW, pp. 377–386 (2006)
19. Buckley, C., Salton, G., Allan, J., Singhal, A.: Automatic query expansion using SMART: TREC 3. In: Overview of the Third Text REtrieval Conference (TREC-3), pp. 69–80 (1995)
20. Fawcett, T.: An introduction to ROC analysis. Pattern Recogn. Lett. **27**(8), 861–874 (2006)

Supporting Exploration of Historical Perspectives Across Collections

Daan Odijk[1]([✉]), Cristina Gârbacea[1], Thomas Schoegje[2], Laura Hollink[2,3],
Victor de Boer[2], Kees Ribbens[4,5], and Jacco van Ossenbruggen[2,3]

[1] ISLA, University of Amsterdam, Amsterdam, The Netherlands
d.odijk@uva.nl
[2] VU University Amsterdam, Amsterdam, The Netherlands
[3] CWI, Amsterdam, The Netherlands
[4] NIOD Institute for War, Holocaust and Genocide Studies,
Amsterdam, The Netherlands
[5] Erasmus School of History, Culture and Communication,
Rotterdam, The Netherlands

Abstract. The ever growing number of textual historical collections calls for methods that can meaningfully connect and explore these. Different collections offer different perspectives, expressing views at the time of writing or even a subjective view of the author. We propose to connect heterogeneous digital collections through temporal references found in documents as well as their textual content. We evaluate our approach and find that it works very well on digital-native collections. Digitized collections pose interesting challenges and with improved preprocessing our approach performs well. We introduce a novel search interface to explore and analyze the connected collections that highlights different perspectives and requires little domain knowledge. In our approach, perspectives are expressed as complex queries. Our approach supports humanity scholars in exploring collections in a novel way and allows for digital collections to be more accessible by adding new connections and new means to access collections.

1 Introduction

A huge amount of digital material has become available to study our recent history, ranging from digitized newspapers and books to, more recently, web pages about people, places and events. Each collection has a different perspective on what happened and this perspective depends partly on the medium, time and location of publication. For example, Lensen [10] analyses two contemporary novels about the second World War and shows that the writers have a different attitude towards the war than previous generations when it comes to issues of perpetratorship, assignation of blame and guilt. We aim to connect collections and support exploration of these different perspectives.

In this work, we focus on the second World War (WWII), as this is a defining event in our recent history. Different collections tell different stories of events in

© Springer International Publishing Switzerland 2015
S. Kapidakis et al. (Eds.): TPDL 2015, LNCS 9316, pp. 238–251, 2015.
DOI: 10.1007/978-3-319-24592-8_18

WWII and there is a wealth of knowledge to be gained in comparing these. Reading a news article on the liberation of the south of the Netherlands in a newspaper collaborating with the occupiers gives the impression that it is just a minor setback for the occupiers. A very different perspective on the same events emerges from the articles in an illegal newspaper of the resistance, leaving the impression that war is ending soon. To complete the picture, these contemporary perspectives can be compared to the view of an historian who—decades later—wrote fourteen books on WWII in the Netherlands and to the voice of thousands of Wikipedians, all empowered with the benefit of hindsight.

We present an interactive search application that supports researchers (such as historians) in connecting perspectives from multiple heterogeneous collections. We provide tools for selecting, linking and visualizing WWII-related material from collections of the NIOD, the National Library of the Netherlands, and Wikipedia. Our application provides insight into the volume, selection and depth of WWII-related topics across different media, times and locations through an exploratory interface. iod We connect digital collections via implicit events, i.e. if two articles are close in time and similar in content, we consider them to be related. Newspaper articles are associated with a clear point in time, the date they were published. However, not all collections have such a clear temporal association. We therefore infer these associations from temporal references (i.e., references in the text to a specific date). Our novel approach to connecting collections can deal with these extracted temporal references.

We provide two insightful visualizations of the connected collections: (1) an exploratory search interface that provides insight into the volume of data on a particular WWII-related topic and (2) an interactive interface in which a user can select two articles/pages for a detailed comparison of the content. Our aim is to provide historical researchers with the means to explore perspectives, not to analyze these for them. These perspectives are expressed in our application through what we call *query contrasts*. These contrasts are, in essence, sets of filters over one or more collections, that are added to search queries and consistently visualized throughout the interface. Example uses of such contrast are comparing newspaper articles versus Wikipedia articles (across collections) or local newspaper versus national newspapers (within a collection).

We focus on events and collections related to WWII. However, as our approach and application can be applied to other digital collections and topics, our work has broader implications for digital libraries. Our main contributions are twofold: (1) we propose and evaluate an approach to connect digital collections through implicit events, and (2) we demonstrate how these connections can be used to explore and analyze perspectives in a working application. Our work is based entirely on open data and open-source technology. We release the extracted temporal references as Linked Open Data and our application as open-source software, for anybody to reuse or re-purpose.

The remainder of this paper is organized as follows. We discuss related work in Sect. 2. Next, we describe our approach to connecting digital collections in Sect. 3. Section 4 describes our exploratory and comparative interfaces. We provide a worked example in Sect. 5, after which we conclude in Sect. 6.

2   Related Work

We discuss related work on studying historical perspectives, on connecting digital collections and on exploring and comparing collections.

Historical Perspectives. Historical news has often been used to study public opinion. For example, Van Vree [18] studied the Dutch public opinion on Germany in the period 1930–1939 based on articles from four newspapers, selected to represent distinct population groups. More recently, Lensen [10] showed how two contemporary novels exhibit new perspectives on WWII. Typically, scholars from the humanities study these perspectives on a small scale. Au Yeung and Jatowt [2] studied how the past is remembered on a large scale. They find references to countries and years in over 2.4 million news articles written in English and study how these are referred to. Using topic modeling they find significant years and topics and compute similarities between countries. Our work provides scholars with the means to explore their familiar research questions on different perspectives on a large scale, without doing this analysis for them.

Connecting Digital Collections. With more collections becoming digitally available, researchers have increasingly attempted to find connections between collections. Common is to link items based on their metadata. When items are annotated with concepts from a thesaurus or ontology, ontology alignment can be used to infer links between items. An example is MultimediaN E-culture, where artworks from museums were connected based on alignments between thesauri used to annotate the collections [16]. An approach that does not rely on the presence of metadata is to infer links based on textual overlap of items. For example, Bron et al. [3] study how to create connections between a newspaper archive and a video archive. Using document enrichment and term selection they link documents that cover same or related events. Similarly, in the PoliMedia project [9] links between political debates and newspapers articles are inferred based on topical overlap. As in our approach, both used publication date to filter documents in the linking process. However, how to score matches using these dates and combine them with temporal references is an open problem. Alonso et al. [1] survey trends in temporal information retrieval and identify open challenges, that include how to measure temporal similarity and how to combine scores for textual and temporal queries.

Exploratory Search. To support scholars in studying historical perspectives, we propose an exploratory search interface. In exploratory search [11], users interactively and iteratively explore interesting parts of a collection. Many exploratory search systems have been proposed; we discuss a few that are closely related to our work. Odijk et al. [14] proposed an exploratory search interface to support historians in selecting documents for qualitative analysis. Their approach addresses questions that might otherwise be raised about the representativeness, reproducibility and rigidness of the document set. de Rooij et al. [8] collected social media content from four distinct groups: politicians, journalist, lobbyist and the public and proposed a search interface to explore this grouped content. Through studying the temporal context and volume of discussion over time, one could answer the questions of who put an issue on the agenda.

Comparing Collections. To support media studies researchers, Bron et al. [4] propose a subjunctive search interface that shows two search queries side-by-side. They study how this fits into the research cycle of media studies researchers. They find that when using the proposed interface, the researcher explore more diverse topics and formulate more specific research questions. ManyPedia [12] allows users to explore different points of view by showing two Wikipedia articles from different languages side-by-side. A similar approach was used to synchronize cross-lingual content [13] on Wikipedia. Similar to the work described above, we provide exploratory search tools that emphasizes different perspectives. Our work differs in that our application provides an end-to-end solution, from connecting multiple collections to exploring and comparing.

3 Connecting Digital Collections

In this section, we describe how we connect multiple heterogeneous digital collections. Simply put, we connect documents from different collections via implicit events using time and content. If we consider two newspaper articles from different newspapers, but published on the same day and with considerable overlap in content, we can infer that it is likely that they cover the same event. However, not all our sources are as neatly dated as the newspapers are. For these, we need to extract the relevant dates based on the document content (detailed in Sect. 3.2). We present and validate our approach to connecting collections in Sects. 3.4 and 3.5 respectively. First, we describe our collections.

3.1 Collections

We connect three heterogeneous collections, each representing a different kind of data source: (1) a digitized collection of around 100 million Dutch newspapers articles, spanning four centuries, (2) the encyclopedic articles of the Dutch Wikipedia, (3) the digitized book series *Het Koninkrijk der Nederlanden in de Tweede Wereldoorlog*[1] by historian Loe de Jong. Table 1 provides an overview of the size and characteristics of each collection. For conciseness, we will refer to these three collections as the *Newspapers, Wikipedia* and *Loe de Jong* collections respectively in the remainder of this paper.

The newspaper archive of the National Library of the Netherlands consists of around 100 million digitized newspaper articles, processed using optical character recognition[2]. Each newspaper article has consistent metadata, including publication date. To focus on the relevant articles we filter out articles published before 1933 or after 1949. The books of Loe de Jong are a standard reference

---

[1] In English: *The Kingdom of the Netherlands during WWII.*
[2] http://delpher.nl/kranten

work about WWII in the Netherlands, consisting of 14 volumes and published between 1969–1988 in 29 parts, all recently digitized [7]. Early parts focus on chronicling events in specific years whereas some later ones focus on specific themes. Each section was treated as a new document. Note that the documents are still substantially longer than those in the two other collections.

3.2   Extracting Temporal References

We connect the three heterogeneous collections presented above via content and time. For the digitized newspaper collection, we use the publication date of an article. This metadata is clean and checked by the National Library. However, for Wikipedia articles and the books of Loe de Jong no clear dates are present. We extract the dates that these articles refer to through a process of temporal tagging, i.e. we extract references to dates from the article content. For this, we use a custom pipeline for xTAS[3], an extendable toolkit for large-scale text analysis. Concretely, our approach for extracting dates consists of three steps: (1) *preprocessing*, (2) *temporal tagging* and (3) *aggregating*.

In the preprocessing step, we normalize the text and prevent common errors we encountered in the subsequent temporal tagging. For Wikipedia articles, we remove all special syntax, used for formatting and creating tables. We extract the textual content of a Wikipedia article using a MediaWiki syntax parser[4]. For the Loe de Jong collection, we remove XML tags and keep only the textual content. This textual content has been obtained from book scans using optical character recognition (OCR). Therefore, it can contain errors in the recognition of terms. We process this digitized text to remedy common OCR errors that we encountered, in particular errors that influence temporal tagging. For example, the numbers 0, 1 and 5 are commonly confused for °, I and S.

For both collections, we also use simple textual replacement rules to prevent common errors we found after an initial evaluation of temporal tagging on our data. A common short-hand way of referring to years is to use an apostrophe followed by only the last to digits: the period '40-'45. As this gives no information on the century being referred to, such a reference is typically ignored by a temporal tagger. However, these references often refer to the 1900s and given that the topic of most of our document collection (WWII), we resolve a reference as above to the period 1940–1945.

After preprocessing, we analyze the content of each Wikipedia article and each document in the Loe de Jong collection using the multilingual cross-domain temporal tagger Heideltime [17]. The aim of a temporal tagger is to find all mentions of time and dates and to pinpoint these as exact as possible to a specific point in time. The output of Heideltime is a set of temporal references normalized according to the TIMEX3 annotation standard [15]. Temporal tagging of historical documents is particularly challenging due to the fact that temporal expressions are often ambiguous and under-specified. For example, *"in the*

---

[3] http://xtas.net
[4] http://github.com/pediapress/mwlib

1930s" refers to the time interval 1930–1939, while *"august 1945"* refers to the entire month of August, 1945. For most of these challenges, Heideltime is able to extract and normalize temporal references, even if they are under-specified.

The final step to extracting dates is aggregating all temporal references to the document level. We separate each temporal reference based on the annotation granularity (i.e. exact day, specific month or only year). We store and treat them differently both in connecting collections (see Sect. 3.4) and in the exploratory visualizations (see Sect. 4.1).

Table 1. Statistics for the collections and for temporal reference extraction.

	Newspapers	Wikipedia	Loe de Jong
Number of documents	21,456,471	2,699,044	1,600
Average number of terms per document	105	91	1776
Number of annotated documents		50	20
Total number of unique / individual annotated date references		609 / 834	469 / 713

3.3 Evaluating Temporal Reference Extraction

We validate our approach to extracting date references with an experiment. We take 50 random documents from the Wikipedia collection and 20 from the Loe de Jong collection. Five judges annotate all date references within the documents. Table 1 details statistics on the annotated documents. The narrative structure of the books of Loe de Jong leads to less frequent temporal references than the encyclopedic style of Wikipedia. We compute inter-annotator agreement over five doubly annotated Wikipedia documents and three documents from the Loe de Jong collection. We observe 97 % agreement on the set of unique dates referenced, signaling excellent agreement among the human annotators. We measure the accuracy of the extracted temporal references by comparing the automatically annotated date references to those annotated by the judges.

On the Wikipedia collection, we observe a mean precision of 98.27 % on the set of all automatically extracted dates, with recall at 86.72 % of unique annotated dates. These scores are comparable to what is reported on standard datasets [17] and signals that the task of extracting dates from Wikipedia articles is well suited to be done automatically.

On the Loe de Jong collection, we obtain substantially lower precision of 63.86 % and recall of 68.04 %. The sections of these books pose two distinct challenges for temporal tagging. First, as the books are digitized using OCR, there are errors in detected terms, including in parts of dates. Our preprocessing approach to remedy some of the common errors has doubled both precision and recall (up from 36.86 % and 30.48 % respectively). The second challenge is more difficult. The books of Loe de Jong are written in a narrative style, where temporal references are often (partially) implicit. For example, a section on famine in the winter of 1944–1945 (referred to as the "hunger winter") often only refers only to days in these winter months, without referring to a year. Given the topic, a reader knows that January 15th refers to January 15th, 1945, but for an automatic approach, this is rather difficult to infer. In fact, half of the fourteen books indicate in the title that they cover only a specific period.

Improving the accuracy of temporal reference extraction on such a collection poses interesting future work for information extraction researchers. Given the length of the documents and thus large number of temporal references, the level of accuracy we obtain after preprocessing is sufficient for ours and similar applications. The extracted temporal references for the Loe de Jong collection are published as Linked Open Data[5]. This enrichment allows for new types of temporal analysis of the collection.

This approach for temporal reference extraction is the first step in connecting these three heterogeneous collections. Using the temporal references for Wikipedia articles and sections of the books of Loe de Jong, combined with the publication dates of newspaper articles, we can find subsets of documents for a specific time period that were either published within that period or refer to a point in time within that period. This provides the researcher with the valuable means to find similar subsets across different collections. However, this does not yet mean that all the documents in the subsets are topically related. For this, we also need to look at the content of the document.

3.4   Combining Temporal and Textual Similarity

To estimate whether two documents refer to the same implicit event, we combine textual similarity with temporal similarity. We measure textual similarity as Manhattan distance over document terms in a TF.IDF weighted vector space. Concretely, we take the subset of maximally 25 terms from a source document, that have the highest TF.IDF score. We then select documents that match at least 30 % of these terms and compute similarity as the sum of TF.IDF scores over the terms. More matching terms thus lead to a higher similarity, as does matching a less common term than a more common term.

We measure temporal similarity using a Gaussian decay function. If two documents are from the same date, they are completely temporally similar. The further the two documents are apart in time, the lower the similarity score. In case we are matching two documents based on temporal references, we multiply the scores we obtain for each temporal reference match. The overall similarity between two documents is then computed by multiplying the temporal similarity with the textual similarity. In this way, temporal similarity functions in a similar matter as a temporal document prior would work, giving preference to documents from a specific period.

---

[5] The exported RDF triples are ingested in the "Verrijkt Koninkrijk" triple store. The updated triple store can be found at http://semanticweb.cs.vu.nl/verrijktkoninkrijk/.

3.5   Evaluating Related Article Finding

We evaluate our approach to measuring similarity using a retrieval experiment to find related documents within the Wikipedia collection. The task is to find documents related to a source Wikipedia article within the Wikipedia collection. We compare two approaches for finding related documents: using only textual similarity and combining temporal and textual similarity.

We sample ten Wikipedia articles out of the 18,361 articles that link to an article with WWII in the title (*"Tweede Wereldoorlog"* in Dutch). We pool the top ten results based on textual similarity and have annotators judge the relatedness of two documents side-by-side on a four-point scale, label from bad to perfect. We obtain the relatedness labels via a crowdsourcing platform. To ensure good quality judgments, we manually create a set of gold standard judgments for twelve document pairs that pilot judges agreed entirely on. Our crowdsourcing judges need to obtain an agreement of over 70 % with the gold standard to start judging. During judging, gold standard pairs are intertwined with unjudged pairs. If the judges do not maintain this agreement on the gold standard, they cannot continue and their judgments are not included. We obtain at least three judgments per document pair, more if the three judges do not agree. We obtain 812 judgments (including the judgments for the gold standard) and measure a mean agreement of 69.5 % over all document pairs. We compute a final rating for a document pair as the mean rating over all judges for that pair.

In our application, related documents are presented to find interesting alternative perspectives from different collections. The related documents are presented as a ranked list, very similar to a standard information retrieval setting. Given this setting and the annotations on a four-point scale, we choose nDCG@10 as our evaluation metric. The nDCG metric can incorporate graded relevance and gives more importance to results higher in the ranked list. We compute nDCG only on the top ten results, as we expect that lower documents are unlikely to be inspected by a user. An nDCG score of 1 indicates that documents are ranked perfectly in order of their relevance score and a score of 0 would mean that all retrieved documents have the lowest relevance score.

Using only textual similarity, we measure an nDCG value of 0.861 and an average rating in the top ten of 2.6 on a scale from 1 to 4. This suggests that the retrieved documents are already of good quality and ranked in a reasonable order. By combining textual and temporal similarity we improve the nDCG score with 3.8 % to 0.894. A detailed look at each of the ten source documents shows improvements in nDCG scores up to 25 %, but also decreased scores up to 5 %. The results suggest that we effectively retrieve related documents and that combining textual and temporal similarity improves effectiveness over only using textual similarity. There are interesting challenges for future research in new approaches for incorporating temporal similarity. For example, intuitively, a match on year references is less strong a match than one on day references. One could therefore weigh matches based on the least fine-grained granularity.

4   Search Interface to Explore Perspectives

We described how we connect collections and their content in Sect. 3 as our first main contribution. Our second contribution is a novel search and analysis application[6]. We provide new means of exploring and analyzing perspectives in these connected collections. Our architecture is modeled in Fig. 1. At the core of the application we use proven open-source technology such as xTAS and Elasticsearch. For each collection, we build a separate index that is exposed to the user interfaces as a combined index.

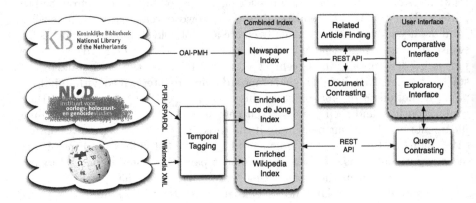

Fig. 1. Architecture to support exploration of perspectives in collections. The interface (right) interacts with a combined index (center) collected from three sources (left).

A researcher interacts with our application through two separate, but connected interfaces: (1) an exploratory search interface for a broad overview, and (2) a comparative interface for detailed analysis. We will discuss the flow between these interfaces with a worked example of a researcher interacting with the application in Sect. 5. First, we will describe each of the two interfaces in more detail below. The comparative interface communicates directly with the combined index, supported by related article finding and document contrasting services. For the exploratory interface, all requests to the index are processed through the query contrasting system.

We provide a researcher with the means to explore perspectives in our application via contrasting. In the comparative interface, two documents are contrasted in detail in a side-by-side comparison. In the exploratory interface, a researcher can combine a keyword query with predefined "query contrasts". A query contrast can be seen as a set of filters that each define a collection subset. A single filter functions in a similar way as facet filters. Such a query contrast filter can simply be contrasting different collections (e.g., newspaper versus Wikipedia articles), or different sources (e.g., a collaborating newspaper versus one run by

---

[6] The fully functional application can be accessed at http://qhp.science.uva.nl.

the resistance) or different locations of publishing. Using a set of these filters (what we call a *query contrast*), what is expressed as a simple keyword query turns into a contrasting comparison between different perspectives.

4.1 Exploratory Interface

To allow researchers to explore the three connected collections, we build an exploratory interface as part of our application. This interface is sketched in Fig. 2a.

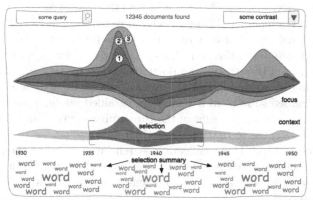

Sketch of the exploratory interface with search on top, a streamgraph to visualize document volume and word clouds to summarize each collection.

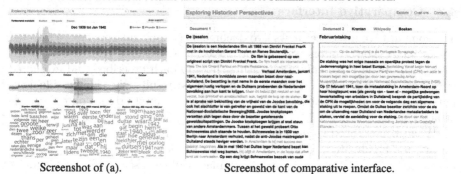

Screenshot of (a).   Screenshot of comparative interface.

Fig. 2. Sketch (a) and screenshot of exploratory (b) and comparative (c) interfaces.

Based on a keyword query and a contrast in the search bar on top, an overview of the search results is presented. Central in the exploratory interface is a visualization that shows the distribution of the volume of documents across time. We visualize this distribution as a streamgraph [5], that can be seen as a streamlined version of a stacked bar chart. A researcher can select a time period of interest while maintaining overview through a Focus+Context interaction design [6]. This allows researchers to focus on a specific period, while at the same time getting an impression of entire time period.

The streams in the context visualization are defined by the selected query contrast and consistently color coded based on this. In the simplest case, each represent one of the three connected collections: newspaper articles, encyclopedic articles and sections of the reference books. For each stream, we show a word cloud representing the most significant terms in the documents in each stream for the selected time period. This provides the researcher with a quick overview of what topics these documents cover.

In Sect. 3.2, we described how we extract temporal references that can have a granularity of a day, month or even a year. If a stream is based on extracted date references, we distinguish in the focus stream graph between different temporal reference granularity. A stream is then split into three substreams: (1) day references, (2) month references and (3) year references (marked in Fig. 2a). The color coding is kept consistent, but opacity decreases as temporal references become less fine-grained. Similarly, the more fine-grained day references are positioned closer to the center of the stream. If a document refers to a specific day, it refers to that month and year as well. In the visualization, we do not count a reference for a less fine-grained substream if a more fine-grained reference occurs for that day. This way, the combined height of the three substreams at any point of time is equal to the references to that day, month or year.

Not depicted in Fig. 2 is the collection search interface, that shows a simple ranked list of documents within any of the three collections. From this search interface, a researcher can select a document to study in more detail in the comparative interface.

4.2   Comparative Interface

The comparative interface shows two documents side-by-side. At first, a selected document is shown on one side, while the other side shows related documents from each of the three collections using the approach described and evaluated in Sect. 3. When selecting a document from these results, the side-by-side comparison is shown. A researcher can return to the related articles on either side at any time.

When comparing two documents side-by-side, interesting parts of the document are highlighted. Using an approach similar to the textual similarity described in Sect. 3.4, we compute the similarity of each sentence in a document to the document on the other side. Sentences with a high similarity are shown clearly, whereas sentences with a low similarity are shown partially transparent. This dimming effect draws the attention of the researcher to the interesting parts of a document in the context of another.

5   A Worked Example

We describe a worked example of how our application can be used to study different perspectives on a specific event in WWII. A full-blown evaluation of our application in a user study with historians is planned for future work.

We go back to an event in Amsterdam, early 1941, as described on Wikipedia. During the winter months of '40/'41, oppression of Jewish citizens of Amsterdam is rising. This leads to open street fights between mobs of both sides. The tensions culminated on February 19th in an ice cream parlor called Koco, where a fight broke out between a German patrol and a mob of regular customers set out to defend the shop. Several arrest were made and the Jewish-German owners where arrested and deported. After roundups in other parts of Amsterdam, the tensions finally lead to the "February strike", the only massive public protest against the persecution of Jews in occupied Europe.

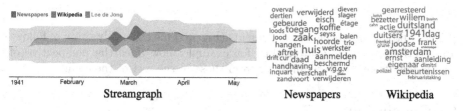

Streamgraph   **Newspapers**   **Wikipedia**

Fig. 3. Screenshots for query "ijssalon koco" from October 1941 until March 1942.

Figure 3 shows screenshots the exploratory search interface, when searching for the name of the shop, contrasting the three distinct collections. From the streamgraph, the historian can clearly see the most documents that mention the shop are from or refer to early 1941. Focusing on this period of interest, the streamgraph depicted in Fig. 3a shows some references to the shop in the buildup towards this event with the bulk in early 1941. A detailed look at the word clouds for newspapers (Fig. 3b), shows that emphasis is given to the perspective of the police. The significant terms include: robbery, thieves, enforcement, gain access, removed and case[7]. On the other hand, the Wikipedia articles referring to this event focus more on the human interest and broader perspectives. The word cloud in Fig. 3c shows the names of the owners and terms as cause, events, February strike, arrested and owner[8].

Diving deeper into the different perspectives, the historian searches for articles related to the Sect. 8.2 of Loe de Jong's fourth book, part II, that covers the events around Koco. He finds Wikipedia articles covering the February strike and related events, but decided to have a more detailed look at the article on the movie "The Ice Cream Parlour". Figure 4 shows a screenshot of the comparison of the content of this article in comparison with the section written by Loe de Jong. The sentences that focus mostly on the movie are faded out, drawing attention to the parts of the article that describe the events in February 1941.

This worked example illustrates how each collection has different perspectives on an important event in WWII, both in comparing subsets of the collection

---

[7] In Dutch: *overval, dieven, handhaving, toegang, verschaft, verwijderd, zaak.*
[8] In Dutch: *aanleiding, gebeurtenissen, Februaristaking, gearresteerd, eigenaar.*

De ijssalon is een Nederlandse film uit 1985 van Dimitri Frenkel Frank met in de hoofdrollen Gerard Thoolen en Renee Soutendijk. Het
camerawerk is van Theo van de Sande. De film is gebaseerd op een origineel script van Dimitri Frenkel Frank. De film heeft als internationale
titels The Ice Cream Parlour en Private Resistance. Er kwamen 35.000 bezoekers naar de De ijssalon in de bioscoop. **Verhaal Amsterdam,
januari 1941, Nederland is inmiddels zeven maanden bezet door nazi-Duitsland.** De bezetting is met name in de eerste maanden over
het algemeen rustig verlopen en de Duitsers probeerden de Nederlandse bevolking aan hun kant te krijgen. Maar dit beleid lijkt mislukt en het
verzet, hoe primitief en amateuristisch ook, begint de kop op de steken. Er is al sprake van beknotting van de vrijheid van de Joodse
bevolking, die ook het slachtoffer is van getreiter en geweld van de kant van de Nationaal-Socialistische Beweging|NSB. **Joodse
knokploegen verzetten zich tegen deze door de bezetter getolereerde geweldsuitspattingen. De Joodse knokploegen krijgen al snel
steun van andere Amsterdammers.** Tussen al het geweld probeert Otto Schneeweiss zich staande te houden. Schneeweiss is in 1939 van
Berlijn naar Amsterdam verhuisd, nadat de anti-Joodse maatregelen in Duitsland steeds heviger werden.

Fig. 4. Screenshot showing the first sentences of the Dutch Wikipedia article on the movie "The Ice Cream Parlour" compared to Loe de Jong's article on the events around that parlour.

(Fig. 3) and in comparing two documents (Fig. 4). One can easily think of follow-up questions to explore after this, for example: how does the perspective of newspapers from Amsterdam differ from those in the rest of the country?

6   Conclusion

We connected multiple heterogeneous collections through implicit events, via time and content. We show that we can extract temporal references with satisfactory accuracy and that we can use these references for related article finding. For future work, we identified interesting challenges in extracting temporal references from historical narratives, such as the books of Loe de Jong. Furthermore, we consider our proposed approach for using extracted temporal references to improve related article finding as just a first attempt. Similar problems exist outside the historical domain.

We presented a novel search interface that supports researchers from the humanities in studying different perspectives on WWII. We showed the value of our application through a worked example of how to study different perspectives on an event in WWII. A full-blown evaluation of our application in a user study with historians is planned for future work. While we focused in this work on events and collections related to WWII, our software and approaches can be applied to any kind of digital collections. We release our work as open data and open-source software[9] to foster future research and applications for digital libraries.

Acknowledgements. This research was supported by Amsterdam Data Science and the Dutch national program COMMIT.

---

[9] The source code is available on https://bitbucket.org/qhp.

References

1. Alonso, O., Strötgen, J., Baeza-Yates, R., Gertz, M.: Temporal information retrieval: challenges and opportunities. In: TWAW Workshop, WWW (2011)
2. Au Yeung, C.-M., Jatowt, A.: Studying how the past is remembered: towards computational history through large scale text mining. In: CIKM 2011, pp. 1231–1240. ACM, New York (2011)
3. Bron, M., Huurnink, B., de Rijke, M.: Linking archives using document enrichment and term selection. In: Gradmann, S., Borri, F., Meghini, C., Schuldt, H. (eds.) TPDL 2011. LNCS, vol. 6966, pp. 360–371. Springer, Heidelberg (2011)
4. Bron, M., Van Gorp, J., Nack, F., de Rijke, M., Vishneuski, A., de Leeuw, S.: A subjunctive exploratory search interface to support media studies researchers. In: SIGIR 2011, pp. 425–434. ACM (2012)
5. Byron, L., Wattenberg, M.: Stacked graphs-geometry & aesthetics. IEEE Trans. Visual Comput. Graphics **14**(6), 1245–1252 (2008)
6. Card, S.K., Mackinlay, J.D., Shneiderman, B.: Readings in Information Visualization: Using Vision to Think. Morgan Kaufmann, San Francisco (1999)
7. de Boer, V., van Doornik, J., Buitinck, L., Marx, M., Veken, T., Ribbens, K.: Linking the kingdom: enriched access to a historiographical text. In: K-CAP 2013, pp. 17–24. ACM, New York (2013)
8. de Rooij, O., Odijk, D., de Rijke, M.: Themestreams: visualizing the stream of themes discussed in politics. In: SIGIR 2013, pp. 1077–1078. ACM (2013)
9. Kleppe, M., Hollink, L., Kemman, M., Juric, D., Beunders, H., Blom, J., Oomen, J., Houben, G.J.: Polimedia analysing media coverage of political debates by automatically generated links to radio & newspaper items. In: Proceedings of the LinkedUp Veni Competition
10. Lensen, J.: De zoektocht naar het midden: nieuwe perspectieven op de herinnering aan de Tweede Wereldoorlog in Vlaanderen en Duitsland. Internationale neerlandistiek **52**(2), 113–133 (2014)
11. Marchionini, G.: Exploratory search: from finding to understanding. Commun. ACM **49**(4), 41–46 (2006)
12. Massa, P., Scrinzi, F.: Manypedia: comparing language points of view of wikipedia communities. In: WikiSym 2012, p. 21. ACM (2012)
13. Monz, C., Nastase, V., Negri, M., Fahrni, A., Mehdad, Y., Strube, M.: Cosyne: a framework for multilingual content synchronization of wikis. In: WikiSym 2011, pp. 217–218. ACM (2011)
14. Odijk, D., de Rooij, O., Peetz, M.-H., Pieters, T., de Rijke, M., Snelders, S.: Semantic document selection. In: Zaphiris, P., Buchanan, G., Rasmussen, E., Loizides, F. (eds.) TPDL 2012. LNCS, vol. 7489, pp. 215–221. Springer, Heidelberg (2012)
15. Pustejovsky, J., Castano, J.M., Ingria, R., Sauri, R., Gaizauskas, R.J., Setzer, A., Katz, G., Radev, D.R.: Timeml: robust specification of event and temporal expressions in text. New Dir. Question Answering **3**, 28–34 (2003)
16. Schreiber, G., et al.: MultimediaN E-Culture demonstrator. In: Cruz, I., Decker, S., Allemang, D., Preist, C., Schwabe, D., Mika, P., Uschold, M., Aroyo, L.M. (eds.) ISWC 2006. LNCS, vol. 4273, pp. 951–958. Springer, Heidelberg (2006)
17. Strötgen, J., Gertz, M.: Multilingual and cross-domain temporal tagging. Lang. Resour. Eval. **47**(2), 269–298 (2013)
18. Van Vree, F.: De Nederlandse Pers en Duitsland, 1930 [-] 1939: Een studie over de vorming van de publieke opinie (1989)

Impact Analysis of OCR Quality on Research Tasks in Digital Archives

Myriam C. Traub[⊠], Jacco van Ossenbruggen, and Lynda Hardman

Centrum Wiskunde & Informatica, Amsterdam, The Netherlands
{Myriam.Traub,Jacco.van.Ossenbruggen,Lynda.Hardman}@cwi.nl

Abstract. Humanities scholars increasingly rely on digital archives for their research instead of time-consuming visits to physical archives. This shift in research method has the hidden cost of working with digitally processed historical documents: how much trust can a scholar place in noisy representations of source texts? In a series of interviews with historians about their use of digital archives, we found that scholars are aware that optical character recognition (OCR) errors may bias their results. They were, however, unable to quantify this bias or to indicate what information they would need to estimate it. This, however, would be important to assess whether the results are publishable. Based on the interviews and a literature study, we provide a classification of scholarly research tasks that gives account of their susceptibility to specific OCR-induced biases and the data required for uncertainty estimations. We conducted a use case study on a national newspaper archive with example research tasks. From this we learned what data is typically available in digital archives and how it could be used to reduce and/or assess the uncertainty in result sets. We conclude that the current knowledge situation on the users' side as well as on the tool makers' and data providers' side is insufficient and needs to be improved.

Keywords: OCR quality · Digital libraries · Digital humanities

1 Introduction

Humanities scholars use the growing numbers of documents available in digital archives not only because they are more easily accessible but also because they support new research tasks, such as pattern mining and trend analysis. Especially for old documents, the results of OCR processing are far from perfect. While improvements in pre-/post-processing and in the OCR technology itself lead to lower error rates, the results are still not error-free. Scholars need to assess whether the trends they find in the data represent real phenomena or result from tool-induced bias. It is unclear to what extent current tools support this assessment task. To our knowledge, no research has investigated how scholars can be supported in assessing the data quality for their specific research tasks.

In order to find out what research tasks scholars typically carry out on a digital newspaper archive (*RQ1*) and to what extent scholars experienced OCR

© Springer International Publishing Switzerland 2015
S. Kapidakis et al. (Eds.): TPDL 2015, LNCS 9316, pp. 252–263, 2015.
DOI: 10.1007/978-3-319-24592-8_19

quality to be an obstacle in their research, we conducted interviews with humanities scholars (Sect. 2). From the information gained in the interviews, we were able to classify the research tasks and describe potential impact of OCR quality on these tasks (*RQ2*). With a literature study, we investigated, how digitization processes in archives influence the OCR quality, how Information Retrieval (IR) copes with error-prone data and what workarounds scholars use to correct for potential biases (Sect. 3). Finally, we report on insights we gained from our use case study on the digitization process within a large newspaper archive (Sect. 4) and we give examples of what data scholars need to be able to estimate the quality indicators for different task categories (*RQ3*).

2 Interviews: Usage of Digital Archives by Historians

We originally started our series of interviews to find out what research tasks humanities scholars typically perform on digital archives, and what innovative additions they would like to see implemented in order to provide (better) support for these research tasks. We were especially interested in new ways of supporting quantitative analysis, pattern identification and other forms of distant reading. We chose our interviewees based on their prior involvement in research projects that made use of digital newspaper archives and/or on their involvement in publications about digital humanities research. We stopped after interviewing only four scholars, for reasons we describe below. Our chosen methodology was a combination of a structured personal account and a time line interview as applied by [4,5]. The former was used to stimulate scholars to report on their research and the latter to stimulate reflection on differences in tasks used for different phases of research. The interviews were recorded either during a personal meeting (*P1, P2, P4*) or during a Skype call (*P3*), transcribed and summarized. We sent the summaries to the interviewees to make sure that we covered the interviews correctly.

We interviewed four experts. (*P1*) is a Dutch cultural historian with an interest in representations of World War II in contemporary media. (*P2*) is a Dutch scholar specializing in modern European Jewish history with an interest in the implications of digital humanities on research practices in general. (*P3*) is a cultural historian from the UK, whose focus is the cultural history of the nineteenth century. (*P4*) is a Dutch contemporary historian who reported to have a strong interest in exploring new research opportunities enabled by the digital humanities.

All interviewees reported to use digital archives, but mainly in the early phases of their research. In the exploration phase the archives were used to get an overview of a topic, to find interesting research questions and relevant data for further exploration. In case they had never used an archive before, they would first explore the content the archive can provide for a particular topic (see Table 1, *E9*). At later stages, more specific searches are performed to find material about a certain time period or event. The retrieved items would later be used for close reading. For example, *P1* is interested in the representations of Anne

Frank in post-war newspapers and tried to collect as many relevant newspaper articles as possible *E1*. *P3* reports on studies of introductions of new words into the vocabulary *E8*. Three of the interviewees (*P1, P3, P4*) mentioned that low OCR quality is a serious obstacle, an issue that is also reflected extensively in the literature [3,6,14]. For some research tasks, the interviewees reported to have come up with workarounds. *P1* sometimes manages to find the desired items by narrowing down search to newspaper articles from a specific time period, instead of using keyword search. However, this strategy is not applicable to all tasks.

Table 1. Categorization of the examples for research tasks mentioned in the interviews. Task type *T1* aims to find the first mention of a concept. Tasks of type *T2* aim to find a subset with relevant documents. *T3* includes tasks investigating quantitative results over time and *T4* describes tasks using external tools on archive data.

ID	Interview	Example	Category
E1	P1	Representation of Anne Frank in post-war media	T2
E3	P1	Contextualizing LDJ with sources used	T4
E4	P2	Comparisons of two digitized editions of a book to find differences in word use	T4
E5	P3	Tracing jokes through time and across newspapers	T3
E6	P3	Plot ngrams frequencies to investigate how ideas and words enter a culture	T1/T3
E7	P3	Sophisticated analysis of language in newspapers	T4
E8	P3	First mention of a newly introduced word	T1
E9	P3 /P4	Getting an overview of the archive's contents	T2
E11	P4	Finding newspaper articles on a particular event	T2

Due to the higher error rate in old material and the absence of quality measures, they find it hard to judge whether a striking pattern in the data represents an interesting finding or whether it is a result of a systematic error in the technology. According to *P1*, the print quality of illegal newspapers from the WWII period is significantly worse than the quality of legal newspapers because of the conditions under which they were produced. As a consequence, it is very likely that they will suffer from a higher error rate in the digital archive, which in turn may cause a bias in search results. When asked how this uncertainty is dealt with, *P4* reported to try to explain it in the publications. The absence of error measures and information about possible preconceptions of the used search engine, however, made this very difficult. *P3* reported to have manually collected data for a publication to generate graphs tracing words and jokes over time (see *E5, E6* in Table 1) as the archive did not provide this functionality. Today, *P3* would not trust the numbers enough to use them for publications again.

P2 and *P3* stated that they would be interested in using the data for analysis independently from the archive's interfaces. Tools for text analysis, such as Voyant[1], were mentioned by both scholars (*E3*, *E4*, *E7*). The scholars could not indicate how such tools would be influenced by OCR errors. We asked the scholars whether they could point out what requirements should be met in order to better facilitate research tasks in digital archives. *P3* thought it would be impossible to find universal methodological requirements, as the requirements vary largely between scholars of different fields and their tasks.

We classified the tasks that were mentioned by the scholars in the interviews according to their similarities and requirements towards OCR quality. The first mention of a concept, such as a new word or concept would fall into category *T1*. *T2* comprises tasks that aim to create a subcollection of the archive's data, e.g. to get to know the content of the archive or to select items for close reading. Tasks that relate word occurrences to a time period or make comparisons over different sources or queries are summarized in *T3*. Some archives allow the extraction of (subsets of) the collection data. This allows the use of specialized tools, which constitutes the last category *T4*.

We asked *P1*, *P2* and *P4* about the possibilities of more quantitative tools on top of the current digital archive, and in all cases the interviewees' response was that no matter what tools were added by the archive, they were unlikely to trust any quantitative results derived from processing erroneous OCRed text. *P2* explicitly stated that while he did publish results based on quantitative methods in the past, he would not use the same methods again due to the potential of technology-induced bias.

None of our interviews turned out to be useful with respect to our quest into innovative analysis tools. The reason for this was the perceived low OCR quality, and the not well-understood susceptibility of the interviewees' research tasks to OCR errors. Therefore, we decided to change the topic of our study to better understanding the impact of OCR errors on specific research tasks. We stopped our series of interviews and continued with a literature study on the impact of OCR quality on specific research tasks.

3 Literature Study: Impact of OCR Quality on Scholarly Research

To find out how the concerns of the scholars are addressed by data custodians and by research in the field of computer science, we reviewed available literature.

The importance of OCR in the digitization process of large digital libraries is a well-researched topic [9,12,18,19]. However, these studies are from the point of view of the collection owner, and not from the perspective of the scholar using the library or archive. User-centric studies on digital libraries typically focus on user interface design and other usability issues [8,20,21]. To make the entry barrier to the digital archive as low as possible, interfaces often try to hide

---

[1] http://voyant-tools.org/.

technical details of the underlying tool chain as much as possible. While this makes it easier for scholars to use the archive, it also denies them the possibility to investigate potential tool-induced bias.

There is ample research into how to reduce the error rates of OCRed text in a post-processing phase. For example, removing common errors, such as the "long s"-to-f confusion or the soft-hyphen splitting of word tokens, has shown to improve Named Entity Recognition. This, however, did not increase the overall quality to a sufficient extent as it addressed only 12 % of the errors in the chosen sample [2]. Focusing on overall tool performance or performance on representative samples of the entire collection, such studies provide little information on the impact of OCR errors on specific queries carried out on specific subsets of a collection. It is this specific type of information we need, however, to be able to estimate the impact on our interviewees' research questions. We found only one study that aimed at generating high-quality OCR data and evaluating the impact of its quality on a specific set of research questions [15]. The researchers found that the impact of OCR errors is not substantial for a task that compares two subsets of the corpus. For a different task, the retrieval of a list of the most significant words (in this case, describing moral judgement), however, recall and precision were considered too low.

Another line of research focuses on how to improve OCR tools or on using separate tools for improving OCR output in a post-processing step [11], for example by using input from the public [10]. Unfortunately, the actual extent, to which this crowdsourcing initiative has contributed to a higher accuracy has not been measured. While effective use of such studies may reduce the error rate, they do not help to better estimate the impact of the remaining errors on specific cases. Even worse, since such tools (and especially human input) add another layer of complexity and potential errors, they may also add more uncertainty to these estimates. Most studies on the impact of OCR errors are in the area of ad-hoc IR, where the consensus is that for long texts and noisy OCR errors, retrieval performance remains remarkably good for relatively high error rates [17]. On short texts, however, the retrieval effectiveness drops significantly [7,13]. In contrast, information extraction tools suffer significantly when applied to OCR output with high error rates [16]. Studies carried out on unreliable OCR data sets often leave the OCR bias implicit. Some studies explicitly protect themselves from OCR issues and other technological bias by averaging over large sets of different queries and by comparing patterns found for a specific query set to those of other queries sets [1]. This method, however, is not applicable to the examples given by our interviewees, since many of their research questions are centered around a single or small number of terms.

Many approaches aiming at improving the data quality in digital archives have in common that they partially reduce the error rate, either by improving overall quality, or by eliminating certain error types. None of these approaches, however, can remove all errors. Therefore, even when applying all of these steps to their data, scholars still need to be able to quantify the remaining errors and assess their impact on their research tasks.

4 Use case: OCR Impact on Research Tasks in a Newspaper Archive

To study OCR impact on specific scholarly tasks in more detail, we investigated OCR-related issues of concrete queries on a specific digital archive: the historic newspaper archive[2] of the National Library of The Netherlands (KB). It contains over 10 million Dutch newspaper pages from the period 1618 to 1995, which are openly available via the Web. For each item, the library publishes the scanned images, the OCR-ed texts and the metadata records. Its easy access and rich content make the archive an extremely rich resource for research projects[3].

4.1 Task: First Mention of a Concept

One of the tasks often mentioned during our interviews was finding the first mention of a term (task *T1* in Sect. 2). For this task, scholars can typically deal with a substantial lack of precision caused by OCR errors, since they can detect false positives by manually checking the matches. The key requirement is recall. Scholars want to be sure that the document with the first mention was not missed due to OCR errors. This requires a 100 % recall score, which is unrealistic for large digital archives. As a second best, they need to minimize the risk of missing the first mention to a level that is acceptable in their research field. The question remains how to establish this level, and to what extent archives support achieving this level. To understand how a scholar could assess the reliability of their results with currently available data, we aim to find the first mention of "Amsterdam" in the KB newspaper archive. A naive first approach is to simply order the results on the query "Amsterdam" by publication date. This returned a newspaper dated October 25, 1642 as the earliest mention. We then explore different methods to assess the reliability of this result. We first tried to better understand the corpus and the way it was produced, then we tried to estimate the impact of the OCR errors based on the confidence values reported by the OCR engine, and finally we tried to improve our results by incremental improvement our search strategy.

Understanding the Digitization Pipeline. We started by obtaining more information on the archive's digitization pipeline, in particular details about the OCR process, and potential post-processing steps.

Unfortunately, little information about the pipeline is given on the KB website. The website warns users that the OCR text contains errors[4], and as an example mentions the known problem of the "long s" in historic documents (see Fig. 1), which causes OCR software to mistake the's' for an'f'. The page does not provide quantitative information on OCR error rates.

---

[2] www.delpher.nl/kranten.
[3] See http://lab.kbresearch.nl for examples.
[4] http://www.delpher.nl/nl/platform/pages/?title=kwaliteit+(ocr).

Fig. 1. Confusing the "long s" for an "f" is a common OCR error in historic texts.

After contacting library personnel, we learned that formal evaluation on OCR error rates or on precision/recall scores of the archive's search engine had not been performed so far. The digitization had been a project spanning multiple years, and many people directly involved no longer worked for the library. Parts of the process had been outsourced to a third party company, and not all details of this process are known to the library. We believe this practice is typical for many archives. We further learned that article headings had been manually corrected for the entire archive, and that no additional error correction or other post-processing had been performed. We concluded that for the first mention task, our inquires provided insufficient information to be directly helpful.

Uncertainty Estimation: Using Confidence Values. Next, we tried to use the confidence values reported by the OCR engine to assess the reliability of our result. The ALTO XML[5] files used to publish the OCR texts do not only contain the text as it was output by the OCR engine, they also contain confidence values generated by the OCR software for each page, word and character. For example, this page[6], contains:

<Page ID="P2"   ...   PC="0.507">

Here, PC is a confidence value between 0 (low) and 1 (high confidence). Similar values are available for every word and character in the archive:

<String ID="P2_ST00800"   ...   CONTENT="AM"   ...
    SUBS_CONTENT="AMSTERDAM."   WC="0.45"   CC="594"/>
<String ID="P2_ST00801"   ...   CONTENT="STERDAM."   ...
    SUBS_CONTENT="AMSTERDAM."   WC="0.30"   CC="46778973"/>

Here, WC is the word-level confidence, again expressed as a value between 0 and 1. CC is the character-level confidence, expressed as a string of values between 0–9, with one digit for each character. In this case, 0 indicates high, and 9 indicates low confidence. This is an example for a word that was split by a hyphen. The representation of its two parts as "subcontent" of "AMSTERDAM" assures its retrieval by the search engine of delpher.

<String ID="P2_ST00766"   ...   CONTENT="Amfterdam,"
    WC="0.36"   CC="0866869771"/>

For the last example, this means the software has lower confidence in the correct "m", than in the incorrect "f". Note that since the above XML data

---

[5] http://www.loc.gov/standards/alto/.
[6] http://resolver.kb.nl/resolve?urn=ddd:010633906:mpeg21:p002:alto.

is available for each individual word, it is a huge dataset in absolute size, that could, potentially, provide uncertainty information on a very fine-grained level. For this, we need to find out what these values mean and/or how they have been computed. However, the archive's website provides no information about how the confidence values have been calculated.

Again, from the experts in the library, we learned that the default word level confidence scores were increased if the word was found in a given list with correct Dutch words. Later, this was improved by replacing the list with contemporary Dutch words by a list with historic spelling. Unfortunately, it is not possible to reproduce which word lists have been used on what part of the archive.

Another limitation is that even if we could calibrate the OCR confidence values to meaningful estimates, they could only be used to estimate how many of the matches found are likely false positives. They provide little or no information on the false negatives, since all confidence values related to characters that were considered as potential alternatives to the character chosen by the OCR engine have not been preserved in the output and are lost forever. For this research task, this is the information we would need to estimate or improve recall. We thus conclude that we failed in using the confidence values to estimate the likelihood that our result indeed represented the first mention of "Amsterdam" in the archive. We summarized our output in Table 2, where for *T1* we indicate that using the confusion matrix is impractical, using the out confidence values (CV output) is not helpful, and using the confidence values of the alternatives (CV alternatives) could have improved recall, but we do not have the data.

Table 2. The different types of tasks require different levels of quality. Quality indicators can be used to generate better estimates of the quality and also (to some extent) to compensate low quality. x stands for an abstract concept that is the focus of interest in the research task.

Category available for:	Confusion matrix sample only	CV output full corpus	CV alternatives not available
T1. 1st mention of x	find all queries for x, impractical	estimated precision not helpful	improve recall
T2. Selecting subset relevant to x	as above	estimated precision, requires improved UI	improve recall
T3. Pattern over time x	pattern summarized over set of alt. queries	estimates of corrected precision	estimates of corrected recall
T3.a. Compare x_1 and x_2	warn for diff. susceptibility to errors	as above, warn for diff. distribution of CVs	as above
T3.b. Compare $corpus_1$ and $corpus_2$	as above	as above	as above

Incremental Improvement of the Search Strategy. We observed that the "long s" warning given on the archive's website is directly applicable to our query. Therefore, to improve on our original query, we also queried for "Amfterdam". This indeed results in an earlier mention: July 27, 1624. This result, however, is based on our anecdotal knowledge about the "long s problem". It illustrates the need for a more systematic approach to deal with spelling variants. While the archive provides a feature to do query expansion based on historic spelling variants, it provides no suggestions for "Amsterdam". Querying for known spelling variants mentioned on the Dutch history of Amsterdam Wikipedia page also did result in earlier mentions.

To see what other OCR-induced misspellings of Amsterdam we should query for, we compared a ground truth data set with the associated OCR texts. For this, we used the dataset[7] created in the context of the European IMPACT project. It includes a sample of 1024 newspaper pages, but these had not been completely finished by end of the project. This explains why this data has not been used in a evaluation of the archive's OCR quality. Because of changes in the identifier scheme used, we could only map 265 ground truth pages to the corresponding OCR text in the archive. For these, we manually corrected the ground truth for 134 pages, and used these to compute a confusion table[8]. This matrix could be used to generate a set of alternative queries based on all OCR errors that occur in the ground truth dataset. Our matrix contains a relatively small number of frequent errors, and it seems doable to use them to manually generate a query set that would cover the majority of errors. We decided to look at the top ten confusions and use the ones applicable to our query. All combinations of confusions resulted in 23 alternative spelling variations of "Amsterdam". When we queried for the misspellings, we found hits for all variations, except one, "Amfcordam". None, however, yielded an earlier result than our previous query.

This method could, however, be implemented as a feature in the user interface, the same way as historic spelling variants are supported[9]. Again, the issue is that for a specific case, it is hard to predict whether such a future would help, or merely provide more false positives.

Our matrix also contains a very long tail with infrequent errors, and for this specific task, it is essential to take all of them into account. This makes our query set very large and while this may not be a technical problem for many state of the art search engines, the current user interface of the archive does not support such queries. More importantly, the long tail also implies that we need to assume that our ground truth does not cover all OCR errors that are relevant for our task.

We conclude that while the use of a confusion matrix does not guarantee finding the first mention of a term, it would be useful to publish such a matrix on each digital archive's website. Just using the most frequent confusions can

---

[7] http://lab.kbresearch.nl/static/html/impact.html.
[8] available on http://dx.doi.org/10.6084/m9.figshare.1448810.
[9] http://www.delpher.nl/nl/platform/pages/?title=zoekhulp.

already help user to avoid the most frequent errors, even in a manual setting. Systematic queries for all known variants would require more advanced backend support.

Fortunately, it lies in the nature of our task that with every earlier mention we can confirm, we can also narrow the search space by defining a new upper bound. In our example, the dataset with pages published before our 1624 upper bound is sufficiently small to allow manual inspection. The first page in the archive of the same title as the 1624 page, is published in 1619, and has a mention of "Amsterdam". It is on the very bottom of the page in a sentence that is completely missing in the OCR text. This explains why our earlier strategy has missed it. The very earliest page in the archive at the time of writing is from June 1618. Its OCR text contains "Amfterftam". Our earlier searches missed this one because it is a very rare variant which did not occur in the ground truth data. While we now have found our first mention in the archive with 100 % certainty, we found it by manual, not automatic means. Our strategy would not have worked when the remaining dataset would have been too large to allow manual inspection.

4.2   Analysis of Other Tasks

We also analyzed the other tasks in the same way. For brevity, we only report our findings to the extent they are different from task $T1$. For $T2$, selecting a subset on a topic for close reading, the problem is that a single random OCR error might cause the scholar to miss a single important document as in $T1$. In addition, a systematic error might result in a biased selection of the sources chosen for close reading, which might be an even bigger problem. Unfortunately, using the confusion matrix is again not practical. The CV output could be useful to improve precision for research topics where the archive contains too many relevant hits, and selecting only hits above a certain confidence threshold might be useful. This requires, however, the user interface to support filtering on confidence values. For the CV alternatives, they again could be used to improve recall, but it is unclear against what precision.

For task $T3$, plotting frequencies of a term over time, the issue is no longer whether or not the system can find the right documents, as in $T1$ and $T2$, but if the system can provide the right counts of term occurrences despite the OCR errors. Here, the long tail of the confusion matrix might be less of a problem, as we may choose to only query for the most common mistakes, assuming that the pattern in the total counts will not be affected much by the infrequent ones. CV output could be used to lower counts for low precision results, while CV alternatives could be used to increase counts for low recall matches. For $T3.a$, a variant of $T3$ where the occurrence over time of one term is compared to another, the confusion matrix could also be used to warn scholars if one term is more susceptible to OCR errors than the other. Likewise, a different distribution of the CV output for the two terms might be flagged in the interface to warn scholars about potential bias. For $T3.b$, a variant where the occurrence of a term in different newspapers is analyzed, the CV values could likely be used to indicate different distributions in the sources, for example to warn for systematic errors caused by differences in print quality or fonts between the two newspapers.

For task $T4$ (not in the table), the use of OCRed texts in other tools, our findings are also mainly negative. Very few text analysis tools can, for example, deal with different confidence values in their input, apart from the extensive standardization these would require for the input/output formats and interpretation of these values. Additionally, many tools suffer from the same limitation that only their overall performance on a representative sample of the data has been evaluated, and little is known about their performance on a specific use case outside that sample. By stacking this uncertainty on top of the uncertain OCR errors, predicting its behavior for a specific case will be even harder.

5 Conclusions

Through interviews we conducted with scholars, we learned that while the uncertain quality of OCRed text in archives is seen as a serious obstacle to wider adaption of digital methods in the humanities, few scholars can quantify the impact of OCR errors on their own research tasks. We collected concrete examples of research tasks, and classified them into categories. We analyzed the categories for their susceptibility to OCR errors, and illustrated the issues with an example attempt to assess and reduce the impact of OCR errors on a specific research task. From our literature study, we conclude that while OCR quality is a widely studied topic, this is typically done in terms of tool performance. We claim to be the first to have addressed the topic from the perspective of impact on specific research tasks of humanity scholars.

Our analysis shows that for many research tasks, the problem cannot be solved with better but still imperfect OCR software. Assessing the impact of the imperfections on a specific use case remains important.

To improve upon the current situation, we think the communities involved should begin to approach the problem from the user perspective. This starts with understanding better how digital archives are used for specific tasks, by better documenting the details of the digitization process and by preserving all data that is created during the process. Finally, humanity scholars need to transfer their valuable tradition of source criticism into the digital realm, and more openly criticize the potential limitations and biases of the digital tools we provide them with.

Acknowledgements. We would like to thank our interviewees for their contributions, the National Library of The Netherlands for their support and the reviewers for their helpful feedback. This research is funded by the Dutch COMMIT/ program.

References

1. Acerbi, A., Lampos, V., Garnett, P., Bentley, R.A.: The expression of emotions in 20th century books. PLoS ONE **8**(3), e59030 (2013)

2. Alex, B., Grover, C., Klein, E., Tobin, R.: Digitised historical text: does it have to be mediOCRe? In: Jancsary, J. (ed.) Proceedings of KONVENS 2012, LThist 2012 Workshop, pp. 401–409. ÖGAI, September 2012

3. Bingham, A.: The digitization of newspaper archives: opportunities and challenges for historians. Twentieth Century Br. Hist. **21**(2), 225–231 (2010)

4. Bron, M.; Exploration and contextualization through interaction and concepts. Ph.D. Thesis (2013)

5. Brown, C.D.: Straddling the humanities and social sciences: the research process of music scholars. Libr. Inf. Sci. Res. **24**(1), 73–94 (2002)

6. Cohen, D.J., Rosenzweig, R.: Digital History: A Guide to Gathering, Preserving, and Presenting the Past on the Web, vol. 28. University of Pennsylvania Press, Philadelphia (2006)

7. Croft, W.B., Harding, S., Taghva, K., Borsack, J.: An evaluation of information retrieval accuracy with simulated OCR output. Technical report, Amherst, MA, USA (1993)

8. Fuhr, N., Hansen, P., Mabe, M., Micsik, A., Sølvberg, I.T.: Digital libraries: a generic classification and evaluation scheme. In: Constantopoulos, P., Sølvberg, I.T. (eds.) ECDL 2001. LNCS, vol. 2163, pp. 187–199. Springer, Heidelberg (2001)

9. Holley, R.: How good can it get? Analysing and improving OCR accuracy in large scale historic newspaper digitisation programs. D-Lib Mag. **15**(3/4) (2009)

10. Holley, R.: Many hands make light work: public collaborative OCR text correction in Australian Historic Newspapers. Technical report, National Library of Australia, March 2009

11. Kettunen, K., Honkela, T., Lindén, K., Kauppinen, P., Pääkkönen, T., Kervinen, J. et al.: Analyzing and improving the quality of a historical news collection using language technology and statistical machine learning methods. In: Proceedings of the 80th IFLA General Conference and Assembly, IFLA World Library and Information Congress (2014)

12. Klijn, E.: The current state-of-art in newspaper digitization a market perspective. D-Lib Mag. **14**, January 2008

13. Mittendorf, E., Schäuble, P.: Information retrieval can cope with many errors. Inf. Retr. **3**(3), 189–216 (2000)

14. Nicholson, B.: Counting culture; or, how to read Victorian newspapers from a distance. J. Victorian Cult. **17**(2), 238–246 (2012)

15. Strange, C., McNamara, D., Wodak, J., Wood, I.: Mining for the meanings of a murder: the impact of OCR quality on the use of digitized historical newspapers. Digital Humanit. Q. 8(1) (2014)

16. Taghva, K., Beckley, R., Coombs, J.: The effects of OCR error on the extraction of private information. In: Bunke, H., Spitz, A.L. (eds.) DAS 2006. LNCS, vol. 3872, pp. 348–357. Springer, Heidelberg (2006)

17. Taghva, K., Borsack, J., Condit, A., Erva, S.: The effects of noisy data on text retrieval. J. Am. Soc. Inf. Sci. **45**(1), 50–58 (1994)

18. Tanner, S., Muñoz, T., Ros, P.H.: Measuring mass text digitization quality and usefulness. D-Lib Mag. **15**(7/8), 1082–9873 (2009)

19. Weymann, A., Luna Orozco, R.A., Mueller, C., Nickolay, B., Schneider, J., Barzik, K.: Einführung in die Digitalisierung von gedrucktem Kulturgut - Ein Handbuch für Einsteiger. Ibero-American Institute (Berlin) (2010)

20. Xie, H.I.: Evaluation of digital libraries: criteria and problems from users' perspectives. Libr. Inf. Sci. Res. **28**(3), 433–452 (2006)

21. Xie, H.I.: Users' evaluation of digital libraries (DLs): Their uses, their criteria, and their assessment. Inf. Process. Manage. **44**(3), 1346–1373 (2008)

Social-Technical Perspectives of Digital Information

Characteristics of Social Media Stories

Yasmin AlNoamany(✉), Michele C. Weigle, and Michael L. Nelson

Department of Computer Science, Old Dominion University,
Norfolk, VA 23529, USA
{yasmin,mweigle,mln}@cs.odu.edu

Abstract. An emerging trend in social media is for users to create and publish "stories", or curated lists of web resources with the purpose of creating a particular narrative of interest to the user. While some stories on the web are automatically generated, such as Facebook's "Year in Review", one of the most popular storytelling services is "Storify", which provides users with curation tools to select, arrange, and annotate stories with content from social media and the web at large. We would like to use tools like Storify to present automatically created summaries of archival collections. To support automatic story creation, we need to better understand as a baseline the structural characteristics of popular (i.e., receiving the most views) human-generated stories. We investigated 14,568 stories from Storify, comprising 1,251,160 individual resources, and found that popular stories (i.e., top 25 % of views normalized by time available on the web) have the following characteristics: 2/28/1950 elements (min/median/max), a median of 12 multimedia resources (e.g., images, video), 38 % receive continuing edits, and 11 % of the elements are missing from the live web.

Keywords: Stories · Storify · Storytelling · Social media · Curation

1 Introduction

Storify is a social networking service launched in 2010 that allows users to create a "story" of their own choosing, consisting of manually chosen web resources, arranged with a visually attractive interface, clustered together with a single URI and suitable for sharing. It provides a graphical interface for selecting URIs of web resources and arranging the resulting snippets and previews (see Fig. 1), with a special emphasis on social media (e.g., Twitter, Facebook, Youtube, Instagram). We call these previews of web resources "web elements", and the annotations Storify allows on these previews we call "text elements".

We would like to use Storify to present automaticaly created summaries of collections of archived web pages in a social media interface that is more familiar users (as opposed to custom interfaces for summaries, e.g. [11]). Since the stories in Storify are created by humans, we model the structural characteristics of these stories, with particular emphasis on "popular" stories (i.e., the top 25 % of views, normalized by time available on the web).

© Springer International Publishing Switzerland 2015
S. Kapidakis et al. (Eds.): TPDL 2015, LNCS 9316, pp. 267–279, 2015.
DOI: 10.1007/978-3-319-24592-8_20

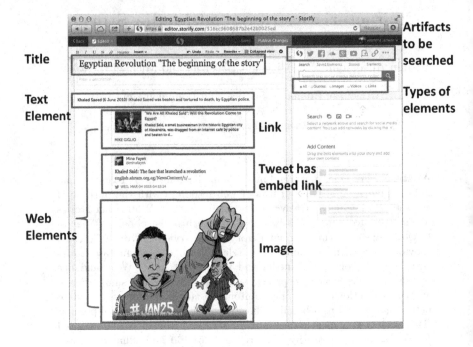

Fig. 1. Example of creating a story on Storify.

In this paper, we will build a baseline for what human-generated stories look like and specify the characteristics of the popular stories. We answer the following questions: What is the length of the human-generated stories? What are the types of resources used in these stories? What are the most frequently used domains in the stories? What is the timespan (editing time) of the stories? Is there a relation between the timespan and the features of the story? Is there a relation between the popularity of the stories and the number of elements? Is there a relation between the popularity and the number of subscribers of the authors? What differentiates the popular stories? How many of the resources in these stories disappear every year? Can we find these missing resources in the archives?

To answer these questions, we analyzed 14,568 stories from Storify comprising 1,251,160 elements. We found that popular stories have a min/median/max value of 2/28/1950 elements, with the unpopular stories having 2/21/2216. Popular stories have a median of 12 multimedia resources (the unpopular stories have a median of 7), 38 % receiving continuing edits (as opposed to 35 %), and only 11 % of web elements are missing on the live web (as opposed to 13 %). The authors of popular stories have min/median/max value of 0/16/1,726,143 subscribers, while the authors of unpopular stories have 0/2/2469 subscribers. We found that there is a nearly linear relation between the timespan of the story and the number of web elements. We also found that only 11 % of the missing resources could be found in public web archives.

2   Related Work

There have been many studies on how the social media is being used in social curation [4,10,12,19,20]. Seitzinger defined social curation as the discovery, selection, collection and sharing of digital artifacts by an individual for a social purpose such as learning, collaboration, identity expression or community participation [15].

Duh et al. [4] studied how Togetter, a popular curation service in Japan, was being used for the social curation of microblogs, such as tweets. They studied the motivation of the curator through defining the topics being curated. They found that there are a diverse number of topics and a variety of social purposes for the content curation, such as summarizing an event and discussing TV shows.

Zhong et al. [20] studied why and how people curate using data sets of three in January 2013 for Pinterest, and over the month of December 2012 for Last.fm. They found that curation tends to focus on items that may not be high ranked in popularity and search rankings, which slightly contradicts our finding in Sect. 4.3. They also found that curation tends to be a personal activity more than being social.

Storify has been used in many studies by journalists [16] and also to explore how curation works in the classroom [8,9]. Cohen et al. believe that Storify can be used to encourage students to become empowered storytellers and researchers [3]. Laire et al. [8] used Storify to study the effect of social media on teaching practices and writing activities.

Stanoevska-Slabeva et al. [16] sampled 450 stories from Storify about the Arab Spring from December 2010 to the end of August 2011. They found that social media curation is done by professionals as well as amateurs. They also found that the longer coverage stories use more resources. They also found that the stories created by both professionals and amateurs presented the primary gatewatching characteristics.

Kieu et al. [6] proposed a method for predicting the popularity of social curation content based on a data set from Storify. They used a machine learning approach based on curator and curation features (for example, the number of followers, the number of stories for the users, and the time that the user started using Storify) from stories. They found that the curator features perform well for detecting the popularity. In this paper, we also investigate if there is relation between the number of views and the features of the story, such as the number of elements.

3   Constructing the Data Set

We created the data set by querying the Storify Search API[1] with the most frequent 1000 English keywords issued to Yahoo[2]. We retrieved 400 results for each keyword, resulting in a total of 145,682 stories downloaded in JavaScript

---

[1] http://dev.storify.com/api.

[2] http://webscope.sandbox.yahoo.com/catalog.php?datatype=l.

Object Notation (JSON). We created the data set in February, 2015 and only considered stories authored in 2014 or earlier, resulting in 37,486 stories. We eliminated stories with only zero or one elements or zero views, resulting in only 14,568 unique stories authored by 10,199 unique users and containing a total of 1,251,160 web and text elements.

4   Characteristics of Human-Generated Stories

Table 1 contains the distribution of story views, web and text elements, and number of subscribers for the story authors. We show the distribution percentiles instead of averages because the distribution of the data is long-tailed. The timespan is the time interval (in hours) in which users edit their stories and is calculated by taking the difference between the story creation-date and last-modified date. The median for all stories is 23 web elements and 1 text element, and 44 % of the stories have no text elements at all. Due to the large range of values we believe median is a better indicator of "typical values" rather than mean.

Table 1. Distribution of the features of the stories in the data set. Timespan is measured in hours.

Features	Views	Web elements	Text elements	Subscribers	Timespan
25th percentile	14	10	0	0	0.18
50th percentile	51	23	1	4	3
75th percentile	268	69	9	21	120
90th percentile	1949	210	19	85	1747
Maximum	11,284,896	2,216	559	1,726,143	36,111

4.1   What Kind of Resources are in Stories?

Using the Storify-defined categories reflected in the UI (Fig. 1), the 1,251,160 elements consist of: 70.8 % links, 18.4 % images, 8.1 % text, 2.0 % videos, and 0.7 % quotes. Text elements are relatively rare, meaning that few users choose to annotate the web elements in their story.

4.2   What Domains are Used in Stories?

To analyze the distribution of domains in stories, we canonicalized the domains (e.g., www.cnn.com → cnn.com) and dereferenced all shortened URIs (e.g., t.co, bit.ly) to the URIs of the final locations. This resulted in 25,947 unique domains in the 14,568 unique stories.

Table 2 contains the top 25 domains of the resources ordered by their frequency. The list of top 25 hosts represents 92.3 % of all the resources. The table

Table 2. The top 25 hosts that are used in human-generated stories. The percentage is the frequency of the host out of 1,150,399.

Host	Frequency	Percentage	Alexa Global Rank as of 2015-03	Category
twitter.com	943,859	82.05 %	8	Social media
instagram.com	45,188	3.93 %	25	Photos
youtube.com	22,076	1.92 %	3	Videos
facebook.com	13,930	1.21 %	2	Social media
flickr.com	7,317	0.64 %	126	Photos
patch.com	5,783	0.50 %	2,096	News
plus.google.com	3,413	0.30 %	1	Social media
tumblr.com	3,066	0.27 %	31	Blogs
blogspot.com	1,857	0.16 %	18	Blogs
imgur.com	1,756	0.15 %	36	Photos
coolpile.com	1,706	0.15 %	149,281	Entertainment
wordpress.com	1,615	0.14 %	33	Blogs
giphy.com	1,055	0.09 %	1,604	Photos
bbc.com	966	0.08 %	156	News
lastampa.it	927	0.08 %	2,440	News
pinterest.com	892	0.08 %	32	Photos
softandapps.info	861	0.07 %	160,980	News
photobucket.com	768	0.07 %	341	Photos
nytimes.com	744	0.06 %	97	News
soundcloud.com	736	0.06 %	167	Audio
wikipedia.org	736	0.06 %	7	Encyclopedia
repubblica.it	682	0.06 %	439	News
theguardian.com	588	0.05 %	157	News
huffingtonpost.com	572	0.05 %	93	News
punto-informatico.it	570	0.05 %	42,955	News

also contains the global rank of the domains according to Alexa as of March 2015. Note that plus.google.com has rank one because Alexa does not differentiate plus.google.com from google.com. We manually categorized these domains in a more fine-grained manner than Storify provides with its "links, images, text, videos, quotes" descriptions (Sect. 4.1).

Although the top 25 list of domains appearing in the stories is dominated by globally popular web sites (e.g., Twitter, Instagram, Youtube, Facebook), the long-tailed distribution results in the presence of many globaly lesser known sites. In Sect. 4.3 we investigate the correlation between Alexa global rank and rank within Storify.

The Embedded Resources of twitter.com. Since Twitter is the most popular domain ($>82\%$ of web elements), we investigate if the tweets have embedded resources of their own. This captures the behavior of users including tweets in the stories because the tweets are surrogates for embedded content. We sampled 5% from Twitter resources (47,512 URIs). Of sampled tweets in the stories, 32% (15,217) have embedded resources, of which there are 14,616 unique URIs. Of the 15,217, 46% are photos from twitter.com (hosted at twimg.com). Table 3 contains the most frequent 10 domains for the embedded resources, which represent 61.6% of the all the URIs embedded in tweets. Note that some Storify stories (0.49%) point to other stories in Storify.

Table 3. The most frequent 10 hosts in the embedded resources of the tweets.

Domain	Percentage	Category
twimg.com	46.17%	Images
instagram.com	4.28%	Images
youtube.com	2.82%	Videos
linkis.com	2.04%	Media sharing
facebook.com	1.40%	Social Media
wordpress.com	0.61%	Blogs
vine.co	0.53%	Videos
blogspot.com	0.52%	Blogs
storify.com	0.49%	Social Network
bbc.com	0.44%	News

4.3 Correlation of Global and Storify Popularity

We calculate the correlation between the frequency of the domains and their Alexa global traffic ranking. Table 4 shows Kendall's Tau τ correlation coefficient for the most frequent n domains. Statistically significant ($p < 0.05$) correlations are bolded. The highest correlation is 0.45 for list of 15 domains. From the results we notice that most of the time the highly ranked real-world resources, such as twitter.com, are correspondingly the most used in human-generated stories. This is interestingly in contrast with [20], which found that the most frequent sites on Pinterest had low Alexa Global Ranking. That possibly returns to the different nature of the usage of both sites. In Pinterest, users pin photos or videos of interest to create theme-based image/video collections such as hobbies, fashion, events. The most used subject areas that are being used by Pinterest users are food and drinks, décor and design, and Apparel and Accessories [5]. Most of the pins on Pinterest come from blogs, or uploaded by users. In Storify, the people tend to use social media and web resources to create their narratives about events, or something of interest.

Table 4. The correlation between the most frequent n domains in the stories and their global Alexa Ranking.

n	10	15	25	50	100
Kendall' τ	0.1555	**0.4476**	**0.3372**	**0.3194**	**0.2485**

4.4 What is the Average Timespan for Stories?

Table 5 shows the percentage of the stories in each time interval. The table also shows the corresponding features of the stories which are divided by their timespan. We normalized the number of views by the age of the story (the time of existence of the story on the live web). The first two intervals represent the stories that were created and modified, then published with no continuing edits.

We see that the majority of the stories in the data set were created and edited in the span of one day. There are 14 % of Storify users who update their stories over a long period of time, with the longest timespan in our data set covering more than four years and with more than 13,000 views. Curiously, it had only 33 web elements and 51 total elements. Although the story with the longest timespan did not have the largest number of elements, from Table 5 we can see that based on the median number of elements in each interval there is nearly linear relation between the time length of the story and the number of elements.

Table 5. The percentage of the stories based on the editing interval along with the median of web elements, text elements, and views. The percentage is out of 15,568 stories.

Intervals	Percentage	Median web elements	Median text elements	Median views
0–60 seconds	14.0 %	15	0	23
1–60 minutes	26.7 %	19	0	53
1–24 hours	23.4 %	25	5	110
1–7 days	13.5 %	26	7	78
1–4 weeks	8.4 %	26	9	80
1–12 months	10.9 %	38	2	129
1–4 years	3.1 %	56	15	156

5 Decay of Elements in Stories

Resources on the web are known to disappear quickly [7,14,18]. In this section we investigate how many resources in the stories are missing from the live web and how many are available in public web archives. We checked the live web and public web archives for 265,181 URIs (202,452 URIs from story web elements +

47,512 randomly sampled tweet URIs + 15,217 URIs of embedded resources in those tweets), in which there are 253,978 unique URIs. We examined the results of the five most frequent domains in the stories (twitter.com, instagram.com, youtube.com, facebook.com, flickr.com).

5.1   Existence on the Live Web

From all the web resources, we checked the existence of the 253,978 unique URIs on the live web. We also checked the pages that give "soft 404s", which return HTTP 200, but do not actually exist [2]. The left two columns of Table 6 contain the results of checking the status of the web pages on the live web. Of all the unique URIs, 11.8 % are missing on the live web. The table also contains the results of the five most frequent domains and all other URIs. We also included the results of checking the existence of Twitter embedded resources at the bottom of the table. From the table, we conclude that the decay rate of social media content is lower than the decay rate of the regular web content and websites.

Table 6. The existence of the resources on the live web (on the left) and in the archives (on the right). Available represents the requests which ultimately return "HTTP 200", while missing represents the requests that return HTTP 4xx, HTTP 5xx, HTTP 3xx to others except 200, timeouts, and soft 404s. Total is the total unique URIs from each domain.

Resources	Existence on live web			Found in archives		
	Available	Missing	Total	Of the available	Of the missing	Total
Twitter	95.5 %	4.5 %	47,385	0.9 %	3.4 %	477
Instagram	86.6 %	13.4 %	43,396	0.3 %	0.07 %	103
Youtube	99.3 %	0.7 %	19,809	16.0 %	0.75 %	3,140
Facebook	95.2 %	4.8 %	12,793	0.6 %	0.49 %	80
Flickr	95.6 %	4.4 %	6,859	0.4 %	0.0 %	25
others	82.1 %	17.9 %	109,120	26.8 %	15.5 %	27,033
Twitter resources	90.1 %	9.9 %	14,616	8.0 %	14.1 %	1,257

5.2   Existence on the Live Web as a Function of Time

We measured the decay of the resources of Storify stories in time by measuring the percentage of the missing resources in the stories over time. For this experiment, we used the 249,964 (all the URIs excluding twitter embedded resources) resources in 14,513 stories to check the rate of the decay in the stories.

We found that 40.8 % of the stories contain missing resources with an average value of 10.3 % per story. Figure 2 contains the distribution of the creation date of stories in our data set in each year and the percentage of the missing resources in each corresponding year. From the graph, we can infer a nearly linear decay rate of resources through time. This finding is very similar to the findings by SalahEldeen and Nelson [13], in which they found that resources linked to from social media resources disappeared at rate of 11 % the first year and 7 % for each following year.

5.3   Existence in the Archives

We checked the 253,978 pages for existence in general web archives in March 2015. The existence in the web archives was tested by querying Memento proxies and aggregator [17].

The right-most columns of Table 6 contain the percentage of the URIs found in the web archives out of the missing and the available URIs on the live web. In total, 12.6 % of the URIs were found in the public web archives. Of the missing resources (29,964), 11 % were found in public web archives. From the table we notice that the social media is not well-archived like the regular web [1]. Facebook uses robots.txt to block web archiving by the Internet Archive, but the other sites do not have this restriction.

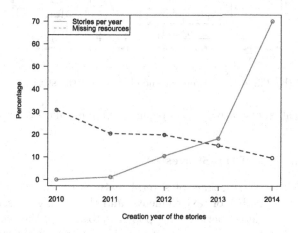

Fig. 2. The distribution of the stories per year and the decay rate of the resources in these stories through time.

6   What Does a Popular Story Look Like?

In this section, we establish structural features for what differentiates popular stories from normal stories for building a baseline for the stories we will automatically create from the archives. We divided the stories into popular and unpopular stories based on their number of views, normalized by the amount of time they were available on the web. We took the top 25 % of stories (3,642 stories) that have the most views and consider those as the popular stories. The 75th percentile of the views that we separated the data based on is 377 views/year.

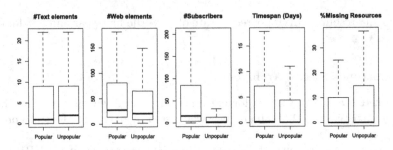

(a) The distributions for the features of the stories.

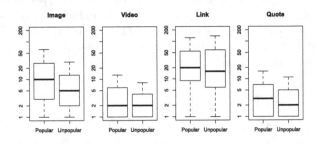

(b) The distributions for the elements of the stories.

Fig. 3. Characteristics of popular and unpopular stories.

6.1   The Features of the Stories

We considered the distributions of several features of the stories: number of web elements, the number of text elements, and the editing timespan. We also check if there is a relation between the popular stories and the relative number of subscribers. Furthermore, we test if popular stories are different from the unpopular stories using Kruskal-Wallis test, which allows comparing two or more samples that are independent and have different sample sizes.

We found that at the $p \leq 0.05$ significance level, the popular and the unpopular stories are different in terms of most of the features: number of web elements, text elements, timespan, and subscribers. Figure 3(a) shows that popular stories tend to have more web elements (medians of 28 vs. 21) and a longer timespan (5 hours vs. 2 hours) than the unpopular stories. The number of elements in the popular stories is between 2 to 1950 web elements with $median = 28$ and text elements from 0 to 559 with $median = 1$. The popular stories tend to have longer editing time intervals than the unpopular stories. For the popular stories, 38 % have an editing timespan of at least one day, while 35 % of the unpopular stories have this feature. The maximum editing time in the popular stories is 4.1 years, while it is 3.5 years for unpopular stories.

6.2 The Type of Elements

Figure 3(b) shows the distributions for the popular and the unpopular stories for each element type. The figure shows that the distribution of the images in popular stories is higher than the distribution for the images in the unpopular stories. The median number of images in popular stories is 10, while it is 5 in the unpopular stories. For the videos the median is 2 for both popular and unpopular. Although the unpopular stories tend to use links more than the popular stories, the median of the links in popular stories (20 links) is higher than the unpopular stories (16 links). We also test if popular stories are different from the unpopular stories using the Kruskal-Wallis test, based on the elements and found $p \leq 0.05$ for all tests.

6.3 Do Popular Stories Have a Lower Decay Rate?

We checked the existence of the missing resources of the popular stories and the unpopular stories to investigate if there is a correlation between popularity and lower decay rate. We found that for the popular stories, 11.0 % of the resources were missing, while 12.8 % of the resources were missing for unpopular stories. Figure 3(a) contains the distribution of the percentage of missing resources per story in popular and unpopular stories. It shows that the resources of the popular stories tends to stay longer than the resources of the unpopular. The 75th percentile of decay rate per popular story is 10 % of the resources, while it is 15 % in the unpopular stories.

7 Conclusions and Future Work

In this paper, we presented the structural characteristics of the human-generated stories on Storify, with particular emphasis on "popular" stories (i.e., the top 25 % of views, normalized by time available on the web). Upon analyzing 14,568 stories, the popular stories have median value of 28 elements, while the unpopular stories have 21. The median value of multimedia elements in popular stories is 12, with only 7 in unpopular stories. Of the popular stories, 38 % receive continuing edits (as opposed to 35 %), and only 11 % of web elements are missing on the live web (as opposed to 13 %). We found that there is nearly a linear relation between the timespan of the story and the number of web elements. There were 11.8 % of the resources missing from the live web, in which 11 % were found in the archives. The percentage of the missing resources is proportional with the age of the stories.

Future work will include investigating if these structural characteristics of stories hold for other social media storytelling services, such as paper.li, scoop.it, and pinterest.com. This study also will inform our future work of automatically creating stories to summarize collections of archived web pages. Using the structural characteristics of human-generated stories, such as number of elements, timespan, and a distribution of domains and types of resources, will provide us with a template with which to evaluate our automatically created stories.

References

1. Ainsworth, S.G., AlSum, A., SalahEldeen, H., Weigle, M.C., Nelson, M.L.: How much of the web is archived? In: Proceedings of the 11th ACM/IEEE-CS Joint Conference on Digital Libraries, JCDL 2011, pp. 133–136. ACM Press (2011)
2. Bar-Yossef, Z., Broder, A.Z., Kumar, R., Tomkins, A.: Sic transit gloria telae: towards an understanding of the web's decay. In: Proceedings of the 13th International Conference on World Wide Web, WWW 2004, pp. 328–337 (2004)
3. Cohen, J., Mihailidis, P.: Storify and news curation: teaching and learning about digital storytelling. In: Second Annual Social Media Technology Conference & Workshop, vol. 1, pp. 27–31 (2012)
4. Duh, K., Hirao, T., Kimura, A., Ishiguro, K., Iwata, T., Yeung, C.M.A.: Creating stories: social curation of twitter messages. In: Proceedings of the 6th International AAAI Conference on Weblogs and Social Media, ICWSM 2012 (2012)
5. Hall, C., Zarro, M.: Social curation on the website pinterest.com. Am. Soc. Inf. Sci. Technol. 49(1), 1–9 (2012)
6. Kieu, B.T., Ichise, R., Pham, S.B.: Predicting the popularity of social curation. In: Nguyen, V.-H., Le, A.-C., Huynh, V.N. (eds.) Knowledge and Systems Engineering. AISC, vol. 326, pp. 419–434. Springer, Heidelberg (2015)
7. Klein, M., Nelson, M.L.: Find, new, copy, web, page - tagging for the (re-)discovery of web pages. In: Gradmann, S., Borri, F., Meghini, C., Schuldt, H. (eds.) TPDL 2011. LNCS, vol. 6966, pp. 27–39. Springer, Heidelberg (2011)
8. Laire, D., Casteleyn, J., Mottart, A.: Social media's learning outcomes within writing instruction in the EFL classroom: exploring, implementing and analyzing storify. Procedia-Soc. Behav. Sci. 69, 442–448 (2012)
9. Mihailidis, P., Cohen, J.N.: Exploring curation as a core competency in digital and media literacy education. J. Interact. Media Educ. 2013(1), 2 (2013). http://dx.doi.org/10.5334/2013-02
10. Ottoni, R., Las Casas, D., Pesce, J.P., Meira Jr., W., Wilson, C., Mislove, A., Almeida, V.: Of pins and tweets: investigating how users behave across image- and text-based social networks. In: Proceedings of the 8th International AAAI Conference on Weblogs and Social Media, ICWSM 2014, pp. 386–395 (2014)
11. Padia, K., AlNoamany, Y., Weigle, M.C.: Visualizing digital collections at archive-it. In: Proceedings of the 12th Annual International ACM/IEEE Joint Conference on Digital Libraries, JCDL 2012, pp. 437–438 (2012)
12. Palomo, B., Palomo, B.: New information narratives: the case of storify. Hipertext.net 12 (2014)
13. SalahEldeen, H.M., Nelson, M.L.: Losing my revolution: how many resources shared on social media have been lost? In: Zaphiris, P., Buchanan, G., Rasmussen, E., Loizides, F. (eds.) TPDL 2012. LNCS, vol. 7489, pp. 125–137. Springer, Heidelberg (2012)
14. SalahEldeen, H.M., Nelson, M.L.: Carbon dating the web: estimating the age of web resources. In: Proceedings of 3rd Temporal Web Analytics Workshop, TempWeb 2013, pp. 1075–1082 (2013)
15. Seitzinger, J.: Curate me! exploring online identity through social curation in networked learning. In: Proceedings of the 9th International Conference on Networked Learning, pp. 7–9 (2014)
16. Stanoevska-Slabeva, K., Sacco, V., Giardina, M.: Content curation : a new form of gatewatching for social media? In: Proceedings of the 12th International Symposium on Online Journalism (2012)

17. Van de Sompel, H., Nelson, M.L., Sanderson, R.: RFC 7089 - HTTP framework for time-based access to resource states - Memento (2013). http://tools.ietf.org/html/rfc7089

18. Weiss, R.: On the Web, Research Work Proves Ephemeral (2003). http://stevereads.com/cache/ephemeral_web_pages.html

19. Zhong, C., Salehi, M., Shah, S., Cobzarenco, M., Sastry, N., Cha, M.: Social bootstrapping: how pinterest and last.fm social communities benefit by borrowing links from facebook. In: Proceedings of the 23rd International Conference on World Wide Web, WWW 2014, pp. 305–314. ACM (2014)

20. Zhong, C., Shah, S., Sundaravadivelan, K., Sastry, N.: Sharing the loves: understanding the how and why of online content curation. In: Proceedings of the 7th International AAAI Conference on Weblogs and Social Media, ICWSM 2013 (2013)

Tyranny of Distance: Understanding Academic Library Browsing by Refining the Neighbour Effect

Dana McKay[1](✉), George Buchanan[2], and Shanton Chang[1]

[1] University of Melbourne, Parkville, VIC 3010, Australia
dmckay1@student.unimelb.edu.au,
shanton.chang@unimelb.edu.au
[2] Centre for HCI Design, City University London, London EC1V 0HB, UK
george.buchanan.1@city.ac.uk

Abstract. Browsing is a part of book seeking that is important to readers, poorly understood, and ill supported in digital libraries. In earlier work, we attempted to understand the impact of browsing on book borrowing by examining whether books near other loaned books were more likely to be loaned themselves, a phenomenon we termed the neighbour effect. In this paper we further examine the neighbour effect, looking specifically at size, interaction with search and topic boundaries, increasing our understanding of browsing behaviour.

Keywords: Browsing · Books · Libraries · Information seeking · Classification systems · Log analysis

1 Introduction

Participants in a 2007 study comparing physical and digital libraries noted there is no digital analogue of library shelves for book seeking, particularly in terms of serendipitous discovery [1]; that same year Rowlands noted the dearth of literature on book selection [2]. Despite later research on book selection, reader behaviour at the library shelves remains largely mysterious.

Key models of information seeking include elements that focus on browsing and exploration [3, 4]. The literature on book seeking also shows that readers consider browsing an important part of book seeking [1, 5, 6]. Readers in both academic [5, 7] and public [8] libraries note lack of browsing support as a reason for avoiding the use of ebooks. Libraries increasingly offer ebooks alongside or in place of print, so avoiding them is likely to negatively affect readers.

Despite readers' insistence that browsing is important, until last year there was little literature on its impact on book use. Our prior work [9] leveraged two established digital library techniques—examining physical libraries [1, 6, 7] and transaction log analysis [10, 11]—to determine whether physical layout of library shelves affects book borrowing patterns. In that work we specifically examined whether proximity to a loaned book increases the chance a book will itself be loaned: we found a strongly

© Springer International Publishing Switzerland 2015
S. Kapidakis et al. (Eds.): TPDL 2015, LNCS 9316, pp. 280–294, 2015.
DOI: 10.1007/978-3-319-24592-8_21

significant increase in this likelihood that we termed the neighbour effect. This paper refines the neighbour effect by examining day-of-week variance, the distance between relatively nearby books borrowed on the same day (co-borrowed books), the impact of topic boundaries, and the distance between co-borrowed books in search results. Given the prevalence of complaints about digital library browsing systems [1, 6, 12], developing a better understanding of browsing in is a vital step to improving users' experience of DLs. This paper thus aims to increase and improve our understanding of browsing. Section 2 presents the background literature, Sect. 3 our methodology. Section 4 presents the results, which are compared to the literature in Sect. 5. We finally draw conclusions and suggest future work in Sect. 6.

2 Background Literature

In this section we will first present the work on browsing in the context of information seeking behaviour, then cover browsing and browsing technologies. Next we will cover the literature on book selection; finally we summarise our previous work that this paper extends.

2.1 Browsing in Human Information Seeking Behaviour

Browsing is a central activity in many major models of information seeking (e.g. [3, 4]). The common models are primarily linear, with elements that address search and triage. Browsing in these models is closely interleaved with search and used for exploration and triage. Beyond these models, serendipity—the opportunity to discover information one otherwise would not—is a key attribute of browsing. Savvy information seekers leverage their surroundings to increase the likelihood of serendipity [13], a strategy also used at the library shelves [1, 6, 14].

2.2 Browsing and Browsing Technologies

Bates—who wrote the seminal work on browsing in electronic environments [15]—also provides a working definition of browsing grounded in a broad literature [16]. She described it as the process of glimpsing a 'scene' (a large collection of potentially interesting objects) sequentially examining objects of interest, and retaining or discarding these. Given this definition, library shelves are, as noted in earlier work on libraries [14], ideally suited to browsing.

More than 20 years since Bates' seminal paper [15], online browsing systems remain under-researched. It may be argued that search results [17], and faceted search in particular [18], re-present opportunities to browse online, however both the limited number of results presented, and the need to navigate away from the 'scene' (or search results) to examine items in depth contravene Bates' definition of browsing [16]. These limitations also negatively affect the close interleaving of search and browse found in information seeking models [3, 4]. The need to prime search-based systems with query terms—a difficult task with imprecise information needs [19]—also limits their usefulness [20].

The literature on browsing-specific systems is limited. A 1995 paper presents an early shelf metaphor system intended for children [10], and a 2004 paper presents three browsing tools within the Greenstone DL system [21]. These tools are, however, only proofs-of-concept. Book browsing systems have been more common in recent research: e.g. exploiting a shelf metaphor [14]; using the non-bibliographic book features readers say they use in decision making [22], and Pearce's iFish tool [23], which creates browsing sets based on user-specified preferences. Whichbook (http://www.openingthebook.com/whichbook/) is one of a number of increasingly common commercial systems. None of these tools however, is rooted in a detailed understanding of browsing behaviour, and none has yet been the subject of rigorous evaluation. The emergence of commercial systems both reinforces the need for research-based approaches, and renders the topic of browsing relevant and timely.

2.3   The Book Selection Process

Rowlands' 2007 critique of the limited research on book selection [2] has been followed by a steady growth in that literature. The process of choosing a book in a library can be divided into 5 components: identifying a need for a book; searching (this step may be skipped); locating books of interest; choosing among them; and reading or otherwise using selected books. Catalogue book search, while it has major usability problems [11, 24], is well-studied elsewhere and not the focus of this work; we will not discuss it further. Similarly, the nature of reading—while an interesting open research question [25]—will not be further discussed here.

There is some work on how readers identify both fiction [8, 26] and non-fiction [27, 28] books of interest. Personal recommendations and shelf browsing are frequently reported, while search appears rare. There is a limited but growing literature on how readers identify and select books at the shelves. Research on children [29, 30] notes that they focus on eye-level shelves, and that shelf order affects the books they select. Adults are also affected by shelf height, though less than children are [7]. Like children [30], they struggle with some aspects of shelf layout in libraries [20], but adults also exploit it, using librarian created displays or recent returns [7, 8, 26] as information resources. Savvy library users value the shelves as a finding aid and relevance cue, and note there is no online equivalent [1]. Participants in numerous studies have said they value the opportunity to browse shelves, even giving it as a reason for avoiding the use of ebooks [1, 5, 7].

Decision-making at the shelves is progressive—looking, then looking more closely, then taking books off the shelf [7, 29]. Readers flip pages, and use index, blurb, and images to determine relevance [6, 7]. These cues are also used in ebooks, but other non-bibliographic cues that are not replicated online—such as dust, book size, and location—are also used [1, 6].

The process of examining shelves and choosing books mirrors Bates' definition of browsing [16]; indeed it seems likely that libraries, classification schemes and shelves have evolved to create a physical browsing engine [31]. The importance to users of browsing is supported by numerous small studies, including [1, 6, 7]. A larger study from 1993 demonstrates that browsing also affects the books readers select: over half of those who located one book identified by searching borrowed at least one further book [32]; a contemporaneous study noted that books near each other on the shelves were likely to be borrowed together [33]. Given the age of these studies and the ascendance of search in the intervening time, it seems reasonable to ask whether browsing still holds such sway: this paper investigates the impact of browsing on book loans.

2.4 Our Previous Work

The only recent study of the impact of browsing—or more specifically shelf location—on loans within academic libraries is our own [9]. We used a large publicly available circulation dataset from the OCLC [34], and selected six libraries based on a set of criteria to ensure their broad similarity. We created a shelf-sorted book set including circulation data for each library.

Circulation data only records the most recent loan for each book; given this limitation we used two tests to look for a neighbour effect. We compared the number of loans among the ten nearest neighbours (five either side) of loaned and unloaned books on the final date recorded in the data, and for randomly selected loans we compared the number of loans among the nearest neighbours on the loan date and on the day before. Both tests showed a strongly significant neighbour effect, supporting our hypothesis that browsing influences loan patterns. One significant limitation remains: without patron data, we cannot prove, for any co-borrowing, that it is the result of use by a single patron. There is a preponderance of evidence, though, in the form of observation [1, 6, 32], log analysis [33] and user self-reporting [5, 8, 26, 27] that makes browsing a logical explanation for a significant proportion of this activity.

The tests we used in this early work were a fairly blunt instrument: they examined a fixed number of books, and did not take search, day-of-week effects or topic boundaries into account. This work aims to refines our method, deepening our understanding of browsing along these axes.

3 Methodology

This paper extends the methodology of our previous work [9]. We will briefly review our dataset, then describe four new tests employed to examine the patterns of co-borrowing.

3.1 Dataset

We used the same six sample libraries as our previous work: Cedarville, Dennison, Case Western Reserve University (CWRU), Oberlin, Ohio Northern (ON) and Ohio State University (OSU). For this study, we omitted Oberlin's Dewey collection, though we retained Oberlin's LC collection: the former has a fragmented floor layout[1] limiting browsing opportunities. We mitigated against overwritten loans (circulation data stores only the most recent loan) by using data from the end of the dataset's collection period, in many cases the final week. The final week of data for OSU, however, had less than half the annual average number of loans per week (650, in comparison with 1592) so we used the penultimate week's data (1071 loans).

3.2   Tests

We conducted four separate tests on our datasets. Our tests primarily involve the ten closest books to an individual text: five to its left and five to the right (duplicates are discarded—they form < 3 % of any collection and usually < 1 %). We refer to this set, which we also used in our previous work, as N_{10}. The only study similar to ours used a width of two [33], or N_4.

Shape of the Neighbour Effect: Books within the N_{10} set will usually be on the same shelf as a target book: 10 books is simply not enough to see the books above and below a borrowed book, nor is it wide enough to tell us when the neighbour effect disappears. Using the final week sets described above, we calculated for each borrowed book the distance to its nearest borrowed neighbour, and the number of books borrowed at each distance out to 150 books either side. We examined this data to see how rapidly the neighbour effect falls away, and whether there are secondary co-borrowing peaks that could account for books above and below a target book.

Day of the Week Effects: There is a known 'day of the week' effect on search behaviour in information retrieval [35, 36], representing changes in user behaviour based on context. The same effects in browsing data would reinforce our impression that the neighbour effect is a product of intentional human behaviour, and not random co-borrowing. We compared the strength of the neighbour effect at N_{10} on the busiest and least busy days of the week (based on the final year of loans data), and compared weekdays with weekends. To ensure that we had a representative time sample we ran this comparison over all complete weeks in 2008 (the final year of data collection—data ceases for all collections in late April or early May).

Comparing Search and Browsing: One explanation for co-borrowing is that users may identify books individually in search and select them independently of one another. While early studies [32, 33] do not support this, the increasing dominance of search [23, 28] suggests that search could be responsible for these results. We created a stop-worded log-rule search index [37] for the titles of the books (the only metadata we had, but data that is frequently used in book search [11, 24, 38]).

We then collated all the possible pairs of books borrowed on the same day in the final week. Those pairs within their N_{10} shelf-set were identified first to create the shelf co-borrowing set. We did two searches against the titles index: a title search, and a

---

[1] See http://www.oberlin.edu/library/main/2.html.

keyword search on their shared title words. Where one book and another loaned on the same day both appeared in the top ten search results (the only results most users look at [37]), they were considered to be search co-borrowings. This created both a title search set and a shared title keyword set of co-borrowings. The difference in result ranking of two results was deemed to be their search distance. For any book pairs that had both shelf (within N_{10}) and search proximity, we determined which distance was shorter (and hence a better explanation for the co-borrowing).

Topic Boundaries: Earlier work suggests that shared topic is almost as important as shared shelf location, even where books are not co-located [33]. Topic and shelf location are intertwined in academic libraries [31]: to investigate this relationship we examined loans near topic boundaries. Topic boundaries were categorised as 3, 2, 1 or 0; rank 3 indicated a difference in the first digit of the call number, 2 the second, 1 the third and 0 none before this. We identified the borrowings near topic boundaries, then did a closer analysis where there was sufficient data.

4   Results

This section reports results for each of the four individual tests described in Sect. 3. We subsequently summarize our findings and point to limitations of our study.

4.1   Shape of the Neighbour Effect

To examine the width of the neighbour effect, we first calculated the distance to nearest co-borrowed book for all books loaned in the final week set. Not every visit to the library will result in co-borrowings, and borrowing rates per book are low (*annual circulation ranged from 0.08 to 0.18 loans per book for all collections* [9]). The data shows a log-log distribution often seen in information behaviour [37]. Nearest neighbour data is shown in Table 1, below:

Table 1. Shelf distance to nearest loan in number of books showing first three quartiles

	Median	Mean	Std.dev	Q1	Q2	Q3
Cedarville	571	1118	178	89	181	640
CWRU	2930	5437	1300	343	1297	3471
Dennison	1507	2177	898	189	919	2063
Oberlin	4917	9833	1910	189	1858	6198
ON	2351	3343	1056	170	992	2972
OSU	4281	8140	1795	471	1773	4585

The distribution of loans suggests a neighbour effect: many loans are close to their nearest borrowed neighbour. Browsing likely accounts for many of the close-by loans, but is unlikely to account for loans hundreds of books distant. There is one possible exception: examining nearest neighbour loans in groups of 50, four libraries (Dennison, Oberlin, ON and OSU) show unexpected peaks between 650 and 700 or 700 and 750, rather than the expected long flat tail. One possible explanation is borrowing from the ends of shelves (as seen in [7]), though this is speculative in the absence of detailed information about physical library layout.

After looking only at nearest neighbour information, we considered the data with respect to all co-borrowings at the micro level (within N_{20}) and the macro level (up to N_{300}).

Examining all neighbours at the micro-level reveals that the rate of co-borrowings falls rapidly with distance for four libraries, but Cedarville and OSU show flatter patterns (Fig. 1).

The macro-level view allows us to test for borrowing from above and below a target book. Of course the number of books per shelf will vary, even within libraries, due to variations in both shelf and book width. Averaging over a large enough sample, however, may reveal a secondary peak representing the books above/below an index book. These books, while closely co-located, are likely to represent greater topic difference than the books beside an index book.

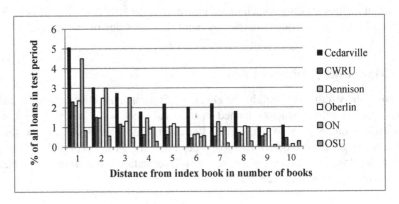

Fig. 1. All co borrowings within N_{20}

Studies of recommender systems show that users will choose items that fall a little outside what they expect [39]; and vertical neighbours on library shelves may have a similar effect. We examined all neighbouring loans (not just the nearest) as we didn't want to exclude patterns where users borrowed books both above or below an index book *and* on either side.

Figure 2 shows co-borrowing over 150 neighbouring books. Beyond 150, loans enter a long flat tail. As expected, there are secondary peaks (in all collections): most have two, one around the 50–70 mark, the next at 100–120. This suggests shelf wrap is c. 60 books, and supports previous evidence [5] that users browse 'three shelves' above and below a target book.

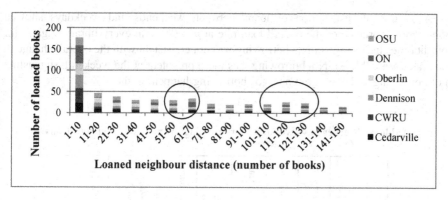

Fig. 2. Total borrowed books for all collections. Circles represent possible vertical browsing.

4.2   Day of the Week Effect

The literature on both search [35, 36] and academic library usage [40] evidences changes in usage between weekdays and weekends; we suspected that the same might hold true for browsing. In view of this we examined both the total number of loans on each day of the week for busy and quiet days (see Table 2), and compared weekdays to weekends (see Table 3). We also compared the prevalence of the neighbour effect in N_{10} against busy days and weekends.

In a χ^2 test of distribution there was clear non-random variance across days of the week at $p < 0.001$ for all libraries. For four libraries, the number of N_{10} loans was not consistent with the overall change in loan rate according to day (see Table 2). In each of those libraries neighbour borrowing was more prevalent on quiet days than we would expect, showing a difference in borrowing behaviour: browsing is more likely on quiet days.

Table 2. Busy and quiet days, frequency of loans of the N_{10} neighbours of each borrowed book

	Busiest day	% of loans	Busy day N_{10} loans	Busy day loan totals	Quietest open day	% of loans	Quiet day N_{10} loans	Quiet day loans	P (Busy vs Quiet N_{10})	χ^2 (df=1)
Cedarville	Mon	21.0	374	2466	Sat	6.0	128	834	<0.0001	80.49
CWRU	Wed	20.6	207	2846	Sat	4.8	93	668	<0.0001	26.85
Dennison	Mon	21.0	36	847	Sat	2.4	11	103	0.0110	6.46
Oberlin	Wed	17.8	121	14239	Sun	9.1	71	7329	0.4257	0.63
ON	Tue	21.5	43	4348	Sat	3.6	10	720	0.4424	0.59
OSU	Wed	19.3	317	47603	Sat	5.2	141	12719	<0.0001	25.06

A χ^2 test of distribution of loans between weekends and weekdays showed non-random variance of the overall loan rate at $p < 0.001$ in every library. Again, for four libraries the change in N_{10} borrowing was not consistent with the overall change in loan rate (see Table 3): N_{10} borrowing was more prevalent at the weekend, suggesting more browsing and less 'grab and go' borrowing happening then.

Table 3. Weekends vs weekdays for browsing: overall and neighbour loans

	W/day%	w/end N_{10} loans	w/end loans	w/day N_{10} loans	w/day loans	P (W/day vs w/end N_{10})	χ^2 (df = 1)
Cedarville	93.9	128	834	1484	10902	0.2063	1.60
CWRU	90.0	199	1364	1079	12329	<0.0001	44.63
Dennison	91.7	34	351	151	3871	<0.0001	23.39
Oberlin	75.3	194	1992	487	6073	<0.0001	816.77
ON	90.3	22	198	155	1846	0.2664	1.23
OSU	89.2	281	2641	1356	21887	<0.0001	880.79

Every library studied shows variance in N_{10} borrowing by day of the week that is inconsistent with overall change in loan rate. In each case this is an increase in N_{10} borrowing on a day that is otherwise 'quiet'. This change suggests altered behaviour at the shelves—readers are more likely to browse during quiet times, perhaps because they have more time to do so.

4.3   The Impact of Search

One key argument against the neighbour effect is that search can account for much of it: users find items that are shelved close to each other during search, rather than from browsing at the shelf. As described in Sect. 3.2, we assess the overlap between the shelf and search co-borrowing pairs (both title and shared keyword searches), and determined for each of these whether books were closer using search or browse (Table 4, left hand side).

While title and keyword search will not account for all possible search scenarios, they are the most common methods [24, 38] seen in libraries. In our data, search only accounts for a small proportion of shelf-based co-borrowings. The argument that search causes co-borrowings is therefore unsustainable. Shelf browsing remains the simplest explanation, accounting for more neighbouring and more total loans, both by rate and by closeness of the pairs.

To further test whether search or shelf location was responsible for co-borrowing, we again took N_{10} sets and allocated any borrowed neighbours to search or browse,

Table 4. Shelf vs loan co-borrowing; search versus browse influence on co-borrowing

	Browse pairs	Title Search pairs	Keyword search pairs	Overlap	Shelf closer	Search closer	Ties	Neighbour loans explained by browse	Neighbour loans explained by search	Total loans	P (browse vs search)	χ^2 (df = 1)
Cedarville	374	212	97	95	54	24	17	333	189	1283	<0.0001	40.00
CWRU	140	71	54	36	28	7	1	132	71	1132	<0.0001	17.89
Dennison	68	32	40	25	12	5	8	51	22	475	0.0005	10.82
Oberlin	128	31	37	10	6	3	1	124	51	765	<0.0001	21.12
ON	46	24	15	11	8	1	2	43	14	200	<0.0001	14.00
OSU	46	14	6	1	1	0	0	46	13	1071	<0.0001	17.40

whichever formed the nearer pair. To maintain independent sets, ties were discarded. We then tested the search versus browse sets using a single-tailed χ^2 test (Table 4 right hand side). The data shows browsing having much more influence on co-borrowing than search. Given readers' penchant for convenience [5] this is unsurprising: books co-located on a shelf one is visiting require little extra effort, while going to another section of shelving takes additional time and effort [37].

4.4 Topic Boundaries

We assessed the impact of topic boundaries on neighbouring loans to understand the interaction between topic and shelf layout. As described in Sect. 3, we defined three levels of topic boundary: from 3 (at the top level of classification), to 0. Our first discovery was that, notably, no book within 5 books of a top-level topic boundary had ever been loaned in any library. Loan numbers were also low when examining level 2 boundaries. We tested this distribution against the nominal likelihoods for each boundary, accounting for the number of topic boundaries and loan rates in each collection, using a two-factor chi-squared test: see the left of Table 5.

Topic boundaries appear to discourage loans: this may be related to how books are shelved close to boundaries: e.g. a level 3 boundary will often begin on a new shelf; given that users look less at top and bottom shelves [7, 29] it is not surprising that these are disadvantaged.

Even at the lowest topic boundaries, loans are limited. Dennison and ON have such small numbers of nearby loans—43 and 74 respectively—that we excluded them from further analysis. For the remaining four libraries we created samples of 100 loans for which N_{10} contained either a level-2 or level-1 boundary, then counted the number of co-borrowed books in each set (i.e. the total sample book count of each type was 1000). A two-tailed Fishers exact test was used to compare sets that included a level 1 boundary with sets with level 0 boundaries to see whether a topic boundary affects neighbouring loans (Table 5, right); results were significant.

Table 5. Number of loans occurring near classification boundaries at any time

	Loans near L2 boundary	Remaining loans	p	χ2 (df = 2)	L1 Neighbour Loans	L0 Neighbour loans	p (L1 vs Lo)
Cedarville	52	92092	0	19288.220	6	20	0.00089
CWRU	19	248644	0	6340.184	2	18	0.0004
Oberlin	28	236920	0	4532.100	1	9	0.0212
OSU	78	304510	0	11795.026	0	12	0.0005
Dennison	11	57830	0	1209.737	Discounted due to lack of data		
ON	71	36448	0	567.963	Discounted due to lack of data		

Clearly topic, as defined by shelf classification, is an important factor in co-borrowing: browsers are more likely to co-borrow within topic boundaries than across them, and are less likely to borrow near topic boundaries than in the rest of the collection.

4.5 Summary and Limitations

Clearly this study shares some of the limitations of our earlier work: we cannot, from the data we have, prove that a single user is responsible for any co-borrowed pair. We are also not privy to users' motivations or behaviour at the shelf. Furthermore we do not have all the metadata required to fully address search; having only title data we cannot address author or other metadata. Nonetheless keyword searching and title searching are the dominant strategies in book search behaviour, and titles represent a linguistically user-friendly corpus [19].

We have examined different aspects of the neighbour effect to obtain new insights into browsing behaviour. Nearest neighbour loans have an overall power-law distribution. Secondary peaks in neighbouring loans were found at distances of 50–60 items. Search accounts for some co-borrowings on the shelf, but not most, and it plausibly generates its own co-borrowings. Across all our data, shelf-browsing accounts for more co-borrowings than title search. The amount of browsing varies by day of the week in all libraries. Finally, borrowing tends to occur within topic boundaries, suggesting that shelf layouts and topic divisions influence borrowing patterns. Each of these findings is novel, and represents a new understanding of human behaviour that could improve digital library design.

5 Discussion

Our findings can be divided into two major themes: the interaction of search, topic, and shelf, and the implications of reader context for browsing.

5.1 Shelf, Topic and Search

The dominant information discovery tool in every existing digital library system is search. Our data demonstrates, though, that search alone cannot fully meet the needs of information seekers. Search explains some co-borrowings in our data, so we would argue that it still plays a valuable role, but many more borrowings are poorly explained by search and well-explained by browsing. Search is extremely well covered by the DL literature (for a starting point see [37]); in contrast there is very little work on browsing. The existing work on browsing is very limited in terms of proposed solutions, and even where these have been proposed (for example [14, 20, 22, 23]) they have not been adequately assessed for their effectiveness. Search effectiveness is assessed with standard metrics [37]; similar metrics do not yet exist for browsing. Our work demonstrates clearly that DLs need to support browsing, most currently do not.

Our data not only shows a need for browsing systems, it also suggests certain key characteristics of such systems. Earlier work suggests browsers check 'three shelves before and after' a target book—this is something in the range of 200–500 books. This is considerably more books than are typically shown in search results, and an order of magnitude more than is shown in some nascent browsing systems [23]. Research on recommender systems has shown that a few 'unexpected' recommendations improve user experience [39]. It is therefore likely that both volume and variety play a role in browsing.

Topic is another clearly important feature of books: our data shows that co-borrowing occurs primarily, though not exclusively, within topic boundaries. Conversely, both our data and the literature [7, 26] suggest that readers occasionally like to see 'distant' books. Physical shelves cannot be rearranged to meet individual user needs, yet we know rearranging shelves affects book selection [41]. Electronic shelves could—and should—leverage topic clustering that occurs across the boundaries of traditional classification schemes to 'rearrange the shelves' to aid discovery; the literature suggests this is likely to be useful [33]. The underlying data already exists within classification schemes [31], but no DL system has yet offered readers these options. Similarly DL systems could leverage topic classification schemes to offer readers a very few different-but-interesting books based on topic data: this approach would mimic the physical shelves, but, being data driven, would offer a higher chance of success.

5.2   Reader Context

It is clear from our data that browsing, like search [35], is affected by day of the week. The way in which co-borrowing is affected by day of the week—increasing on quiet days, weekends, or both—suggests that reader context has a significant impact on browsing behaviour. This dovetails with earlier work: students report staying longer in the library when they do not have to rush to class [40], and academics report searching the catalogue inside the library when they are looking for inspiration, and outside the library when they are in a rush [6]. Similarly browsing the shelves [1, 6, 7] and serendipitous discovery [13] are activities that require time and attention, which, when a searcher is meeting an urgent information need, may not be available. This context-dependent approach to browsing has significant implications for DL design: users need to be able to 'grab and go' [7] when it suits them, but browsing facilities should be a visibly tempting way for users to spend any extra time they have.

6 Conclusions and Future Work

Not all who wander in libraries are lost: the library shelves afford browsing and serendipitous discovery in ways that simply do not exist in current DL systems. Our data demonstrates clearly that shelf arrangement and reader context—and thus browsing—have a clear impact on borrowing that cannot be explained by search. To be truly effective information resources, DL systems need to facilitate browsing; the literature shows that DLs are not meeting readers' needs in this space. Our data further points to clear design implications: it is not enough to offer users a small number of books (as, for example in search results). The neighbour effect extends above and below a target book, as well as to the left and right—a range of over 200 books. Browsing systems also need to be optional: browsing happens under different circumstances to search, and users must be able to engage in the most appropriate information seeking strategy for their context. If they have time, they should be visibly tempted to linger and browse; if they do not, 'grab and go' should be an option.

Browsing is not just a necessity for DL systems, it is also an opportunity. DL systems have the potential to offer features that cannot exist in the library shelves: shelves can be rearranged to reflect cross-classification topic clustering, for example. DL systems have the power to offer readers new and exciting paths to wander, and must leverage that if they are to provide adequate user experience.

Our study, of course, does not answer all (or even most) of the questions about browsing. To determine how much browsing occurs that does not result in loans, and what users motivations are requires a different kind of study; these questions remain future work.

Acknowledgements. This paper contains information from OhioLINK Circulation Data (http:// www.oclc.org/research/activities/ohiolink/circulation.htm), made available by OCLC Online Computer Library Center, Inc. and OhioLINK under the ODC Attribution License (http://www. oclc.org/research/activities/ohiolink/odcby.htm). Travel to the conference was partially funded by a Google Travel Scholarship administered by the Department of Computing and Information Systems at the University of Melbourne.

References

1. Makri, S., Blandford, A., Gow, J., Rimmer, J., Warwick, C., Buchanan, G.: A library or just another information resource? A case study of users' mental models of traditional and digital libraries. JASIST **58**, 433–445 (2007)
2. Rowlands, I., Nicholas, D., Jamali, H.R., Huntington, P.: What do faculty and students really think about e-books? Aslib Proc. **59**, 489–511 (2007)
3. Kuhlthau, C.C.: Inside the search process: information seeking from the user's perspective. JASIST **42**, 361–371 (1999)
4. Marchionini, G.: Information Seeking in Electronic Environments. Cambridge University Press, Cambridge (1995)
5. McKay, D.: Gotta keep 'em separated: why the single search box may not be right for libraries. In: CHINZ 2011, pp. 109–112. ACM (2011)

6. Stelmaszewska, H., Blandford, A.: From physical to digital: a case study of computer scientists' behaviour in physical libraries. IJDL **4**, 82–92 (2004)
7. Hinze, A., McKay, D., Vanderschantz, N., Timpany, C., Cunningham, S.J.: Book selection behavior in the physical library: implications for ebook collections. In: JCDL 2012, pp. 305–314. ACM (2012)
8. Ooi, K.: How Adult Fiction Readers Select Fiction Books in Public Libraries: A Study of Information Seeking in Context. Master of Library and Information Studies. Victoria University of Wellington, Wellington (2008)
9. McKay, D., Smith, W., Chang, S.: Lend me some sugar: borrowing rates of neighbouring books as evidence for browsing. In: DL 2014, pp. 145–154. IEEE (2014)
10. Borgman, C.L., Hirsh, S.G., Walter, V.A., Gallagher, A.L.: Children's searching behavior on browsing and keyword online catalogs: the science library catalog project. JASIS **46**, 663–684 (1995)
11. Lau, E.P., Goh, D.H.-L.: In search of query patterns: a case study of a university OPAC. Inform Process Manag **42**, 1316–1329 (2006)
12. McKay, D., Hinze, A., Heese, R., Vanderschantz, N., Timpany, C., Cunningham, S.J.: An exploration of ebook selection behavior in academic library collections. In: Zaphiris, P., Buchanan, G., Rasmussen, E., Loizides, F. (eds.) TPDL 2012. LNCS, vol. 7489, pp. 13–24. Springer, Heidelberg (2012)
13. Makri, S., Blandford, A., Woods, M., Sharples, S., Maxwell, D.: "Making my own luck": Serendipity strategies and how to support them in digital information environments. JASIST **65**, 2179–2194 (2014)
14. Kleiner, E., Rädle, R., Reiterer, H.: Blended shelf: reality-based presentation and exploration of library collections. In: CHI 2013, pp. 577–582. ACM (2013)
15. Bates, M.J.: The design of browsing and berrypicking techniques for the online search interface. Online Inform Rev. **13**, 407–424 (1993)
16. Bates, M.J.: What is browsing–really? A model drawing from behavioural science research. Inform Res **12**, 330 (2007)
17. Oksanen, S., Vakkari, P.: Emphasis on examining results in fiction searches contributes to finding good novels. In: JCDL 2012, pp. 199–202. ACM, 2232855 (2012)
18. Kules, B., Capra, R., Banta, M., Sierra, T.: What do exploratory searchers look at in a faceted search interface? In: JCDL 2009, pp. 313–322. ACM (2009)
19. Borgman, C.L.: Why are online catalogs *still* hard to use? JASIS **47**, 493–503 (1996)
20. McKay, D., Conyers, B.: Where the streets have no name: how library users get lost in the stacks. In: CHINZ 2010, pp. 77–80. ACM (2010)
21. McKay, D., Shukla, P., Hunt, R., Cunningham, S.J.: Enhanced browsing in digital libraries: three new approaches to browsing in Greenstone. IJDL **4**, 283–297 (2004)
22. Thudt, A., Hinrichs, U., Carpendale, S.: The bohemian bookshelf: supporting serendipitous book discoveries through information visualization. In: CHI 2012, pp. 1461–1470. ACM (2012)
23. Pearce, J., Chang, S.: Exploration without keywords: the bookfish case. In: OzCHI 2014, pp. 76–79. ACM (2014)
24. McKay, D., Buchanan, G.: One of these things is not like the others: how users search different information resources. In: Gradmann, S., Borri, F., Meghini, C., Schuldt, H. (eds.) TPDL 2011. LNCS, vol. 6966, pp. 260–271. Springer, Heidelberg (2011)
25. Marshall, C.C.: Reading and Writing the Electronic Book. Morgan & Claypool, Chapel Hill (2010)
26. Saarinen, K., Vakkari, P.: A sign of a good book: readers' methods of accessing fiction in the public library. J. Doc. **69**, 736–754 (2013)

27. Rowlands, I., Nicholas, D.: Understanding information behaviour: how do students and faculty find books? J. Acad. Libr. **34**, 3–15 (2008)

28. Tenopir, C., King, D.W., Edwards, S., Wu, L.: Electronic journals and changes in scholarly article seeking and reading patterns. Aslib Proc. **61**, 5–32 (2009)

29. Reutzel, D.R., Gali, K.: The art of children's book selection: a labyrinth unexplored. Read. Psychol. **19**, 3–50 (1998)

30. Moore, P.: Information problem solving: a wider view of library skills. Contemp. Educ. Psychol. **20**, 1–31 (1995)

31. Svenonius, E.: The Intellectual Foundation of Information Organization. MIT Press, Boston (2000)

32. Hancock-Beaulieu, M.: Evaluating the impact of an online library catalogue on subject searching at the catalogue and at the shelves. J. Doc. **46**, 318–338 (1993)

33. Losee, R.M.: The relative shelf location of circulated books: a study of classification, users, and browsing. Libr. Resour. Tech. Serv. **37**, 197–209 (1993)

34. O'Neill, E.T., Gammon, J.A.: Consortial book circulation patterns: the OCLC-OhioLINK study. C&RL **75**, 791–807 (2014)

35. Sanderson, M., Dumais, S.T.: Examining repetition in user search behavior. In: Amati, G., Carpineto, C., Romano, G. (eds.) ECiR 2007. LNCS, vol. 4425, pp. 597–604. Springer, Heidelberg (2007)

36. Jansen, B.J., Spink, A.: How are we searching the World Wide Web? A comparison of nine search engine transaction logs. Inform. Process. Manag. **42**, 248–263 (2006)

37. Baeza-Yates, R., Ribeiro-Neto, B.: Modern information retrieval. ACM Press, New York (1999)

38. McKay, D., Buchanan, G.: Boxing clever: how searchers use and adapt to a one-box library search. In: OZCHI 2013, pp. 497–506. ACM, 2541031 (2013)

39. Herlocker, J.L., Konstan, J.A., Terveen, L.G., Riedl, J.T.: Evaluating collaborative filtering recommender systems. ToIS **22**, 5–53 (2004)

40. Fried Foster, N., Gibbons, S.: Studying Students: The Undergraduate Research Project at the University of Rochester. Association of College and Research Libraries, Rochester (2007)

41. Saarti, J.: Feeding with the spoon, or the effects of shelf classification of fiction on the loaning of fiction. Inform. Serv. Use. **17**, 159 (1997)

The Influence and Interrelationships Among Chinese Library and Information Science Journals in Taiwan

Ya-Ning Chen[1](✉), Hui-Hsin Yeh[2], and Po-Jui Lai[2]

[1] Department of Information and Library Science, Tamkang University,
New Taipei City, Taiwan
arthur9861@gmail.com
[2] Shou Ray Information Service Co., Ltd., New Taipei City, Taiwan
{vicky,kevin}@sris.com.tw

Abstract. This study aims to investigate the influences and interrelationships between journals of library and information science in Taiwan, in terms of information flow. Eleven Chinese journals and 2,031 articles during from 2001 to 2012 have been selected as subject and an 11×11 matrix was generated to conduct journal-to-journal analysis. Several bibliometric indicators proposed by Xhignesse and Osgood [16] have been examined, including indegree, outdegree, sending-receiving and self-feeding ratios. Degree and betweenness centrality of social network analysis have also employed to investigate the central and brokerage position of eleven journals in terms of network structure. In addition to overall structured analysis of twelve years, this study has furthering separated 12 years into three individual periods of four years to conduct a both synchronic and diachronic journal-to-journal citation analysis. Finally, this study discussed the implications and limitation of this study for Chinese journals of library and information science in Taiwan.

Keywords: Information flow · Journal network · Journal citation analysis

1 Introduction

Scholarly journals and articles are published as a primary means to disseminate the research results and progress of theoretical and practical advancements in most fields. Thus researchers and professionals have used journals to exchange ideas and learn the state-of-the-art information to support their studies and works. Seamlessly scholarly journals and their articles have built up a communication network for researchers and professionals.

Citation analysis is a useful approach to draw an image of scholarly communication network based on intercitations between journals. Since a seminal work published by Cason and Lubotsky [2], journal-to-journal citation study has often been used to investigate the information influences, influenced, interrelationships and their patterns between journals in a specific field, including accounting [11, 15], economy [6], communication [14], distance education [17], genetics [10], information systems [13], Internet research [12], management [3], psychology [4, 8, 16], social work [1],

© Springer International Publishing Switzerland 2015
S. Kapidakis et al. (Eds.): TPDL 2015, LNCS 9316, pp. 295–305, 2015.
DOI: 10.1007/978-3-319-24592-8_22

sociology [5], and water pollution and humanization of labor [7]. To our best knowledge, there is nearly no study has examined the influences and interrelationships of information flow between journals through by synchronic and diachronic citation simultaneously. Therefore, this study will adopt a journal-to-journal citation to analyze the influences and interrelationships embedded in 11 Chinese library and information science (hereafter LIS) journals in Taiwan.

2 Literature Review

In a study of journal network, Xhignesse and Osgood [16] have used concepts of receiver and source to examine the information exchange between psychological journals in 1950 and 1960. They have also offered many fundamental bibliometric indicators to measure the information flow of journal network and inspired many studies of various disciplines, including accounting [11, 15], economy [6], education [17], and genetics [10]. On the other hand, clustering and structural equivalence of social network analysis (hereafter SNA) have been employed to categorize specific field's journals into various subject-based groups and then investigated the interrelationships between categorized groups for selected journals. Such studies include demography [9], psychology [4, 8], social work [1], and sociology [5]. In addition to clustering and structural equivalence, centrality and cliques have also used by other studies to examine the journal's prestige and position of specific discipline within network structure, such as accounting [15] and communication [14]. Furthermore, the aforementioned studies have selected either of synchronic or diachronic journal-to-journal citation to make a comparison between periods or an overview image of a longitudinal period. Therefore, it needs a study to conduct a journal-to-journal citation analysis to examine more in-depth interactions and characteristics between journals for scholarly communication.

3 Methodology

This study selected Chinese LIS journals in Taiwan as subject, because no study has focused on Chinese journal-to-journal citation analysis for LIS in Taiwan. There are about 20 LIS journals in Taiwan, but not all of them were qualified for this study. Few LIS journals in Taiwan have renamed their journal titles to continue in publishing articles, and this study regarded the above journals as one rather than two different journals. Next, some LIS journals in Taiwan are covered by a research report for LIS journals ranking, and indexed by Taiwan Academic Citation Index (http://award. libraryandbook.net/taci/) and Taiwan Citation Index: Humanities and Social Sciences (http://tci.ncl.edu.tw). Thus some LIS journals were excluded out form this study if they are not covered by the aforementioned report and citation indexes. Third, only scholarly articles were eligibly selected as subject, and reviews, interviews, visiting reports and news of library activities were excluded out from this study. Totally 11 Chinese LIS journals and 2,031 articles (see Table 1) during 2001 and 2012 were selected as subject. As a result, an 11 × 11 matrix was generated as a basis for journal-to-journal citation analysis. Fourth, this study also conducted verification to correct errors of bibliographic description of cited references, including journal titles and publishing year. Fifth, although cited journal appears more than two times in article's reference of citing journal, this study has counted them as once rather than as many times as they were cited (see Table 2). Sixth, in addition to 12 years (2001 −2012), this study has also separated 12 years into three individual periods of 4 years to take an in-depth comparative study for journal-to-journal citation analysis. Lastly, this study employed indegree, outdegree, sending-receiving and self-feeding ratios, as well as centrality (including degree and between) of SNA by using UCINet 6 software to examine the influences and interrelationships between Chinese LIS journals in Taiwan in terms of information exchange.

Table 1. Objects of 11 Chinese LIS Journals and Articles for Analysis

Journal Name	JRN Code	Frequency	Publication Year (20-)												
			01	02	03	04	05	06	07	08	09	10	11	12	Total
Journal of Library and Information Science Research	JRN01	Semi-annual	20	24	28	25	41	6	12	10	10	10	10	10	206
Journal of Library and Information Science	JRN02	Semi-annual	19	14	16	24	18	16	20	14	15	13	18	12	199
Journal of Educational Media & Library Sciences	JRN03	Quarterly	29	32	40	39	31	24	23	21	20	20	21	20	320
Journal of Library and Information Studies	JRN04	Semi-annual	9	7	16	18	12	5	5	5	5	12	12	12	118
National Central Library Bulletin	JRN05	Semi-annual	20	20	21	20	18	12	13	14	11	12	12	13	186
University Library Journal	JRN06	Semi-annual	20	19	18	18	18	17	16	16	14	15	17	16	204
Bulletin of Library and Information Science	JRN07	Semi-annual	31	33	25	28	26	27	26	27	23	20	19	11	296
Journal of Cultural Enterprise and Management	JRN08	Semi-annual	0	0	0	4	0	7	5	5	11	9	14	12	67
Information Management for Buddhist Libraries	JRN09	Semi-annual	23	24	21	23	21	20	18	27	20	9	16	17	239
Journal of Information, Communication, and Library Science	JRN10	Quarterly	40	33	10	6	7	11	0	0	0	0	0	0	107
National Cheng Kung University Library Journal	JRN11	Annual	16	11	12	5	9	8	7	6	4	4	2	5	89

4 Analysis

4.1 Indegree Ratio

From 2001 to 2012 (see Table 3), the highest total number of receiving citations (indegree, self-citations excluded) is JRN01 (289), followed by JRN06 (187), JRN03 (177), JRN07 (169) and JRN10 (161). According to standardized citation values, papers published in JRN10 generated more citations per paper (1.50), followed by JRN01 (1.4), JRN06 (0.92), JRN02 (0.66) and JRN04 (0.64). In terms of evolutional change of three periods, JRN01, JRN02, JRN03, JRN05 and JRN11 have increased in receiving citations across over periods. JRN04 and JRN10 have increased in receiving citations from period 1 to 2, and then moved down from period 2 to 3. In contrary,

JRN06 has declined in receiving citations from period 1 to 2, and then moved up from period 2 to 3.

4.2   Outdegree Ratio

In 12 years, the highest total number of sending citations (outdegree, self-citations excluded) is JRN07 (299), followed by JRN06 (226), JRN03 (207), JRN01 (172) and JRN05 (148) (see Table 3). Based on standardized citation values, papers published by JRN06 (1.11) sending most citations per paper, followed by JRN07 (1.01), JRN04 (0.91), JRN01 (0.83) and JRN05 (0.80). In terms of evolutional change of three periods, JRN03 and JRN07 have ascended in sending citations across over periods. JRN01, JRN02 and JRN04 have decreased in sending citation from period 1 to 2, and then went up from period 2 to 3. Contrarily JRN05 and JRN08 have increased in sending citation from period 1 to 2, and then moved down from period 2 to 3. JRN06, JRN09, JRN10 and JRN11 have descended in sending citations over 3 periods.

4.3   Sending-Receiving Ratio

According to Xhignesse and Osgood [16], "journals with high ratios are storers, and those with low ratios are feeders of information in the network". Eagly [6] has furthering proposed value of 1 as threshold to distinguish between storers and feeders, and defined that a storer is a seldom cited by others, whereas a feeder is often cited by others in terms of journal network. During from 2001 to 2012, JRN01, JRN02, JRN09, JRN10 and JRN11 are feeders, and JRN03-07 are storers with exclusion of self-citations (see Table 3). During evolutional change over three periods with exclusion of self-citations, JRN01 and JRN10 were feeders, and JRN03, JRN06 and JRN07 were storer purely. JRN02, JRN09 and JRN11 have gradually changed from storer into feeder. JRN04 and JRN05 have the preference to play a role as a storer than a feeder.

4.4   Self-Feeding Ratio

Based on Eagly's definition [6], journals with higher values than average values are highly specialized. Within 12 years, the average self-feeding ratio is 0.3 in journal network. JRN07, JRN08 and JRN09 were highly specialized and JRN03 and JRN05 were close to highly specialized (see Table 3). On the other hand, the average self-feeding ratios for three periods are 0.20, 0.189 and 0.31 respectively. Three (i.e., JRN03, JRN05 and JRN09) were highly specialized journals across over three periods. Detailed information of highly specialized journals for three periods was listed as below:

- 2001−2004: JRN03, JRN05, JRN07 and JRN09
- 2005−2008: JRN03−07 and JRN09
- 2009−2012: JRN03, JRN05 and JRN08−09

Table 2. Relationship between Citing Journal (Vertical Axis) and Cited Journal (Horizontal Axis)

	JRN01				JRN02				JRN03				JRN04				JRN05				JRN06				JRN07			
Year (20-)	01–04	05–08	09–12	01–12	01–04	05–08	09–12	01–12	01–04	05–08	09–12	01–12	01–04	05–08	09–12	01–12	01–04	05–08	09–12	01–12	01–04	05–08	09–12	01–12	01–04	05–08	09–12	01–12
JRN01	26	12	17	55	9	3	4	16	7	5	5	17	4	7	8	19	11	4	5	20	13	6	5	24	17	5	12	34
JRN02	2	9	10	21	4	2	11	17	8	8	6	22	3	3	7	13	4	5	3	12	6	6	7	19	6	1	8	15
JRN03	16	15	14	45	4	12	15	31	21	22	29	72	0	6	6	12	5	4	4	13	9	7	12	28	10	5	12	27
JRN04	6	3	15	24	4	3	4	11	7	4	11	22	4	6	10	20	3	0	3	6	10	1	6	17	4	1	8	13
JRN05	18	15	8	41	6	7	6	19	8	7	5	20	1	1	1	3	9	9	21	39	5	10	2	17	8	8	4	20
JRN06	24	22	8	54	7	6	6	19	15	11	18	44	9	2	0	11	3	8	7	18	13	15	15	43	17	15	13	45
JRN07	33	28	16	77	6	11	9	26	13	16	14	43	3	3	8	14	6	14	8	28	23	17	18	58	48	22	17	87
JRN08	0	0	2	2	0	0	0	0	0	0	1	1	0	0	0	0	0	0	0	0	1	0	1	0	0	0	0	0
JRN09	10	0	0	10	3	2	0	5	2	0	0	2	1	0	0	1	0	1	1	2	6	0	0	6	4	0	0	4
JRN10	7	0	0	7	2	0	0	2	2	0	0	2	2	0	0	2	0	0	0	0	6	1	0	7	2	1	0	3
JRN11	7	1	0	8	0	1	2	3	1	2	1	4	0	0	1	1	0	0	1	1	8	2	0	10	7	1	0	8
Total^b	149	105	90	344	45	47	57	149	84	75	90	249	27	28	41	96	42	45	52	139	99	66	65	230	123	59	74	256
Total^b*	123	93	73	289	41	45	46	132	63	53	61	177	23	22	31	76	33	36	31	100	86	51	50	187	75	37	57	169

	JRN08				JRN09				JRN10				JRN11				Total^a				Total^a*			
Year (20-)	01–04	05–08	09–12	01–12	01–04	05–08	09–12	01–12	01–04	05–08	09–12	01–12	01–04	05–08	09–12	01–12	01–04	05–08	09–12	01–12	01–04	05–08	09–12	01–12
JRN01	0	0	0	0	1	1	3	5	23	5	0	28	9	0	0	9	120	48	59	227	94	36	42	172
JRN02	0	0	0	0	2	2	1	5	10	2	1	13	1	0	1	2	46	38	55	139	42	36	44	122
JRN03	0	0	0	0	2	1	2	5	19	7	4	30	3	8	5	16	89	87	103	279	68	65	74	207
JRN04	0	0	0	0	0	0	0	0	6	1	4	11	2	1	0	3	46	20	61	127	42	14	51	107
JRN05	0	0	0	0	6	4	1	11	8	5	1	14	1	1	1	3	70	67	50	187	61	58	29	148
JRN06	0	0	0	0	1	2	3	6	18	4	1	23	1	3	2	6	108	88	73	269	95	73	58	226
JRN07	0	0	0	0	2	5	0	7	14	9	7	30	4	8	4	16	152	133	101	386	104	111	84	299
JRN08	0	0	1	1	0	1	0	1	0	0	0	0	0	0	0	0	0	2	4	6	0	2	3	5
JRN09	0	0	0	0	25	26	17	68	9	1	0	10	0	0	0	0	60	30	18	108	35	4	1	40
JRN10	0	0	0	0	0	0	0	0	11	0	0	11	0	0	0	0	32	2	0	34	21	2	0	23
JRN11	0	0	0	0	3	0	0	3	1	1	0	2	3	0	0	3	31	8	4	43	28	8	4	40
Total	0	0	1	1	42	42	27	111	119	35	18	172	24	21	13	58	754	523	528	1805				
Total*	0	0	0	0	17	16	10	43	108	35	18	161	21	21	13	55					590	409	390	1389

a = receiving citations; b = sending citations; * = self-citations excluded.

4.5 Degree and Betweenness Centrality

In terms of normalized indegree centrality of SNA, JRN01 has played a role in being most cited by other journals in LIS journal network, followed by JRN06, JRN03, JRN07 and JRN10 (see Table 4). In terms of evolutional change of three periods, JRN02 have ascended, and JRN01, JRN06, JRN09, JRN10 and JRN11 have descended in indegree centrality across over periods. JRN05 has increased in indegree centrality from period 1 to 2, and then declined from period 2 to 3. JRN03, JRN04, and JRN07 has got contrary pattern to JRN05. JRN08 got 0 of indegree centrality (Fig. 1).

Table 3. Indegree, Outdegree, Sending-receiving and Self-feeding Ratios

Journal	Year (20-)	Indegree	Outdegree	Indegree[a]	Outdegree[a]	Sending-receiving ratio	Self-feeding ratio
JRN01	01-04	123	94	1.27	0.97	0.76	0.17
	05-08	93	36	1.35	0.52	0.39	0.11
	09-12	73	42	1.83	1.05	0.58	0.19
	01-12	289	172	1.40	0.83	0.60	0.16
JRN02	01-04	41	42	0.56	0.58	1.02	0.09
	05-08	45	36	0.66	0.53	0.80	0.04
	09-12	46	44	0.79	0.76	0.96	0.19
	01-12	132	122	0.66	0.61	0.92	0.11
JRN03	01-04	63	68	0.45	0.49	1.08	0.25
	05-08	53	65	0.54	0.66	1.23	0.29
	09-12	61	74	0.75	0.91	1.21	0.32
	01-12	177	207	0.55	0.65	1.17	0.29
JRN04	01-04	23	42	0.46	0.84	1.83	0.15
	05-08	22	14	0.81	0.52	0.64	0.21
	09-12	31	51	0.76	1.24	1.65	0.24
	01-12	76	107	0.64	0.91	1.41	0.21
JRN05	01-04	33	61	0.41	0.75	1.85	0.21
	05-08	36	58	0.63	1.02	1.61	0.20
	09-12	31	29	0.65	0.60	0.94	0.40
	01-12	100	148	0.54	0.80	1.48	0.28
JRN06	01-04	86	95	1.15	1.27	1.10	0.13
	05-08	51	73	0.76	1.09	1.43	0.23
	09-12	50	58	0.81	0.94	1.16	0.23
	01-12	187	226	0.92	1.11	1.21	0.19
JRN07	01-04	75	104	0.64	0.89	1.39	0.39
	05-08	37	111	0.35	1.05	3.00	0.37
	09-12	57	84	0.78	1.15	1.47	0.23
	01-12	169	299	0.57	1.01	1.77	0.34
JRN08	01-04	0	0	0.00	0.00	0.00	0.00
	05-08	0	2	0.00	0.12	0.00	0.00
	09-12	0	3	0.00	0.07	0.00	1.00
	01-12	0	5	0.00	0.07	0.00	1.00
JRN09	01-04	17	35	0.19	0.38	2.06	0.60
	05-08	16	4	0.19	0.05	0.25	0.62
	09-12	10	1	0.16	0.02	0.10	0.63
	01-12	43	40	0.18	0.17	0.93	0.61
JRN10	01-04	108	21	1.21	0.24	0.19	0.09
	05-08	35	2	1.94	0.11	0.06	0.00
	09-12	18	0	0.00	0.00	0.00	0.00

(Continued)

Table 3. (*Continued*)

Journal	Year (20-)	Indegree	Outdegree	Indegree[a]	Outdegree[a]	Sending-receiving ratio	Self-feeding ratio
		161	23	1.50	0.21	0.14	0.06
JRN11	01-04	21	28	0.48	0.64	1.33	0.13
	05-08	21	8	0.70	0.27	0.38	0.00
	09-12	13	4	0.87	0.27	0.31	0.00
	01-12	55	40	0.62	0.45	0.73	0.05

[a]Standardized values based on citations per paper

Table 4. Normalized Degree and Betweenness Centrality of SNA

Journal	Year	Indegree	Outdegree	Betweenness
JRN01	01-04	12.300	9.400	1.624
	05-08	9.300	3.600	1.111
	09-12	7.300	4.200	3.963
	01-12	28.900	17.200	2.669
JRN02	01-04	4.100	4.200	1.280
	05-08	4.500	3.600	3.591
	09-12	4.600	4.400	3.667
	01-12	13.200	12.200	0.910
JRN03	01-04	6.300	6.800	1.280
	05-08	5.300	6.500	1.556
	09-12	6.100	7.400	9.222
	01-12	17.700	20.700	2.669
JRN04	01-04	2.300	4.200	0.741
	05-08	2.200	1.400	0.444
	09-12	3.100	5.100	2.037
	01-12	7.600	10.700	0.529
JRN05	01-04	3.300	6.100	0.661
	05-08	3.600	5.800	3.963
	09-12	3.100	2.900	10.148
	01-12	10.000	14.800	0.317
JRN06	01-04	8.600	9.500	1.624
	05-08	5.100	7.300	12.667
	09-12	5.000	5.800	1.074
	01-12	18.700	22.600	2.669
JRN07	01-04	7.500	10.400	1.624
	05-08	3.700	11.100	5.444

(*Continued*)

Table 4. (*Continued*)

Journal	Year	Indegree	Outdegree	Betweenness
		5.700	8.400	0.815
	01-12	16.900	29.900	0.910
JRN08	01-04	0.000	0.000	0.000
	05-08	0.000	0.200	0.000
	09-12	0.000	0.300	0.000
	01-12	0.000	0.500	0.000
JRN09	01-04	1.700	3.500	0.503
	05-08	1.600	0.400	1.667
	09-12	1.000	0.100	0.000
	01-12	4.300	4.000	1.389
JRN10	01-04	10.800	2.100	0.503
	05-08	3.500	0.200	0.741
	09-12	1.800	0.000	0.000
	01-12	16.100	2.300	0.000
JRN11	01-04	2.100	2.800	0.159
	05-08	2.100	0.800	0.000
	09-12	1.300	0.400	0.185
	01-12	5.500	4.000	0.159

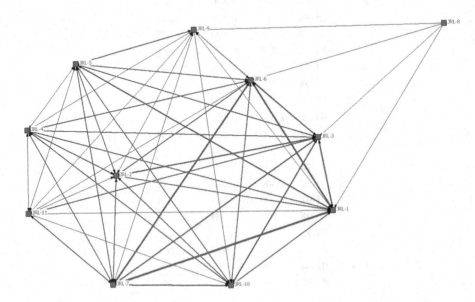

Fig. 1. Sciogram of Interrelationships among Chinese LIS Journals in Taiwan

Based on normalized outdegree centrality of SNA, JRN07 has played a role in most citing other LIS journals, followed by JRN06, JRN03, JRN01 and JRN05. In terms of evolutional change of three periods, JRN05, JRN06, JRN09, JRN10 and JRN11 have descended, whereas JNR08 have ascended in outdegree centrality. JRN01-04 have declined in outdegree centrality from period 1 to 2, and then gone up from period 2 to 3.

On the other hand, in terms of normalized betweenness centrality of SNA, JRN01, JRN03 and JRN06 have played an active role in bridging intercitation between journals, followed by JRN09, JRN02 and JRN07. In terms of evolutional change of three periods, JRN02, JRN03 and JRN05 have ascended in betweenness centrality of SNA across over periods. JRN01, JRN04 and JRN11 have declined in betweenness centrality from period 1 to 2, and then move up from period 2 to 3. Contrarily JRN06, JRN07, JRN09 and JRN10 have increased in betweenness centrality from period 1 to 2, and then moved down from period 2 to 3. JRN08 got 0 of betweenness centrality both of overall of 12 years and 3 individual periods.

5 Discussion

According to top-5 standardized citation values of receiving citations, JRN10 got a significant number of citations from nine other journals, followed by JRN01 and JRN06 from ten journals, and JRN02 and JRN04 from nine journals. These five journals have played an important role in receiving citations from LIS journals in Taiwan. Furthermore, JRN01 and JRN02 have also kept steady growth in receiving citations from other LIS journals across over periods. On the other hand, based on top-5 standardized citation values of sending citations, JRN06 sent a significant number of citations to ten other journals, followed by JRN07 to 10 journals, JRN04 to eight journals, and JRN01 and JRN05 to nine journals. Within evolution of three periods, JRN06 has descended in sending citations, whereas JRN07 has ascended in sending citations to other journals.

In terms of feeders and storer of citations, JRN01 and JRN10 are truly feeders of Taiwan LIS journal network, based on standardized value of indegree is larger than that of outdegree and sending-receiving ratio in terms of overall and periodical analysis. However, the reasons for JRN01 and JRN10 to become a feeder of LIS journal in Taiwan are different. The reason for JRN01 may lie in that JRN01 is one of the oldest and most established LIS journals in Taiwan, and has been selected as one of LIS journals covered by Taiwan Social Citation Index since 2010 in this field. JRN10 has continued in receiving citations from other Taiwan LIS journals over years although it had ceased publication. It reveals that articles published by JRN10 are still valuable to last its influence in attracting many articles' citations from other LIS journals. On the other hand, JRN03, JRN05 and JRN09 are categorized as more specialized journal among eleven LIS journals in Taiwan and the reasons are related to their editorial policy of topic coverage. In addition to LIS, JRN03, JRN05 and JRN09 also cover topics in a specialized fragment of educational technology, sinology and Buddhism studies respectively. Fragmentation of non-LIS specialized topic coverage may lessen citations from other LIS-oriented journals in Taiwan. The results of self-feeding and sending-receiving ratios are a useful reference for journal's policy of subject coverage.

In terms of SNA, degree centrality stands for the prestige of a journal based on citation. JRN01, JRN06, JRN03, JRN07 and JRN10 have played a star role in LIS journal network based on the top-5 highest overall indegree centrality. JRN01, JRN06 and JRN10 gradually have declined their prestige, whereas JRN03 and JRN07 have risen up their prestige according to SNA's indegree centrality over three periods. JRN08, JRN09 and JRN11 are more isolated from other LIS journals in Taiwan. It reveals that some journals (e.g., JRN02 and JRN04) received higher citations, but they have not played a central role in LIS journal network; whereas some journals (e.g., JRN03 and JRN07) have not received higher citations, but they have played a central role in the flow of LIS journal citation. Furthermore, JRN07, JRN06, JRN03, JRN01 and JRN05 have also played a central role in sending citations to other LIS journals based on top-5 outdegree centrality. On the other hand, in terms of SNA's betweenness centrality, JRN01, JRN06, JRN03, JRN09, JRN02 and JRN07 have played a top-5 bridging role to facilitate intercitation between other LIS journals in Taiwan, and JRN08 got zero of betweenness centrality during from 2001 to 2012. In addition to receiving citations, results of SNA centrality also reveal the significance of journal role in LIS journal network, and can be a reference for editor's journal management and author's paper submission.

6 Conclusion

This is an in-depth study on journal-to-journal citation analysis. This study has not only employed typical bibliometric indexes including indegree, outdegree, and sending-receiving and self-feeding ratios, but also adopted degree and betweenness centrality of SNA to investigate the information flow between Chinese LIS journals in Taiwan, both of overall of twelve years and three individual periods of four years citation analysis. Therefore, this study offers a snapshot of connected interrelationships between journals that can be assessed as one indicator of prestige and importance of scholarly journals in LIS field in Taiwan. Furthermore, the citation patterns which embedded in Chinese LIS journals has identified by this study can serve an informative guide to the study, and a useful reference for editor's journal business management and author's paper submission of LIS field in Taiwan.

A limitation of this study is the counting way for journal-to-journal citations and target Chinese LIS journals. The way to count citations as once may lead to different from that counting as many times as they are either of cited or citing. Therefore, this study will conduct another journal-to-journal citation analysis to examine the distinction between two different counting ways for Chinese LIS journals in Taiwan in the near future.

References

1. Baker, D.R.: A structural analysis of social work journal network. J. Soc. Serv. Res. 15(3–4), 153–168 (1992)
2. Cason, H., Lubotsky, M.: The influence and dependence of psychological journals on each other. Psychol. Bull. 33(2), 95–103 (1936)

3. Danell, R.: Stratification among journals in management research: a bibliometric study of interaction between european and american journals. Scientometrics **49**(1), 23–38 (2000)
4. Doreian, P.: Structural equivalence in a psychology journal network. J. Am. Soc. Inform. Sci. **36**(6), 411–417 (1985)
5. Doreian, P., Fararo, T.J.: Structural equivalence in a journal network. J. Am. Soc. Inform. Sci. **36**(1), 28–37 (1985)
6. Eagly, R.V.: Economics journals as a communication network. J. Econ. Lit. **13**(3), 878–888 (1975)
7. Leydesdorff, L.: The development of frames of references. Scientometrics **9**(3–4), 103–125 (1986)
8. Liu, Z.: Scholarly communication in educational psychology: a journal citation analysis. Collection Building **26**(4), 112–118 (2007)
9. Liu, Z., Wang, C.: Mapping interdisciplinarity in demography: a journal network analysis. J. Inform. Sci. **31**(4), 308–316 (2005)
10. McCain, K.W.: Core journal networks and cocitation maps: new bibliometric tools for serials research and management. Libr. Q. **61**(3), 311–336 (1991)
11. McRae, T.W.: A citational analysis of the accounting information network. J. Acc. Res. **12**(1), 80–92 (1974)
12. Peng, T.Q., Wang, Z.-Z.: Network closure, brokerage, and structural influence of journals: a longitudinal study of journal citation network in internet research (2000-2010). Scientometrics **97**(3), 675–693 (2013)
13. Polites, G., Watson, R.T.: Using social network analysis to analyze relationships among IS journals. Journal of the Association for Information Systems **10**(8), 595–636 (2009)
14. Rice, R.E., Borgman, C.L., Reeves, B.: Citation networks of communication journals, 1977 – 1985: cliques and positions, citations made and citation received. Hum. Commun. Res. **15**(2), 256–283 (1988)
15. Wakefield, R.: Networks of accounting research: a citation-based structural and network analysis. Br. Account. Rev. **40**(3), 228–244 (2008)
16. Xhignesse, L.V., Osgood, C.E.: Bibliographical citation characteristics of the psychological journal network in 1950 and in 1960. Am. Psychol. **22**(9), 778–791 (1967)
17. Zawacki-Richter, O., Anderson, T.: The geography of distance education: bibliographic characteristics of a journal network. Distance Educ. **32**(3), 441–456 (2011)

Poster and Demo Papers

An Experimental Evaluation of Collaborative Search Result Division Strategies

Thilo Böhm[1]([✉]), Claus-Peter Klas[2], and Matthias Hemmje[1]

[1] University of Hagen, 58084 Hagen, Germany
{thilo.boehm,matthias.hemmje}@fernuni-hagen.de
[2] GESIS, 50667 Köln, Germany
claus-peter.klas@gesis.org

Abstract. Collaboration during information retrieval has been identified by many empirical studies as a common pattern of teams in everyday work, e.g. [5]. This collaboration is characterized by two or more individuals, who set out together to resolve a shared information need [4]. In this paper, we present an experimental evaluation of different search result division strategies in simulated collaborative search tasks. We compare our proposed approach, which defines optimum collaboration strategies as integer linear problem, with proven principles like, e.g., the PRP.

1 Introduction

When team members collaborate to satisfy the same information need, they often use the same or very similar query terms [3], which is likely to result in highly similar ranked search result-sets returned by the Digital Library. In consequence, their retrieved document-sets can overlap, which may lead to less coverage and less productivity due redundant work. Optimizing the individual's contribution calls for ranking the available documents with respect to the diversity of the team and allocating an appropriate result-set to each single team member.

2 Approach

We propose a criterion (or rule) that helps coordinating team work in a flexible and adjustable manner. To this aim, we introduce a mapping that allocates a specific sub-set of documents $D_i \subset D$ to a team member $\tau_i \in T$ for examination, with D being the union of (potentially overlapping) retrieved document-sets and $M = |D|$. The mapping $a : T \times D \rightarrow \{0,1\}$ allocates document $d_j \in D$ to team member τ_i, if the tuple (i,j) is mapped to 1, otherwise it is mapped to 0.

Each team member is assumed to have only a limited capacity to accept and assess documents, which we denote *assessment capacity* K_i. Let $\rho_{i,j}$ be the user-specific probability of a document d_j being relevant for a user τ_i. The mapping a should allocate sub-sets of documents to the most suited individuals, i.e. documents relevant for the team should be maximize and redundant work minimized, which is described by the following integer linear program (**ILP**):

© Springer International Publishing Switzerland 2015
S. Kapidakis et al. (Eds.): TPDL 2015, LNCS 9316, pp. 309–312, 2015.
DOI: 10.1007/978-3-319-24592-8_23

$$\max \sum_{i=1}^{N} \sum_{j=1}^{M} a_{i,j} \rho_{i,j}$$

$$\text{subject to} \sum_{i=1}^{N} a_{i,j} \leq 1, \forall j \text{ and } \sum_{j=1}^{M} a_{i,j} \leq K_i, \forall i \tag{1}$$

In this criterion, the mapping $a_{i,j}$ represents the unknowns of the ILP to be determined, e.g. using a numeric solver, and $\rho_{i,j}$ represents the user-specific probabilities of relevance of a given document d_j and team member τ_i.

3   Experimental Evaluation

Experimental Setup. Experiments were conducted using the OHSUMED collection, from which we selected topics that have at least 20 relevant documents assigned that have a topic description consisting of at least five terms. This ensured that there were enough relevant documents for examining search result division, i.e. if we only had a few of relevant documents per topic, then it is likely that only one user could accomplish the search task, which would not be particularly interesting. Because we focused on recall-oriented search tasks, we chose the overall recall of the team as measure. We used Apache Lucene as search engine and lp_solve as solver for integer linear programs.

We chose the team size of two as this has been empirically identified as typical size for collaborative search teams [7]. We employed the general simulation procedure used for evaluation by Shah et al. [8], which consisted of the following steps: (1) Each simulated user issued a query. (2) Documents of both query responses were merged into a shared result-set using the CombSUM algorithm [8]. (3) From this shared result-set, each simulated user were provided with a result-page for assessment consisting of K_i documents. Result-pages were extracted from the shared result-set either by applying one of the baseline procedures (see below) or by applying our optimum criterion (**ILP**). Assessment capacities considered were $K_i \in \{10, 20, 40, 60, 80, 100, 120, 140\}$.

We used the query generation approach examined in [1], which modeled a user who selects terms from an imagined ideal relevant document: $P(t|query) = (1 - \lambda)P(t|topic) + \lambda P(t)$. Sometimes, chosen terms will be on topic, $P(t|topic)$, while other times terms will be off topic, $P(t)$. The distribution $P(t|topic)$ describes the occurrences of terms in the ideal relevant document and represents the user's domain knowledge. For estimating $P(t|topic)$, we used the strategy called *Conditional* [1]. Considering that users with profound domain knowledge generally use a more specific vocabulary [9], we chose to simulate a difference in domain knowledge by assigning the following parings of λ parameters that we denote *slight* $\lambda \in \{0.3, 0.7\}$ and *strong* $\lambda \in \{0.1, 0.9\}$ knowledge difference.

Baselines. We used the following two baselines for our experiments. (1) To extract a result-page from the shared result-set, for each user, we re-ranked the whole shared result-set according to the user's knowledge and did a cut-off after

K_i documents. This simulated team members employing search tools designed for individual usage, i.e. results are optimized for one individual. We called this baseline **PRP**. The obvious disadvantage is that the result-pages created for team members are likely to have an overlap. (2) To avoid this overlap, we also used a baseline applied in [8]: The shared result-set is split using a Round Robin procedure, cut-off and re-ranked for the corresponding team member. We called this baseline **RR**. This baseline did both, avoiding redundancy and optimizing search results for an individual.

3.1 Results

Figure 1 depicts the results of our experiments. Each data point in the diagrams is an average of 38 samples (one per topic considered). The PRP baseline resulted in the lowest retrieval performance, which most likely resulted from the overlap between the result-pages. Both of the other approaches, ILP and RR, were successful at improving the performance over PRP in both settings, i.e. slight (Fig. 1a) and strong (Fig. 1b) knowledge difference. An expected result was the performance of the RR baseline that benefits from the avoided overlap between the result-pages. In Fig. 1a, we can see that with this simple division of labor strategy, the RR baseline performed nearly equivalently to the ILP approach. However, this is different in Fig. 1b, where the ILP approach clearly outperforms both baselines. In this setting, i.e. strong knowledge difference, the performance improvement of the ILP approach over the second best baseline (RR) is in a range of 3.85 % and 8.58 %.

Generally, in both considered team-compositions, the ILP approach allowed a team finding more or as many relevant documents, as both baselines did, i.e. it resulted in the best retrieval effectiveness. Additionally, as the diversity of the team increased (measured by the knowledge difference), the performance improvement of the ILP approach increased. Hence, our ILP approach leveraged the diversity within the team.

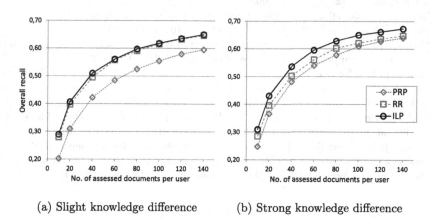

(a) Slight knowledge difference   (b) Strong knowledge difference

Fig. 1. Plot of overall recall as the number of assessed documents per user change

4 Conclusions

We demonstrated the application of an optimum criterion that can be used for determining optimum collaboration strategies in collaborative search sessions. Although parameters of our simulation were well justified, real world user behavior is more complex and our simulation represents an idealized collaborative session. While in practice, such ideal conditions may not apply, the conducted experiments indicate that our optimum criterion provides the potential for more (retrieval) effective collaboration in search sessions. We think of our approach as an enhancement for today's collaboration environments that consist of tools and interfaces designed for individual usage [2, 6]. Extending this environment with a service dedicated team support would provide the potential for an increase of effectivness as indicated by our simulation. Such a service might incorporate our technique in different ways, e.g. document sets estimated by our approach could be recommended to users or highlighted in results presented to users. Our future research will focus on utilizing the presented criterion to implement and evaluate services that enhance tools actually used in today's practice. The work presented in this paper represents a first step towards this goal.

References

1. Azzopardi, L.: Query side evaluation: an empirical analysis of effectiveness and effort. In: Proceedings of the 32nd International ACM SIGIR Conference on Research and Development in Information Retrieval, pp. 556–563. ACM (2009)
2. Böhm, T., Klas, C.-P., Hemmje, M.: Supporting collaborative information seeking and searching in distributed environments. In: Proceedings LWA 2013 Conference (2013)
3. Foley, C., Smeaton, A.F.: Division of labour and sharing of knowledge for synchronous collaborative information retrieval. Inf. Process. Manag. 46(6), 762–772 (2010)
4. Golovchinsky, G., Qvarfordt, P., Pickens, J.: Collaborative information seeking. IEEE Comput. 42(3), 47–51 (2009)
5. Hansen, P., Järvelin, K.: Collaborative information retrieval in an information-intensive domain. Inf. Process. Manag. 41(5), 1101–1119 (2005)
6. Morris, M.R.: Collaborative search revisited. In: Proceedings of the 2013 Conference on Computer Supported Cooperative Work, pp. 1181–1192. ACM (2013)
7. Morris, M.R., Horvitz, E.: Searchtogether: an interface for collaborative web search. In: Proceedings of the 20th Annual ACM Symposium on User Interface Software and Technology, pp. 3–12. ACM (2007)
8. Shah, C., Pickens, J., Golovchinsky, G.: Role-based results redistribution for collaborative information retrieval. Inf. Process. Manag. 46(6), 773–781 (2010)
9. White, R.W., Dumais, S.T., Teevan, J.: Characterizing the influence of domain expertise on web search behavior. In: Proceedings of the Second ACM International Conference on Web Search and Data Mining, pp. 132–141. ACM (2009)

State-of-the-Art of Open Access Textbooks and Their Implications for Information Provision

Ya-Ning Chen[✉]

Department of Information and Library Science, Tamkang University,
New Taipei City, Taiwan
arthur9861@gmail.com

Abstract. The skyrocketing price of textbooks is beyond students' affordability. Educators and teachers have taken an open access approach to textbooks to address this issue. This study aimed to investigate open access textbooks from the perspectives of source of provision, use license, mode of use, file format, and business model in terms of information provision. Eighteen use cases of open access textbooks were selected for systematic review by two researchers. Future suggestions for open access textbooks are also discussed.

Keywords: Open access textbook · Open textbook · Open source textbook

1 Introduction

In education, the price of textbooks has skyrocketed in the same way as that of electronic journals and books. The cost of textbooks is now beyond the reach of many students and has blocked them from course registration [7]. Owing to wider adoption and success of open access journals (OAJ) and books (OAB) in offering researchers access to scholarly information, educators and teachers have regarded open access textbooks (OAT) as a feasible solution for students to acquire textbooks. In terms of information provision, libraries are also concerned about how to deploy OAT to support teaching and learning.

2 Background

Since textbook prices have increased [5, 8], books are not only beyond students' budget [7] but also, based on a survey conducted by Florida Distance Learning Consortium in 2010, "23 % students responded that they had occasionally not registered for a particular course or section because of the high textbook costs" [4].

National governments and regional projects have noticed this issue. For instance, the U.S. Federal Government has taken legislative action by implementing the H.R.4575 Open College Textbook Act of 2010 and the S.1704/H.R. 3583 The Affordability College Textbook Act. Some U.S. states have also taken similar action, including Washington State, Ohio State and Texas State. In Poland, the Polish Ministry

© Springer International Publishing Switzerland 2015
S. Kapidakis et al. (Eds.): TPDL 2015, LNCS 9316, pp. 313–316, 2015.
DOI: 10.1007/978-3-319-24592-8_24

of Education has funded the "Digital School" program to provide access to free and open e-textbooks [2]. The ProjectLATIn, a collaboration of South American countries offers 23 textbooks to students freely [6]. Furthermore, universities have sponsored the Open Textbook Initiative, Alternative Textbook Initiative or Alternate Textbook Project to encourage faculty to develop or adopt OATs to lower dependence on commercial textbooks. Although several projects and cases in the world have implemented OATs, to our best knowledge, no study has examined current OAT development and related issues as they pertain to information provision. Therefore, this study investigated OAT in terms of information provision for libraries.

3 Methodology

This study used systematic review as an approach to overview the development and issues related to OAT. This study was designed according to steps defined by systematic review [1] as follows:

- Framing research questions: including source of provision, use license, mode of use, file format, and business model.
- Identification of relevant use cases: several terms were used to retrieve and identify cases and their literature related to OATs, including open textbook, OTB, open access textbook, OAT, open source textbook, free online textbook, free web-based textbook, free electronic textbook, free e-textbook, free dynamic textbook, and free wiki-based textbook.
- Selection of qualified cases and literature: in addition to cases and their literature related to OAT, these cases had to provide a web-based platform and related documents (such as About, About Us, Description, FAQ, Goals, Introduction, Missions, and conference presentations) to allow researchers in cross-verify the research questions proposed by this study. After the exclusion of the Global Text Project, Open Textbook Initiative, McGraw-Hill and Pearson, 25 cases and their related literature were qualified for this study.
- The 18 cases were the American Institute of Mathematics, Bookboon, Boundless, CK-12, Flat World Knowledge (FWK), IEEE Signal Processing Society, Open Course Library, Open SUNY Textbooks, OpenEd, OpenStax College, O'Reilly Atlas, O'Reilly Open Access Books, Orange Grove Plus, ProjectLATIn, Saylor, STEMWiki Hyerlibrary, WikiBooks, and Writing Spaces.
- Extraction and summary of the evidence-based data: details can be found in the Results section.
- Synthesis and interpretation of the findings: detailed information can be found in the Discussion section.

In this study two researchers conducted the systematic review. The research profiles of the researchers focused on both the strategy and practice of open access study. At the beginning of the systematic review, the researchers held discussions to attain a consistent understanding of the scope of OAT, and defined research parameters and questions. Then the researchers extracted, summarized, synthesized and interpreted the data. Lastly researchers exchanged and discussed their data to achieve an agreement on interpretation and their evidence supporting the defined research questions.

4 Results

According to the characteristics of sponsors, sources of OATs can be categorized as follows: commercial publishers, legislative actions, non-profit organizations, repositories of open educational resources, wiki-based resources, scholarly associations, libraries and others. In terms of type of license, most OAT providers release OAT materials through Creative Commons (CC), but there various CC terms, including BY, BY-SA, BY-ND, BY-NC-SA and BY-NC-ND, and CC BY and CC BY-NC-SA are often used by OAT providers. In addition, GNU Free Documentation License (GFDL), GNU General Public License (GPL) and All Rights Reserved (ARR) are also used. GFDL and GPL license are used by computer science OATs, such as O'Reilly's Altas. On the other hand, some commercial textbooks have rights released by authors or publishers to offer free access to them as OATs, but they are still protected by ARR. In terms of file format, HTML, XML or similar compatible formats (e.g., wiki-based webpage) are the most common for online reading. APK, DAISY, ePub, Mobi and PDF are used for download for offline use, and PDF is the most popularly used format. There are three types of use mode for OAT. The first is to use OAT online, but some providers will limit their use within specified coverage. For example, FWK only offers users access to three chapters of OAT. The second is to download an OAT file for offline use, either to read on screen or print out. The last mode is a combination of the first and second type which allows users to read online or download the OAT for offline use. According to business models proposed by Anderson [1] for the Internet, only freemium, advertising and gift economy models are used by OAT providers. For example, Bookboon claimed that they will offer less than 15 % advertisement space for each piece of OAT content.

5 Discussion

A key characteristic of OATs that contrasts with OAJ and OAB is that users are permitted to flexibly customize contents for various models of teaching and learning. Several OAT providers (e.g., CK-12, FWK, OPEN SUNY Textbooks and Writing Spaces) offer a faculty account to edit the content of OAT materials. However, some have not allowed faculty to make any changes or modifications to existing OAT contents, such as Bookboon. In terms of license, some OAT providers have restricted their contents by using CC ND license to prevent changes or modifications. However, OATs are a kind of OA-based open content. According to Wiley's definition [7], open access content is open to retain, reuse, revise, remix and redistribute. If OAT materials cannot be modified for various teaching models, are OATs still open content? Based on the aforementioned two cases, OAT providers have only agreed on open access, but have not achieved agreement on other principles and IPR requirements of OATs, such as freedom to modify.

On the other hand, openness is very important for libraries to deliver OATs to teachers and students according to consideration of information provision. Standard open file format is useful to download or print for learning without any ICT restrictions. Although CC is mostly adopted by OAT providers to provide an open license, legal stipulations of BY-NC-ND have prevented teachers from modifying or remixing existing OAT content to produce customized teaching materials for students. Furthermore, a workable business model for sustainability to keep long-term availability of OAT is also required for libraries to effectively support teaching and learning.

6 Conclusion

OA has impacted higher education in various ways ranging from scholarly publishing to accessing to scholarly information resources. Based on the aforementioned analysis and discussion, OAT appears to be an embodiment of the golden road of OA. It is valuable as it provides quick access, practical educational materials and high information currency in terms of information provision. This study has addressed OAT applications and their implications in library information services to support teaching and learning.

Acknowledgements. This paper was supported by the Ministry of Science and Technology of Taiwan under MOST Grants: MOST 104-2410-H-032-081.

References

1. Anderson, C.: Free $0.00 is the future of business. Wired **16**(3), 140–149, 194 (2008). http://www.wired.com/images/press/pdf/free.pdf
2. Centrum Cyfrowe: Projekt Polska Works Towards Social Change and Enhancing Citizens (2012). http://centrumcyfrowe.pl/english/
3. Khan, K.S., Kunz, R., Kleijnen, J., Antes, G.: Five steps to conducting a systematic review. J. R. Soc. Med. **96**(3), 118–212 (2003)
4. Morris-Babb, M., Henderson, S.: An experiment in open-access textbook publishing: changing the world one textbook at a time. J. Sch. Publishing **43**(2), 148–155 (2012)
5. National Association of College Stores: Higher Education Retail Market Facts & Figures (2013). http://www.nacs.org/research/industrystatistics/higheredfactsfigures.aspx
6. ProjectLATIn: Products (2011). http://latinproject.org/index.php/en/products
7. Senack, E.: Fixing the Broken Textbook Market: How Students Respond to High Textbook Costs and Demand Alternatives (2014). http://www.uspirg.org/sites/pirg/files/reports/NATIONAL%20Fixing%20Broken%20Textbooks%20Report1.pdf
8. United States Government Accountability Office.: College Textbooks: Enhanced Offerings Appear to Drive Recent Price Increases: Report to Congressional Requesters (2005). http://www.gao.gov/new.items/d05806.pdf
9. Wiley, D.: Defining Open (2009). http://opencontent.org/blog/archives/1123

Adaptive Information Retrieval Support for Multi-session Information Tasks

Daniel Backhausen[1]([⊠]), Claus-Peter Klas[2], and Matthias Hemmje[1]

[1] Distance University in Hagen, Hagen, Germany
Daniel.Backhausen@Fernuni-Hagen.de, Matthias.Hemmje@FernUni-Hagen.de
[2] GESIS - Leibniz Institute for the Social Sciences, Mannheim, Germany
Claus-Peter.Klas@Gesis.org

Abstract. Goals and corresponding tasks are both major drivers of information needs which are satisfied within information behaviours and lead to information tasks and finally information retrieval. Changing goals or a changing information need and task interruption during execution are the two most challenging barriers of people working on complex and longitudinal goals or tasks. People often have problems re-locating already visited pages or recapturing previous queries and their results. Thus the support of task-based information retrieval is still not sufficient today. Due to this we examined new concepts that make task-based information retrieval more efficient and useful.

Keywords: Interactive information retrieval · Task-based information retrieval · Adaptive systems · Personalized search · Task management

1 Introduction

Today researchers agree that work goals and corresponding work tasks are the leading drivers for creating information needs (e.g. [1,2,8]). Each one of us has work goals which are either job-related or may be driven by personal interests. Complex goals like writing a funding proposal, building a house, or the restoration of a classic car, can take hours, days, or even months to complete. To achieve such a goal, it is necessary to execute and complete derived tasks since they define concretely what has to be done. In addition, executing a task requires a particular knowledge which might already exists for routine tasks that are executed regularly. But if a task has never been done before, there is normally a lack of knowledge which hinders its execution and completion. Hence specific information is needed and we enter into an information task, i.e., an information seeking behaviour and within this behaviour we quite often initiate a corresponding information retrieval task to acquire the missing knowledge. During the different stages of such an information task processing, the required knowledge changes from more general up to more specific. This influences how we seek and search for information, i.e., it influences the information retrieval tasks that we are initiating.

© Springer International Publishing Switzerland 2015
S. Kapidakis et al. (Eds.): TPDL 2015, LNCS 9316, pp. 317–320, 2015.
DOI: 10.1007/978-3-319-24592-8_25

In this way, searching for information by means of initiating information or more specifically information retrieval tasks, is an important support action for many work tasks. Project- and task-management tools support users to manage their work on such work goals in work tasks and work sub-tasks. Unfortunately, they do not use this explicitly provided information about the work task to support the related information tasks, i.e., to e.g. support the corresponding information retrieval tasks at hand and to contextualize and in this way personalize our web-based information searches.

On the other hand different web-based information support services are trying to help users to find pages that contain relevant information, but unfortunately they do not have any awareness about their users' work context, i.e., their current work goal or current work task and thus this kind of knowledge is not part of the relevance calculation of the underlying information retrieval mechanism. This is also true for information about the stage or status and the complexity of the work task, the corresponding information tasks, possible sub-tasks or higher level tasks, and information already explored in previous work task contexts and their information tasks.

Our current research activities tackle this problem by proposing new concepts of task-supported information retrieval focusing on complex goals and work tasks that span over multiple sessions and which might be executed in parallel. Through the support of an integrated task tool, we are aiming at enabling users to structure their goals at hand into tasks and sub-tasks and utilize this explicit information as an important part of representing the users' current work and information context in a machine readable way. Assuming that a task is explicitly represented and managed in this way, all implicit and explicit user interaction over several work sessions is stored with reference to this task. Having the task specific context information at hand we are then able to implement system support features and user awareness features like task-based *personalization*, *recommendation*, and *adaptivity* which we will present in the next chapter.

2   Concepts and Prototype

Based on and in extend to previous work like [3–7,9] we want to introduce *three* new concepts to increase task performance and to get a higher usefulness in supporting users within their tasks. To test and evaluate these concepts, we implemented a prototype called TasksTodo, which is an add-on for the web browser Firefox. The implementation as a browser add-on allows us to realize the concept of integrating task management directly into web-based information retrieval activities.

As discussed earlier, realistic work situations can be longitudinal and multifaceted and thus very complex. The current task is a major factor for the situational information need, but beyond that it is often only a partial view on the overall work definition. Hence it is necessary to also take the goal and the overall task structure into account when supporting the user during task completion. This way it is e.g. possible to determine task complexity and predict next steps. For that reason we propose a *first* concept, using the overall information about the work goal and its tasks, not only limiting to the current task. Therefore TasksTodo allows users to create an arbitrary number of goals and within each goal users can define an arbitrary number of tasks on different levels (i.e. a task which contains sub-tasks). This allows grouping of concrete tasks to an abstract goal. For each goal and task it is possible to define its name or title, the urgency and priority, and the due date. Goals and tasks can be flagged as completed, to reduce situational importance (Fig. 1).

Fig. 1. TasksTodo - Search page adaption and task-related history

One of the main problems users have with longitudinal multi-session tasks is task interruption and resumption. Hence, for task efficiency it is important to support users in such situations. People need to get a quick overview on their past activities related to a certain task to easily continue work. For the web these activities are at least the last opened tabs, visited pages, and bookmarks. Our *second* concept enhances the previous research mentioned above. In TasksTodo all information retrieval activities are tracked in relation to the selected task. In this way, users can use the task-related history to review the last visited pages within a certain task. Users can also restore the tabs of the last task session. Besides this, TasksTodo allows to easily manage task-related bookmarks. For example this can be done using the context menu of the browser upon a web page or link. Each bookmark automatically contains the page title, the URL, and the date it has been created. In addition it also takes a screenshot and stores the complete page content for later use. TasksTodo offers support for switching between goals and tasks due to multiple task execution.

The *third* concept we want to introduce here is task-based adaption. Part of our tracking of browser activities is the visit and use of search engines. This allows us to offer support for information search across different services (domains) and enables us to provide further features like the adaption of functionality and content of the retrieval system. TasksTodo visualizes the last five queries for Google, Google Scholar, Bing, Microsoft Academic Search, Yahoo, YouTube, Wikipedia, and the digital library system of Gesis related to the current work task. This way users can easily re-capture and re-execute last search activities. TasksTodo also modifies bookmarked pages by adding a green top bar with a notification to the page. This indicates a page as bookmarked, therefore, being relevant for the current task. Contrariwise it is also possible to flag pages as not being relevant according to the current task resulting in a red bar with a notification on the top of the page. This information will be used in future to adapt the result lists.

The title of the selected task is set as a parameter to the visited URL (e.g. http://example.org?tt_task=my+task+title). Therefore page providers could use this information to adapt their page and content to the users current task. In that or a similar way adaption can also be provided by page providers.

More information about the TasksTodo project and a link to download the software can be found at taskstodo.org.

References

1. Belkin, N.J.: A methodology for taking account of user tasks, goals and behavior for design of computerized library catalogs. SIGCHI Bull. **23**(1), 61–65 (1991)
2. Byström, K., Järvelin, K.: Task complexity affects information seeking and use. Inf. Process. Manage. **31**(2), 191–213 (1995)
3. Kellar, M., Watters, C., Shepherd, M.: The impact of task on the usage of web browser navigation mechanisms. In: Proceedings of Graphics Interface 2006, GI 2006, pp. 235–242. GIPS, Toronto (2006)
4. MacKay, B., Watters, C.: Exploring multi-session web tasks. In: Proceedings of the SIGCHI Conference on Human Factors in Computing Systems, CHI 2008, pp. 1187–1196. ACM, New York (2008)
5. MacKay, B., Watters, C.: Building support for multi-session tasks. In: CHI 2009 Extended Abstracts on Human Factors in Computing Systems, CHI EA 2009, pp. 4273–4278. ACM, New York (2009)
6. MacKay, B., Watters, C.: An examination of multisession web tasks. J. Am. Soc. Inf. Sci. Technol. **63**(6), 1183–1197 (2012)
7. Rajamanickam, M.R., MacKenzie, R., Lam, B., Su, T.: A task-focused approach to support sharing and interruption recovery in web browsers. In: CHI 2010 Extended Abstracts on Human Factors in Computing Systems, CHI EA 2010, pp. 4345–4350. ACM, New York (2010)
8. Vakkari, P.: Task-based information searching. Annu. Rev. Inform. Sci. Technol. **37**(1), 413–464 (2003)
9. Wang, Q., Chang, H.: Multitasking bar: prototype and evaluation of introducing the task concept into a browser. In: Proceedings of the SIGCHI Conference on Human Factors in Computing Systems, CHI 2010, pp. 103–112. ACM, New York (2010)

Transformation of a Library Catalogue into RDA Linked Open Data

Gustavo Candela[✉], Pilar Escobar, Manuel Marco-Such,
and Rafael C. Carrasco

Departamento de Lenguajes y Sistemas Informáticos,
Universidad de Alicante, Alicante, Spain
gustavo.candela@cervantesvirtual.com

Abstract. The 200,000 records in the catalogue of the Biblioteca Virtual Miguel de Cervantes have been migrated to a new relational database whose data model adheres to the FRBR and FRAD specifications. The database content has been later mapped to RDF triples which employ the RDA vocabulary to describe the entities, as well as their properties and relationships. The intermediate relational model —ensuring, for example, referential integrity— provides tighter control over the process and, therefore, enhanced validation of the output. This RDF-based semantic description of the catalogue is now accessible online and supports browsing and searching the information.

Keywords: Linked open data · Bibliographic and authority data · Cultural heritage · Semantic web

1 Introduction

Linked open data are data which others are allowed to use, modify and redistribute, and previously enriched with information about their relations with other data. The metadata which describe such relations are often expressed as binary relations between *Uniform Resource Identifiers* and encoded in RDF (Resource Description Framework) format, a standard for data interchange.[1] Applying the *linked open data* concepts to the cultural heritage domain has become an active and challenging field [3]. Meanwhile, modern standards for cataloguing are emerging as an alternative replacement to the traditional ones. For example, RDA (Resource, Description and Access) follows the concepts and terminology of the Functional Requirements for Bibliographic Records (FRBR) and the Functional Requirements for Authority Data (FRAD) and will also adopt the Functional Requirements for Subject Authority Data (FRSAD), a family of models promoted by the IFLA (FR* models) which define entities, relationships, and attributes that should be used to describe resources.

This paper describes the steps applied for the automation of the migration process from a MARC21 collection of records to a set of RDF triples containing

---

[1] See http://www.w3.org/RDF.

© Springer International Publishing Switzerland 2015
S. Kapidakis et al. (Eds.): TPDL 2015, LNCS 9316, pp. 321–325, 2015.
DOI: 10.1007/978-3-319-24592-8_26

bibliographic metadata in RDA. The process is strongly based on the currently available open-source technology and relies on the creation of a relational database which provides controlled generation of linked data in RDA.

2 An FRBR-FRAD Relational Model for MARC21 Records

Inspired by the IFLA conceptual models, an Entity-Relationship (ER) model —schematically represented in Fig. 1— has been defined to store the Biblioteca Virtual Miguel de Cervantes descriptive metadata. The abstract entity *Creation* generalises the entities in FRBR Group 1 and their relationships with other entities in the model such as *Agent* or *Subject*. Other relationships were defined such as adaptations, translations or reproductions. Some additional elements were incorporated to the model in order to address the catalogue specificities. For example, *Collection* entities were needed to host arbitrary groupings of objects, such as works in a bibliography, items with a common provenance (e.g., a partner library holdings or items in a personal archive), which are not properly creations and usually have no associated descriptive metadata. Since authors are often the subject of a book in a library with a focus on literature, a new type of relationship was introduced to describe creations having agent as *subject*; conversely, agents play different *roles* when contributing to a document —for example, printer, editor or illustrator. Entities for the UDC and Unesco classifiers were also added to the model. The relation *part of* was defined in order to describe nested inclusions, such as journals publishing volumes, made of issues containing articles.

Once the relational database was set up, a number of tools were implemented, in particular, a new catalogue manager assisting the creation of new entries in FRBR and an automatic procedure —schematically depicted in Fig. 2— to migrate the old MARC21 records into the new database. The migration process first loads MARC21 records using the MARC4J[2] library and creates in-memory Java objects representing the FRBR entities according to the Library of Congress guidelines for the MARC to FRBR transformation. Then, a transducer implements the mapping of each object field to the appropriate field in the database. The transducer uses the Hibernate ORM[3] platform which supports object/relational mapping and provides a simple interface to store and retrieve objects.

3 From FRBR to RDA Linked Open Data

Once the descriptive metadata are stored in a relational database with an FRBR model, they can be published as semantic information with the method described below. The publication of RDA in RDF format first loads the fields

---

[2] https://github.com/marc4j.
[3] http://hibernate.org/orm.

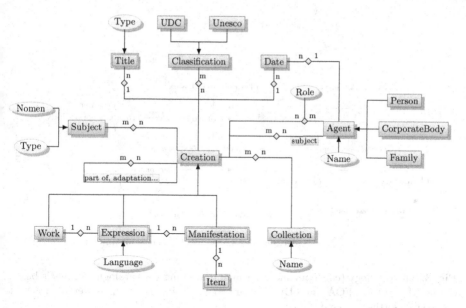

Fig. 1. Simplified diagram of the Entity-Relationship model of the relational database.

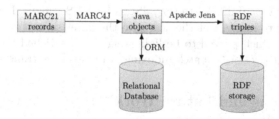

Fig. 2. Schematic representation of the migration and conversion process.

in the database as in-memory Java objects with the Hibernate ORM library. The Apache Jena library is then used to map of the object fields into the RDA components for the RDF graph (nodes and connections). As shown in Fig. 3, the core vocabulary consists of the elements and relationship designators recently approved by the Joint Steering Committee on Development of RDA. The OWL-Time ontology[4] has been used to describe temporal events such as publication years; external content, hosted by partner libraries, was described with FOAF elements [1] and subjects triples were created with the Dublin Core property *dc:subject*. The identifiers are prefixed with their domain and the entity type.

---

4 www.w3.org/TR/owl-time.

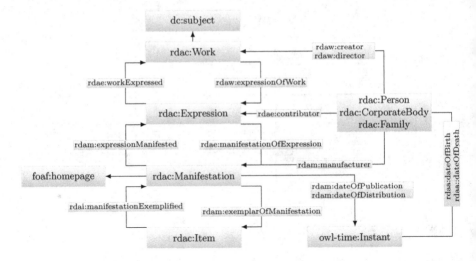

Fig. 3. The ontology (concepts and relations) describing the catalogue entries is based on the RDA, OWL, FOAF and Dublin Core vocabularies. Tag prefixes denote different name-spaces (the source ontology).

Several options to provide SPARQL access to the RDF storage were evaluated, including OpenLink Virtuoso[5], 4Store[6], and Sesame[7]. The last one was selected in order to implement the access to the data, since it is an open-source Java framework which proved to be light-weight and satisfied the requirements supporting full-text queries, batch indexing, and database transactions.[8]

4   Conclusions and Future Work

The semantic content published in the Biblioteca Virtual Miguel de Cervantes linked open data-set includes:

- Near 8 million triples and 90 different types of relationships.
- About 194,000 expressions and 380,000 items.
- About 55,000 persons, 3500 corporate bodies and 9 families.
- Nearly 77,000 dates of publication and 50,000 dates of distribution.

The output can be downloaded from data.cervantesvirtual.com in several formats: RDF, JSON, Dublin Core and ESE (Europeana Semantic Elements). Furthermore, an online demonstrator of the conversion process has been implemented and the user can upload there (bvmcresearch.cervantesvirtual.com/arms) a MARC21 record and obtain the RDA description in RDF format.

---

[5] http://virtuoso.openlinksw.com.
[6] http://4store.org.
[7] http://rdf4j.org.
[8] For an extensive comparative study of platforms, see [2].

References

1. Brickley, D., Miller, L.: FOAF Vocabulary Specification 0.99 (2014). http://xmlns. com/foaf/spec (visited on 03 January 2015)
2. Haslhofer, B., et al.: Europeana RDF Store Report. University of Vienna,Technical Report, Vienna, March 2011
3. Marden, J., et al.: Linked open data for cultural heritage: evolution of an information technology. In: Albers, M.J., Gossett, K. (ed.) Proceedings of the 31st ACM international conference on Design of communication, Greenville, September 30 - October 1, 2013, pp. 107–112. ACM (2013)

Segmenting Oral History Transcripts

Ryan Shaw[(✉)]

School of Information and Library Science,
University of North Carolina at Chapel Hill,
Chapel Hill, USA
ryanshaw@unc.edu
https://aeshin.org

Abstract. Dividing oral histories into topically coherent segments can make them more accessible online. People regularly make judgments about where coherent segments can be extracted from oral histories. But when different people are asked to extract coherent segments from the same oral histories, they often do not agree about where such segments begin and end.

Keywords: Oral history · Discourse segmentation · Natural language processing · Digital libraries

1 Introduction

Oral histories are rich and unique documents of the past and our memory of it. Putting oral histories on the web makes them more accessible, but they remain daunting to consume [4]. It requires a significant time commitment to listen to a one or two hour interview. This is why curators of oral history collections, when creating "exhibits" of their materials for the public, usually select short extracts from longer interviews. Scholars also select extracts from their recordings when presenting their work to a live audience [7, 265]. But are extracts merely subjective judgments, or do they reflect a topical structure about which consensus might be reached? An analysis of 829 judgments about oral history extract boundaries suggests that while these judgments are not purely subjective, consensus about topical structure is weak.

2 Data and Methods

Our corpus consisted of 19 transcripts of oral history interviews conducted by the Southern Oral History Program (SOHP) at the University of North Carolina.[1] In an earlier project, SOHP staff transcribed and selected salient extracts from each of the interviews. Each interview was divided into segments, of which some

---

[1] All of the data and code discussed in this paper are available at https://github.com/contours.

© Springer International Publishing Switzerland 2015
S. Kapidakis et al. (Eds.): TPDL 2015, LNCS 9316, pp. 326–329, 2015.
DOI: 10.1007/978-3-319-24592-8_27

subset (the selected extracts) were judged to be topically coherent. On average, half of the segments were selected as salient extracts. The unselected segments tended to be longer, with a mean length of 70 sentences compared to 44 sentences for the selected extracts. This suggests that the extraction process may not identify some potential topic boundaries (those that appear within the segments of interviews judged to be less salient).

Two non-expert annotators were asked to imagine that they had been tasked with curating an online collection of oral histories and to select "the most important parts" from each transcript. Each transcript was presented as an HTML page showing only the names of the speakers and the transcribed text of their speech. The annotators could click on the text to split it into segments and then indicate which segments were to be selected as extracts. They were instructed that each extract "should cover a single topic or anecdote and should be understandable on its own." To give them a sense of the expected granularity of the extracts, the annotators were told that the length of extracts could vary considerably but would average around 30–50 sentences (the average extract length in the original project).[2] Extracts could not overlap, and not all of the text had to be extracted (i.e. it was permissible to "leave out" parts of the transcript between extracts). Extract boundaries were not limited to speaker changes or paragraph breaks and could be placed between any two adjacent sentences.

3 Comparing Segmentations

The three annotators placed a total of 829 boundaries between sentences of the 19 interview transcripts, creating a total of 886 segments. The distribution of segment lengths appears to be exponential (see Fig. 1). The mean segment length was 50 ± 52.7 sentences ($n = 886$). The shortest and longest segments created were one sentence and 621 sentences long, respectively. The longest segment may be an error, since it is far longer than any other segment in the dataset and far longer than the other segments created by that annotator for that interview. Some of the very short segments seem to be cases of "trimming" behavior, where a short segment is created specifically to be excluded (e.g. an interviewer unsuccessfully trying to interrupt an interviewee with a new question) or highlighted (e.g. a particularly vivid single quote that makes sense in isolation). Different tendencies to "trim" like this may explain some of the variations observed across the annotators' segmentations.

If the annotators had consistently segmented at the same level of granularity, one would expect to see little variance in the segment lengths and a positive correlation between each transcript's length and the number of segments into which it was divided. But segment lengths varied significantly. While there was a positive correlation between transcript length and the number of segments, there was also a positive correlation between transcript length and the median segment length. Thus while the annotators divided longer transcripts into more segments,

---

[2] The complete text of the instructions provided to the annotators is available at https://github.com/contours/segment/blob/5404fce/public/instructions.html.

they also created longer segments for longer transcripts. Longer interview transcripts might reflect the fact that some interviewees are more long-winded than others. If that were the case then one might expect longer segments for longer interviews: each topic takes longer to cover. However it is also possible that longer interview transcripts simply cover more topics than others. If so, then the division of longer interviews into longer segments may have been due to annotator exhaustion, resulting in interviews segmented at different levels of granularity.

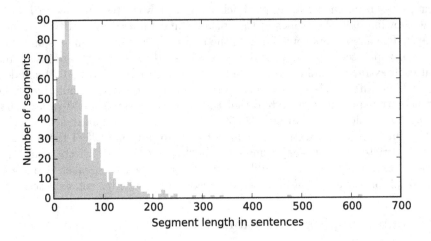

Fig. 1. Lengths in sentences of the manually-created segments.

Segmentation granularity also varied across annotators. The original extractor placed boundaries slightly more frequently than the overall rate, which was about one boundary per 59 potential boundaries. Annotator A placed fewer boundaries (creating longer segments) on average than either the original extractor or annotator B. Annotator B placed more boundaries (creating shorter segments) on average than the other two. This could indicate that annotator B engaged in more "trimming" than the other annotators. These differences in boundary placement rates across annotators might lead one to expect low inter-annotator agreement, and indeed that is the case.

To measure inter-annotator agreement the *boundary edit distance* metric proposed by Fournier [3] was used. Boundaries differing by more than eight sentences were treated as misses, while boundaries differing by eight or less sentences were treated as "near misses", scaled the distance between the boundaries. A boundary edit distance of one indicates perfect agreement. The micro-averaged edit distance between pairs of boundaries placed by two annotators was 0.27 ± 0.0232 (95 % CI, $n = 1270$). This pairwise mean boundary edit distance measures actual agreement; to correct for chance agreement one can calculate Fleiss' π^* coefficient [2], which was also 0.27.

4   Discussion and Future Work

This study examined flat, non-overlapping segmentations of oral histories, but topical structure is generally believed to be hierarchical [6]. Ashplant [1, 107] suggests that this is true of oral histories in his analysis of part of *The Dillen* [5]. He discerns a three-level topical hierarchy, with *anecdotes* grouped into *narrative elements*, which are in turn grouped into broad topical *clusters*. Differing judgments about which of these levels is the appropriate one for selecting extracts from may have been a factor contributing to the varying segmentation granularities found in this study.

Segmentation and segment-level description could make oral histories more accessible. But even though identifying topically coherent segments within interviews is an accepted part of working with oral histories, people often do not agree on the boundaries of those segments. In this study, annotators agreed (exact match or "near miss") on less than half of the boundaries they placed. Further progress will depend on clearer definition of the segmentation task.

Acknowledgments. This work was funded by the Institute of Museum and Library Services. Thanks to the Southern Oral History Program for making the interview transcripts available and Kathy Brennan and Sara Mannheimer for their annotation work.

References

1. Ashplant, T.G.: Anecdote as Narrative Resource in Working-class Life Stories: Parody, Dramatization and Sequence. In: Chamberlain, M., Thompson, P. (eds.) Narrative and Genre, pp. 99–113. Routledge, London (1998)
2. Fleiss, J.L., Nee, J.C., Landis, J.R.: Large sample variance of kappa in the case of different sets of raters. Psychol. Bull. **86**(5), 974–977 (1979). http://dx.doi.org/10.1037/0033-2909.86.5.974
3. Fournier, C.: Evaluating Text Segmentation using Boundary Edit Distance. In: Proceedings of 51st Annual Meeting of the Association for Computational Linguistics, p. to appear. Association for Computational Linguistics, Stroudsburg (2013). http://anthology.aclweb.org/P/P13/P13-1167.pdf
4. Frisch, M.: Oral history and the digital revolution: toward a post-documentary sensibility. In: Perks, R., Thomson, A. (eds.) The Oral History Reader, 2nd edn. Routledge, London (2006)
5. Hewins, G.H., Hewins, A.: The Dillen: Memories of a man of Stratford-Upon-Avon. Elm Tree Books, London (1981)
6. Manning, C.D.: Rethinking Text Segmentation Models: An Information Extraction Case Study. Technical report, University of Sydney (1998). http://nlp.stanford.edu/cmanning/papers/SULTRY-980701.ps
7. Thompson, P.: The Voice of the Past: Oral History, 3rd edn. Oxford University Press, Oxford (2000)

Digital Libraries Unfurled: Supporting the New Zealand Flag Debate

Brandon M. Thomas, Joanna M. Stewart,
David Bainbridge$^{(\boxtimes)}$, David M. Nichols, William J. Rogers,
and Geoff Holmes

Department of Computer Science, University of Waikato,
Hamilton, New Zealand
{bmt11,jms78}@students.waikato.ac.nz,
{davidb,daven,coms0108,geoff}@cs.waikato.ac.nz

Abstract. This article reports on the development of an interactive web environment, backed by a digital library, that supports the creation of new flag designs. Specifically, it supports the user through an iterative design process, guided by principles drawn from the field of Vexillology. The work has been motivated by a legally binding referendum on the issue in New Zealand, planned to occur in late 2015/early 2016.

Keywords: Vexillology · Digital libraries · Iterative design · Image similarity

1 Introduction

What the future flag of New Zealand will look like has become the topic of significant debate with the country's current Prime Minister, John Key, scheduling a two-stage binding referendum on the matter over the years 2015–2016. With a background in digital libraries, human computer interaction, and open source software development the authors of this paper decided to develop an online interactive environment that allows members of the public to try out their ideas for a new flag, which we report on here. We enrich the core design activity with feedback over how similar the flag the user is designing is to other existing flags. We also allow the user to visualise the flag flying in a variety of conditions. This is all achieved in real-time within the user's web browser, with the results of users' designs stored in a digital library.

2 Worked Example

The aim of our work was to create an on-line environment in which users could develop their own flag designs. In particular we wished the environment to embody the principles of good flag design [2]. Figure 1—which shows a sequence of snapshots taken during a session by one user interacting with the developed

© Springer International Publishing Switzerland 2015
S. Kapidakis et al. (Eds.): TPDL 2015, LNCS 9316, pp. 330–333, 2015.
DOI: 10.1007/978-3-319-24592-8_28

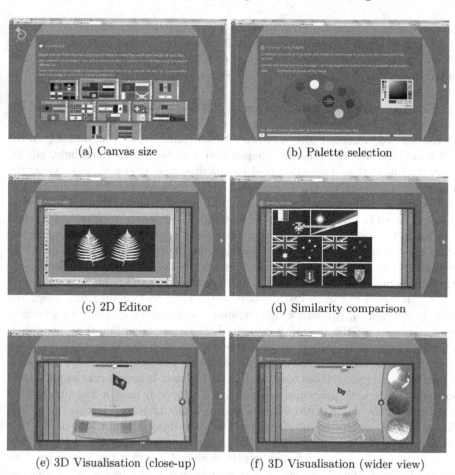

(a) Canvas size

(b) Palette selection

(c) 2D Editor

(d) Similarity comparison

(e) 3D Visualisation (close-up)

(f) 3D Visualisation (wider view)

Fig. 1. The developed iterative flag design process (Color figure online).

web site—demonstrates the solution we devised. After a splash page introducing the site (not shown) the user is presented with a page asking them to select the aspect ratio of their flag (Fig. 1a). The aspect ratios are derived from a digital library collection of the world's flags we have developed. Boxes visually matched to show the various ratios are presented, with the example flags that have those dimensions shown within the respective boxes. The displayed boxes are frequency sorted left to right, showing most common to least common. Clicking on one of the other boxes changes the selection, and hovering over a flag brings up a tooltip which lets the user see the name of the country.

When the user is ready to proceed to the next step, the instructions displayed directs them to click the right arrow on the page, or else perform a right-to-left swipe action. This brings up Fig. 1b, where an artist's palette is the key feature. By default the three colours used in the current New Zealand flag are shown.

Our snapshot was taken a little while later on, after the user interacted with the page, using the slider at the bottom to increase the number of colours up to 7. From this screen the user can continue (click right arrow, or swipe) to the next stage (flag design), or else use the left arrow (left-to-right swipe) to return to the aspect ratio stage.

Figure 1c shows our user went on to the flag design stage. This is the key stage of the environment, where the user is likely to spend most of their time. Here they are presented with a browser-based "paint-style" program. Down the left-hand side are the different shapes that can be drawn (rectangle, ellipse, star, etc.) as well as a selection tool for manipulating drawn shapes. Along the bottom of the paint program area are controls for selecting the current fill and line colour of the shapes being drawn. Through the file menu (top-left) more substantial edits can be effected, such as importing a symbol from a library of shapes. In the case of our user, for their flag design they have imported a fern leaf from the Open Clip Art web site[1] through the file menu, changed its colour to white (it was green initially) and positioned it on the left. They then duplicated the shape, mirrored it in the y-axis, and positioned to the right.

Figure 1 also has some vertical bars to the right (coloured brown, green and blue in the figure), through which the user can access views that provide additional context to the on-going design. Utilising an accordion-type metaphor, clicking on one of these bars squeezes the current view closed to make way for a new view to be display. The remaining snapshots in Fig. 1 (d)–(f), are snapshots taken from accessing these bars. In Fig. 1d the user is shown how similar their current design is to other flags in the world. Hovering over a flag brings up a tooltip with the name of the country and a similarity score—in this case, 75 %. The next bar over in the interface does the same thing with the alternative New Zealand flag designs (not shown). Clicking on the right-most bar brings up the view shown in Fig. 1e, which displays the designed flag flying on top of the New Zealand Parliament Building. This is an interactive 3-dimensional simulation. The user can shift and move around the location to view the flag from other angles, as shown in Fig. 1f. This last snapshot also shows the simulation allows for elements of the weather to be adjusted. The scroll-bar at the top of the viewing area controls the strength of the wind, and through the pull-out area to the right, they can experience what the flag looks like when it is raining and snowing, in addition to when it is sunny. When the user is satisfied with their design they click/swipe on to the final stage (not shown) where they are asked to fill out contact details if they would like their flag to be published in the site's digital library collection of alternative designs.

3   Implementation

In terms of implementation, the core component is a Greenstone 3 digital library [4] consisting of two image collections: one for the world's flags, and the other comprised of alternative designs for the New Zealand flag. We utilise

---

[1] https://openclipart.org/.

the baseline functionality of Greenstone 3, which includes incremental ingest and metadata assignment, to incorporate the new designs. To meet the additional functionality needed by our site, advances in web browser-based standards meant we were able to achieve this without resorting to customized digital library services. While the Greenstone DL architecture allows for extensible services, performing this functionality in the web browser means there is a considerable saving in computational cost in providing the site.

The 3D rendering is accomplished using WebGL.[2] A particle system utilising springs is used to provide a cloth-like simulation [1]. The image similarity calculation is based on an HSV colour quantized histogram calculation [3]. Within the web browser the individual pixels to an image are accessible when rendered to an HTML *canvas* element. We use an off-screen canvas to render the flags at full size for the calculation.

When a user first visits the flag design site, an AJAX call is used to see if computed colour histograms for the flags in the DL already exist. If they do not, then the browser computes these, and sends back the computed values to be stored in the DL as metadata, run-length encoded to save space. If a web browser supports local storage, this information is also stored in the browser to avoid even the need of pulling the computed data over the network on subsequent visits to the site. A comparable technique is used for computing and storing the aspect ratios of the flags. A checksum scheme is used to notice when data on the server has changed, and consequently local information should be discarded.

4   Conclusion

To conclude, in this article we have presented the design and implementation of an interactive web environment, backed by a digital library, that supports the development of new flag designs. Motivation for the work has come from a planned flag referendum to be held in our country. Beyond the core activity of developing a new flag design with a "paint-style" application, our solution provides additional context to support the user in this process: showing the user how similar their design is to other flags; and allowing them to visualise, in 3D, the flag flying in a variety of conditions. The site is accessible via the URL: www. greenstone.org/nz-flag-design.

References

1. Jakobsen, T.: Advanced character physics. In: Game Developers Conference, pp. 383–401 (2001)
2. Kaye, T.: Good Flag, Bad Flag: How to Design a Great Flag. North American Vexillological Association, Boston (2013)
3. Shapiro, L.G., Stockman, G.C.: Computer Vision. Prentice Hall (2003)
4. Witten, I.H., Bainbridge, D., Nichols, D.M.: How to Build a Digital Library, 2nd edn. Morgan Kaufmann, San Francisco (2010)

---

[2] https://get.webgl.org/.

Evaluating Auction Mechanisms for the Preservation of Cost-Aware Digital Objects Under Constrained Digital Preservation Budgets

Jose Antonio Olvera[✉], Paulo Nicolás Carrillo,
and Josep Lluis de la Rosa

TECNIO EASY Innovation Center University of Girona, Campus de Montilivi,
E17071 Girona, Catalonia (EU), Spain
{joc7188,paulocarrillopena}@gmail.com,
peplluis@eia.udg.edu

Abstract. This is a novel approach to managing costs in digital preservation, one that takes advantage of an object-centric approach developed by means of self-preserving digital objects whereby the objects manage their preservation not only by maximizing the chances of avoiding obsolescence but also doing it at a minimum cost. To accomplish this, we assign a budget that the objects manage to achieve the necessary preservation services at a given cost. Several strategies apply, such as maximal preservation service at all costs or burn low even if the preservation is not perfect. We explore optimizing the budget of self-preserving digital objects through micro-negotiations of objects and services, expecting accurate balance of costs and quality of preservation. Specifically, in negotiation, we will explore the price-based algorithms that are the electronic auctions, notably the combinatorial and multi-unit auctions. We compare the expected lifetime of digital objects with the two electronic auction algorithms with the aim of deciding in what conditions these algorithms apply and deliver good results. In all, this work, exploratory in nature, studies a bottom-up approach of cost management in digital preservation, contrary to the prevailing top-down approach of the state of the art, using e-auctions.

Keywords: Digital preservation · Self-preservation · Auctions · Agents

1 Introduction

As was explained in [4], the prevailing paradigm is centralized, top-down, in which institutions are the main players. We propose studying a change in paradigm, to one that is mainly bottom-up, in which the digital objects self-preserve.

Works in this line such as [1–3] leave a path open to digital objects that have incorporated behaviors to combat digital obsolescence with a promising level of success; however, part of the success is also due to the correct management of the budget that digital objects have for their preservation. In this concept, the digital objects become active actors in their own Long Term Digital Preservation (LTDP) by a new definition of the digital object to be preserved, the Cost Aware Digital Object (CADO),

© Springer International Publishing Switzerland 2015
S. Kapidakis et al. (Eds.): TPDL 2015, LNCS 9316, pp. 334–337, 2015.
DOI: 10.1007/978-3-319-24592-8_29

which has a budget to compete with other CADOs and negotiate with digital preservation services in search of their own LTPD.

We will contribute to the digital preservation as an object-centric perspective of recovery and cost management, a new recovery method in itself with a clear focus on costs and efficacy.

2 Cost Aware Digital Objects

In [5], we focused on the optimization of the budget of self-preserving digital objects through micro-negotiations as combinatorial and multi-units of preservation and curation services for the objects to be preserved. This paper expands our previous work, seeking an efficient cost management following our paradigm of self-preserving digital objects. In this paradigm, we rethink the relationship between DOs and digital preservation services, as follows.

2.1 Definitions

The CADO. It is responsible for its own-conservation with its own budget for hiring preservation services, such as risk assessment data loss, data recovery, metadata extraction, migration and new storage formats. We define the state of obsolescence with four attributes: accessibility, readability, integrity and authenticity, the combination of which represents the quality of CADO and digital longevity.

Virtual Currency for Preservation. To extemporize and universalize the price assignment of the LTDP services costs, so that we will talk of the same prices now and in the future regardless of the monetary paradigm of the future, as well as provide a budget for CADO in a transparent, easy, and general way, we define a virtual currency called PRESERVA ($). We define 1$ as the cost of preserving a DO for 100 years. In our experiments, we use the m$ (miliPRESERVA).

Preservation Services. There are specific preservation actions such as format migration, metadata extraction, and new storage. Examples of them were taken from [7], including services that are of relevant types and compatible with a CADO so that they can increase one or more attributes at the same time. We organize them in terms of a generic type of preservation (i.e., general families of services that will be valid 100 years from now, such as storage, migration, etc.) and a particular type (the services themselves), an initial sales price and a minimum one, and an estimation of the increase in percentage of the four attributes that a CADO can experience with this service. The quality percentage of increase at each CADO attribute is calculated by estimates that have been made based on several heuristic criteria, such as family and specific type and the number of downloads. The price is estimated with respect to the general family of the services with a random variation of ± 10 %. We define five general families of service: storage, metadata, migration, integrity, and authenticity. To obtain the price of each family of service, we argue: if the cost for preserving a DO for 100 years is 1$ and we estimate the number of times that DOs will need the different families of preservation services during this time, we can obtain the price of each family of service in miliPRESERVA. We obtained a price of 15.38 m₽ for the storage services and 25 m₽ for the metadata, migration, integrity and authenticity services.

SAW – Software Adoption Wave. This represents a massive upgrade of old software to new software, a wave of change. It occurs at the same pace of the formats update, i.e., 5 years [6]. In each SAW, we activate digital preservation actions to recover the CADO quality loss suffered in these waves. We simulated the SAWs with a percentage decrement of CADO attributes: randomly, 10 % or 20 % decrements are applied.

2.2 Simulations Set Up

5 simulations per scenario are run to have statistical ground, and as was explained above, in each scenario, the prices of the services vary ± 10 % as well as the impact of the quality boost in the CADOs. 20 iterations occur at every run (of 100 years), and a SAW is simulated at every iteration, that is, every 5 years. The simulations run 1000 CADOs and 100 preservation services. Three scenarios are developed according to the initial budget assigned to all CADOs. The scenarios are the following: low budget is 500 m₽, medium budget is 1000 m₽ and high budget is 2000 m₽.

3   Results and Conclusions

Before studying the results obtained with the different e-auctions, the best baseline was chosen. The three baselines that were run are: the best quality, where in each SAW, the CADO chooses the service with more quality among compatible services; the cheapest, where in each SAW, the CADO chooses the cheapest service among compatible services; and the first choice, where in each SAW, the CADO chooses the first compatible service that is found, without considering either the price or the quality of the service. The best baseline strategy for all the scenarios (low, medium and high budget) was buying the digital preservation services with the best quality, and it is compared with the multi-unit and combinatory auctions (both descending and reverse).

As observed in the three simulation scenarios (see Fig. 1), the best auction mechanism, that is, the one that is better able to manage the budget, is the combinatory

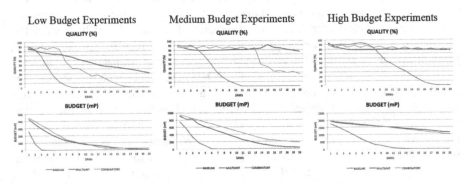

Fig. 1. Low, medium and high budget experiments

auction for high budgets, and the multi-unit auction for lower budgets. It has been found that the higher the budget is the better are the results of the combinatory auction regarding the other two e-auctions.

Finally, this result encourages us to follow the research in this line with experiments of wider scope and refined behavioral auction heuristics.

Acknowledgements. This research is partly funded by the TIN2013-48040-R (QWAVES) Nuevos métodos de automatización de la búsqueda social basados en waves de preguntas, the IPT20120482430000 (MIDPOINT) Nuevos enfoques de preservación digital con mejor gestión de costes que garantizan su sostenibilidad, and VISUAL AD Uso de la Red Social para Monetizar el Contenido Visual, RTC-2014-2566-7 and GEPID Gamificación en la Preservación Digital desplegada sobre las Redes Sociales, RTC-2014-2576-7, the EU DURAFILE num. 605356, FP7-SME-2013, BSG-SME (Research for SMEs) Innovative Digital Preservation using Social Search in Agent Environments, as well as the AGAUR 2012 FI_B00927 awarded to José Antonio Olvera and the grup de recerca consolidat CSI-ref. 2014 SGR 1469.

References

1. Cartledge, C.L., Nelson, M.L.: When Should I Make Preservation Copies of Myself? Digital Libraries 8–12 September 2014, London, UK. Available at http://www.cs.odu.edu/~mln/pubs/jcdl-2014/jcdl-2014-cartledge-copies.pdf
2. de la Rosa, J.L., Olvera, J.A.: First studies on self-preserving digital objects. In: Artificial Intelligence Research and Development, Proceedings of the 15th International Conference of the Catalan Association for Artificial Intelligence, CCIA 2012, vol. 248, pp. 213–222. Alacant (2012)
3. Nelson, M.: Buckets: Smart Objects for Digital Libraries. Ph.D. thesis, Old Dominion Univ (2001)
4. Olvera, J.A.: Digital preservation: a new approach from computational intelligence. In: Joint Conference on Digital Libraries 2013, JCDL Doctoral Consortium 2013, July 22–26, Indianapolis, Indiana, USA. Available at http://www.ieee-tcdl.org/Bulletin/current/papers/olvera.pdf (2013)
5. Olvera, J.A., de la Rosa, J.Ll., Carrillo, P.: Combinatorial and multi-unit auctions applied to digital preservation. In: Museros, L., et al. (eds.) Artificial Intelligence Research and Development, vol. 269, pp. 265–268. IOS Press, in press (2014). doi:10.3233/978-1-61499-452-7-265
6. Rotherberg, J.: Ensuring the longevity of digital documents. Sci. Am. **272**(1), 42–47 (1995)
7. Ruusalep, R., Dobreva, M.: Digital Preservation Services: State of the Art Analysis. Available at www.dc-net.org/getFile.php?id=467

Mobile Annotation of Geo-locations in Digital Books

Annika Hinze$^{(\boxtimes)}$, Haley Littlewood, and David Bainbridge

Department of Computer Science, University of Waikato, Hamilton, New Zealand
{hinze,davidb}@waikato.ac.nz, hml15@students.waikato.ac.nz

Abstract. This demo paper introduces an editor for manual annotation of locations in digital books, using a crowd-sourcing approach. It is the first of its kind and allows book lovers and literary travel enthusiasts to annotate the locations in their digital books on-the-go. We show both a mobile and a desktop version, and briefly explain the linkage to the Digital Library that is holding the digital books.

Keywords: Location · GPS · Geo-location in books · Semantic markup · Semantic annotation · Story lines · Literary tourism · Crowd-sourcing

1 Introduction

The importance of places and locations in literature is highlighted by Malcolm Bradbury's observation that a "very large part of our writing is a story of its roots in a place: a landscape, region, village, city, nation or continent" [3]. As a literary device, the role of places and locations is primarily to help explore aspects of identity [4], and so not only may a location in the narrative be real or imaginary, often the boundaries between these are intentionally blurred to create imaginary spaces [7].

Literary tourism is a type of cultural tourism, which focuses on the real-world locations of fictional works and the lives of the authors. It reflects a common desire in many readers to gain a deeper appreciation and understanding of a book they have read—particularly if a book made a significant impact on them. In many ways, modern literary tourism is the successor of the British *Grand Tour* of Victorian times, in which the upper class would voyage through the cultural centres of Europe [5]. Digital technologies make it easier for the literary tourist of today to engage with geo-information before or after their travel. Furthermore, gaining access to this information *in-situ* during their travels is technically getting easier now that literary tourists may use smart phones and tablets instead of, or in addition to, books and maps. Currently, many resources for literary tours are relatively static (e.g., web sites providing lists of locations sometimes accompanied by a stylized map). More compelling is the dynamic display of location information as a reader travels (e.g., as in Tipple [8]); however, for this to be supported, geo-location annotation of places and locations in literary works is required.

© Springer International Publishing Switzerland 2015
S. Kapidakis et al. (Eds.): TPDL 2015, LNCS 9316, pp. 338–342, 2015.
DOI: 10.1007/978-3-319-24592-8_30

This paper introduces an editor for manual annotation of locations in electronic books, stored in a digital library and, in doing so, provides a tool for crowd-sourcing the location annotations. Previously, we used simplified automatic annotation using a gazetteer [10], and explored more recently the requirements of creating geo-locations for digital books in three case studies. The editor, called *Book Navigator*, offers both a mobile app version and a desktop-based interface for creating geo-location annotations, which are stored in addition to the eBook in Greenstone [12]. The remainder of the paper shows a brief walk-through of the mobile and desktop system (Sect. 2) along with a overview of the system architecture and implementation in Sect. 3. We discuss related approaches in Sect. 4, and conclude with a short summary.

2   Walk-Through

Figures 1 and 2 show screenshots of the mobile *Book Navigator* interface for annotating locations in books *in-situ*. In the example shown, a user visiting the Hamilton Gardens is annotating a book about the Paradise Gardens on their mobile device. Figure 1 is the document view during the annotation of a short paragraph. Navigation through the book is done through up/down scrolling; annotation is initiated through a long-click on a text element. In this screen, one can see that the user is able to select an area or a path (two buttons at the lower end of the screen) to annotate. Figure 2 shows the equivalent map view based on the user's current location. A path is defined through a number of locations; its direction is defined through the order in which the locations are selected. Users can either annotate locations with their current position (based on GPS) or with distant locations selected via the map. The desktop interface is not shown here; in the desktop version all locations have to be explicitly selected on the map for annotation. However, the larger screen makes it possible to show both text and map at the same time.

Fig. 1. Book view

Fig. 2. Map view

3   Architecture and Implementation

Figure 3 shows the architecture of the editor software for annotating eBooks with locations. The map information is imported from Open Street Map, the eBooks are held in a Greenstone Digital Library. The display component allows a user to select an eBook from a collection, and display the text in the Map Display. The user can navigate through the text of the book; via the Annotator component, they can then select a part of the text and assign location markers—see Fig. 4 for the conceptual diagram of the relationships between marker, location markup, text annotation, and book. The annotations (i.e., text positions and location information) are first stored locally on the mobile device (to ensure independence from

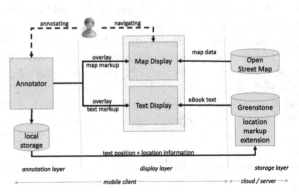

Fig. 3. Architecture of mobile annotator

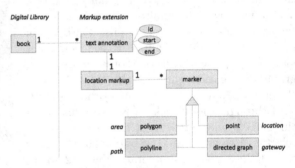

Fig. 4. Conceptual design

a server connection) and can be uploaded later to the digital library running on the server side or to a cloud storage for exchange. This architecture can also be run self-contained on a smart phone using the mobile configuration of Greenstone [2]. We currently support (see Fig. 4) markers for areas (polygon), paths (polyline), single GPS locations (point) and gateways to indicate entries and exists (directed graph). Text elements (e.g., words, phrases or paragraphs) can have several locations assigned. The annotation only stores the beginning and the end of an annotation and does not refer to the text content itself.

4   Related Work

Maps of literary spaces are typically hand-crafted, i.e., the locations are collected through literary analysis and then entered into maps. Examples of such mappings are found in the "Literary Atlas of Europe" [16] and the "Atlas of Literature" [3]. Older projects are "Projekt Historischer Roman" (mapping 6700

German language novels) and "Atlas das Literaturas Regionais do Brasil" (mapping 550 Brazilian literary works); however, neither of these provide open sources nor allow further location annotations [15]. The *Gutenkarte* automatically identifies geo-locations in classic texts from project Gutenberg (http://www.gutenberg.org/) and the MetaCarta API. Its approach is similarly simplistic and limited as our work using a gazetteer of place names [10].

Literature analysis has long focussed on the temporal aspects of a narrative. In recent decades, this has shifted to a focus on spatial aspects [6], which has been referred to as a *spatial turn* [17] and *topographical turn* [18]. Bachti uses the concept of *chronotopoi*—elements of space-time of a story—which form the foundation for "showing and representing" events in a literary text, e.g., on maps or timelines [1].

Piatti et al. [15] note that literary geography is based on individual readings and acknowledge a *double uncertainty* in both the primary material (uncertainty of location information in the text) and in the methodology (location annotation). We embrace his uncertainty and the highly individually biased nature of literary location annotations through our crowd-sourcing approach. Inspiration for our annotation interface was drawn from semantic annotation tools [11]. The native support for semantics in digital libraries (e.g., JeromeDL [13], Greenstone [9]) typically abstracts at the document level and these works did not provide the type of fine grained access required here. In fact, most systems that provide location-based information either use automatically collected coordinates [14] or expert manual location markup [3,16].

5 Summary

This paper introduced our editor prototype for creating mobile annotation of geo-locations in eBooks. By storing the books and suggested annotations in a digital library, we provide an environment that allows the crowd-sourcing of this information. We investigated related work on mapping literature and distinguished our crowd-sourcing approach from the work of literary cartographers.

References

1. Bachtin, M.: Formen der Zeit und des Chronotopos im Roman. In: Kowalski, E., Wegner, M. (eds.) Untersuchung zur Poetik und Theorie des Romans. Aufbau Verlag (1986) (1937/38)
2. Bainbridge, D., Jones, S., McIntosh, S., Witten, I.H., Jones, M.: Beyond the client-server model: self-contained portable digital libraries. In: Buchanan, G., Masoodian, M., Cunningham, S.J. (eds.) ICADL 2008. LNCS, vol. 5362, pp. 294–303. Springer, Heidelberg (2008)
3. Bradbury, M., Ahrens, R.: The atlas of literature. De Agostini Editions London (1996)
4. Brown, P., Irwin, M.: Literature and Place, 1800–2000. Peter Lang (2008)

5. Cunningham, S.J., Hinze, A.: Supporting the reader in the wild: identifying design features for a literary tourism application. In: SIGNZ Human-Computer Interaction (CHINZ) (2013)
6. Fischer-Lichte, E.: The shift of a paradigm: from time to space? introduction. In: XIIth Congress of the International Comparative Literature Association, vol. 5, pp. 15–18 (1988)
7. Fluck, W.: Theories and methods imaginary space; or, space as aesthetic object. Space, Place, Environment **15**, 15 (2004)
8. Hinze, A., Bainbridge, D.: Listen to Tipple: creating a mobile digital library with location-triggered audio books. In: Zaphiris, P., Buchanan, G., Rasmussen, E., Loizides, F. (eds.) TPDL 2012. LNCS, vol. 7489, pp. 51–56. Springer, Heidelberg (2012)
9. Hinze, A., Buchanan, G., Bainbridge, D., Witten, I.: Semantics in Greenstone. In: Kruk, S.R., McDaniel, B. (eds.) Semantic Digital Libraries, pp. 163–176. Springer, Heidelberg (2009)
10. Hinze, A., Gao, X., Bainbridge, D.: The TIP/Greenstone bridge: a service for mobile location-based access to digital libraries. In: Gonzalo, J., Thanos, C., Verdejo, M.F., Carrasco, R.C. (eds.) ECDL 2006. LNCS, vol. 4172, pp. 99–110. Springer, Heidelberg (2006)
11. Hinze, A., Heese, R., Luczak-Rösch, M., Paschke, A.: Semantic enrichment by non-experts: usability of manual annotation tools. In: Cudré-Mauroux, P., Heflin, J., Sirin, E., Tudorache, T., Euzenat, J., Hauswirth, M., Parreira, J.X., Hendler, J., Schreiber, G., Bernstein, A., Blomqvist, E. (eds.) ISWC 2012, Part I. LNCS, vol. 7649, pp. 165–181. Springer, Heidelberg (2012)
12. Witten, I.H., Bainbridge, D., Nichols, D.M.: How to Build a Digital Library. Morgan Kaufman Publishers, San Francisco (2009)
13. Kruk, S., Cygan, M., Gzella, A., Woroniecki, T., Dabrowski, M.: JeromeDL: the social semantic digital library. In: Kruk, S.R., McDaniel, B. (eds.) Semantic Digital Libraries, pp. 139–150. Springer, Heidelberg (2009)
14. Pat, B., Kanza, Y., Naaman, M.: Geosocial search: finding places based on geotagged social-media posts. In: International World Wide Web Conference. ACM (1915)
15. Piatti, B., Reuschel, A.-K., Hurni, L.: Literary geography-or how cartographers open up a new dimension for literary studies. In: International Cartography Conference (2009)
16. Reuschel, A.-K., Piatti, B., Hurni, L.: Mapping literature. the prototype of "A Literary Atlas of Europe". In: Proceedings of the 24th International Cartographic Conference (2009)
17. Soja, E.W.: Thirdspace: Journeys to Los Angeles and other real and imagined places. Trans. Inst. Br. Geogr. **22**(4), 529–540 (1997)
18. Weigel, S.: On the 'topographical turn': concepts of space in cultural studies and Kulturwissenschaften. a cartographic feud. Eur. Rev. **17**(01), 187–201 (2009)

Teaching Machine Learning: A Geometric View of Naïve Bayes

Giorgio Maria Di Nunzio[⊠]

Deptartment of Information Engineering, University of Padua, Padua, Italy
dinunzio@dei.unipd.it
http://www.dei.unipd.it/~dinunzio/

Abstract. In this demo, we present two applications which allow users to 'see' a geometric interpretation of the Bayes' rule and interact with a Naïve Bayes text classifier on a real dataset, namely the Reuters-21578 newswire collection. The main objective of this demo is to show how the pattern recognition capabilities of the human increase the effectiveness of the classifier even when technical details are not known in advance or the user is not an expert in the field. These two applications were developed with the R package Shiny; they have been deployed online and they are freely accessible from the links indicated in the paper.

1 Introduction

When we want to quantify the uncertainty of the outcome of an experiment, we can use Bayesian modelling to build the mathematical model of the experiment. In this context, Bayes' rule is used to compute the posterior probability of a variable given some observed data. Posterior probabilities can be hard to compute; therefore, a "naïve" solution is to make some assumptions that allow for a factorisation of the posterior probability into simple conditional probabilities which are easy to compute [3]. Naïve Bayes (NB) classifiers have been widely used in the literature of Data Mining and Machine Learning since they are easy to train and reach satisfactory results which are comparable to the results of more complex state-of-the-art classification algorithms. However, the optimisation of the parameters of NB classifiers is often not adequate, if not missing at all. Based on the idea of Likelihood Spaces, a two-dimensional representation of probabilities [2], we have developed two Web applications which provide an adequate data and knowledge visualization to teach in a real machine learning setting how parameter optimisation and misclassification costs affect the performance of the classifier. In this demo, we show a geometric interpretation of the Bayes' rule which can be used to teach to non-experts how NB works and how to optimise the parameters of these classifiers in a very intuitive way. In addition, we present a real text classification problem based on the Reuters 21578 collection.[1]

---

[1] http://www.daviddlewis.com/resources/testcollections/reuters21578/.

© Springer International Publishing Switzerland 2015
S. Kapidakis et al. (Eds.): TPDL 2015, LNCS 9316, pp. 343–346, 2015.
DOI: 10.1007/978-3-319-24592-8_31

2   Mathematical Background

Bayes' rule gives a simple but powerful link between prior and posterior probabilities of events. For example, assume that we have two classes c_1 and c_2 and we want to classify objects according to some measurable features of the objects. The probability that an object o belongs to c_1 is:

$$\overbrace{P(c_1|o)}^{posterior} = \frac{\overbrace{P(o|c_1)}^{likelihood}\ \overbrace{P(c_1)}^{prior}}{P(o)} \tag{1}$$

Bayes' rule tells us that, starting from a prior probability on c_1, we may change our idea about the probability of class c_1 after observing the object o, according the 'likelihood' of that object. If o is represented by a set of features $F = \{f_1, f_2, ..., f_j\}$, a Naïve Bayes approach factorises $P(o|c_1)$ as:

$$P(o|c_1) = \prod_j P(f_j|c_1) \tag{2}$$

where features are independent from each other given the class, that is the conditional independence assumption.

3   Demo

The first part of the demo[2] shows the implications of the visual interpretation of the Bayes' rule on a two-dimensional space. In the second part of the demo,[3] we show how this two-dimensional space can be used in a real scenario of text classification using the Reuters-21678 newswire collection.

Bayes' Rule. If we imagine $P(c_1|o)$ and $P(c_2|o)$ as two coordinates of a cartesian space, we can draw objects on the segment with endpoints $(1,0),(0,1)$ (since $P(c_1|o) = 1 - P(c_2|o)$). In Fig. 1, objects are represented by three binary features f_1, f_2, and f_3 (this is called a multivariate Bernoulli NB classifier). Each point in the plot represents an object (three binary features, $2^3 = 8$ objects), when a point is below the line that passes through the origin (i.e. $P(c_1|o) > P(c_2|o)$), it is classified under c_1. The user can change the values of the conditional probability for each feature and class and see the effect in terms of the position of the points.

The demo allows users to study situations that are more realistic in terms of real machine learning problems. First, it is often the case that the normalisation factor $P(o)$ is not computed when classifying the object because it is a constant that does not change the classification. By de-selecting it from the interface, we see how the coordinates change accordingly $(P(c_1|o) \propto P(o|c_1)P(c_1))$. Users can

---

[2] https://gmdn.shinyapps.io/bayes2d/.
[3] https://gmdn.shinyapps.io/shinyK/.

Bayesian 2D

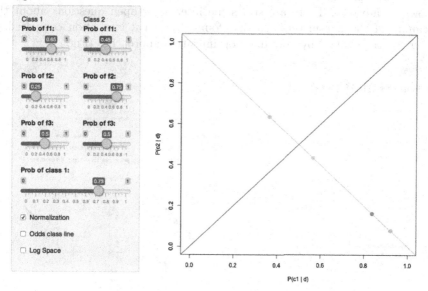

Fig. 1. Teaching how Bayes' rule work on a two-dimensional space.

also adjust the prior $P(c_1)$ and see the effect of an "unbalanced" class situation. When normalisation is not active, we can even choose whether to shift points or rotate the classification line by selecting 'Odds class line' ($P(c_1|o) \propto P(o|c_1)$). In addition, it is possible to show the coordinates of the objects as logarithms of probabilities by selecting 'Log space' on the interface. Therefore we can study the problem in terms of log-likelihood, $\log(\prod P(f_j|c_1)) = \sum \log(P(f_j|c_1))$.

Newswires Classification Task. In the second part of the demo, we show a real text classification problem with a multivariate NB classifier. The interface allows users to: choose one out of ten categories, select the number of training/validation folds and features for training the classifier, smooth probabilities to avoid zero probabilities [5], adjust misclassification costs [4]. In real cases like this one, the logarithm transformation is necessary. Starting from default parameters, the user can interact with the model and find the parameters that produce a good separation between the two classes with little effort and without necessarily being an expert in the field [1].

4   Conclusions

The objective of this demo was two-fold: to introduce a geometric interpretation of Bayes' rule which can be used to teach how NB classifier works in an intuitive way; to show how the pattern recognition capabilities of the human can improve

the effectiveness of the default NB classifier even when technical details are not
known in advance. There are still some interesting open questions about the
meaning of the parameters that are found by visual inspection compared to
other solutions found by automatic optimization approaches (Fig. 2).

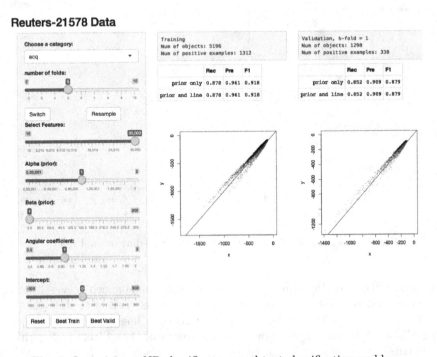

Fig. 2. Optimizing a NB classifier on a real text classification problem.

References

1. Ankerst, M., Ester, M., Kriegel, H.-P.: Towards an effective cooperation of the user
   and the computer for classification. In: Proceedings of the Sixth ACM SIGKDD
   2000, pp. 179–188 (2000)
2. Di Nunzio, G.M.: A new decision to take for cost-sensitive naïve bayes classifiers.
   Inf. Process. Manag. **50**(5), 653–674 (2014)
3. Domingos, P., Pazzani, M.: On the optimality of the simple bayesian classifier under
   zero-one loss. Mach. Learn. **29**(2–3), 103–130 (1997)
4. Elkan, C.: The foundations of cost-sensitive learning. In: Proceedings of the 17th
   International Joint Conference on Artificial Intelligence, IJCAI 2001, vol. 2, pp.
   973–978. Morgan Kaufmann Publishers Inc., San Francisco (2001)
5. Yuan, Q., Cong, G., Thalmann, N.M.: Enhancing naive bayes with various smooth-
   ing methods for short text classification. In: Proceedings of the International Conference on WWW 2012, pp. 645–646. ACM, New York (2012)

Study About the Capes Portal of E-Journals Non-users

Wesley Rodrigo Fernandes$^{(\boxtimes)}$ and Beatriz Valadares Cendón

School of Information Science,
Federal University of Minas Gerais, Belo Horizonte, Brazil
wesleyronline@yahoo.com.br, cendon@eci.ufmg.br

Abstract. This study investigated the non-users of the CAPES Portal of E-Journals, a governmental initiative to offer free access to e-journals to consortiated federal educational and research institutions in Brazil. The research used a, mostly, quantitative research methodology, which collected some qualitative data through a Web Survey. 16.1 % of survey respondents declared that they did not use the Portal. These non-users were asked (1) which were the reasons which led to non-use, (2) if there were other electronic information sources they used, and (3) if they would use the Portal in case the barriers identified for the non-use were remedied. The results show that the non-use of the Portal is caused mainly because respondents lack information about its existence (24.5 % of responses), use other information resources (22.3 % of responses) and prefer printed journals (11.6 % of responses). Another important finding is that 82.1 % of the non-users would become users if the barriers indicated by them as the cause of non-use were solved. The study contributes to the scarce literature on non-users of digital libraries and presents recommendations to improve use based on the results obtained.

Keywords: Study of non-users · Digital libraries · Electronic journals · Capes portal of E-Journals (Brazil)

1 Introduction

The Capes Portal of E-Journals was launched in Brazil in November of 2000 as a governmental initiative. It currently provides, to over 420 higher education or research institutions, access to the full text of over 37,000, mostly international, journals. It also offers some 130 reference databases in addition to technical standards, patents, theses, dissertations and books covering all areas of knowledge [8]. The Portal is freely accessible locally through the IPs of the licensed institutions. Remote access to the full contents of the site is also possible if the partner institution offers this possibility to its community of users.

Because of its importance and relevance, several studies have adopted the Portal as an empirical object of research. Some of this research have identified that part of the potential users of the Portal do not access it [3, 5, 6]. Motivated by this finding, the current study proposed, as a research question, to investigate why a portion of the faculty in universities belonging to the Capes Consortium do not use or underutilize the Capes Portal of E-Journals.

© Springer International Publishing Switzerland 2015
S. Kapidakis et al. (Eds.): TPDL 2015, LNCS 9316, pp. 347–350, 2015.
DOI: 10.1007/978-3-319-24592-8_32

2 Non-users Studies

While there is a vast literature discussing the studies of users of information sources, the study of the non-users has not been given as much attention. Totterdell [9] points out that user studies are very much focused on users and warns that studies must extend this population in the coming years as it is also necessary to know the reasons for non-use. Le Coadic [4] pointed out that all information systems have non-users and these are by far more important than the users.

The advent of the Internet brought to reality the possibility of remote access to the digital format of traditional journals. The access to and use of full-text digital scholarly journals has risen dramatically over the last ten years – and still continues to rise at a substantial rate, leading to a revolution in scholarly communications. A massive and unprecedented migration of research works from the print to the virtual world is happening [7].

Adegbore [1] emphasizes that scholars indicate satisfaction with and acceptance of electronic resources and signal interest in continuing to use them. Given this finding, the author warns that barriers to their use should be reduced. To this end, information elicited from non-users can provide an excellent source of feedback to help identify and diminish the problems, as highlighted by Consonni [2]. Non-users can help to point out service limits and issues, give suggestions for improvement and provide information about what it is generally expected from the digital library. However, information about the non-user is not easily available: there is not a bibliography of studies of non-users of digital libraries, and although the topic is mentioned in a few researches as a challenge, it is never seriously tackled [2].

3 Methodology

Data was collected through a web survey. Initially the researchers sent an invitation letter to 14,763 faculties from 17 universities which had access to the Portal. All universities selected had schools and courses on eight great areas of knowledge and were distributed in the five geographic regions of the country. Faculty who did not respond to the initial message, received two additional invitations, with an interval of 30 days between them, encouraging them to participate in the study. The Web Survey included four questionnaires. The first consisted of questions relating to personal and professional characteristics of respondents and was directed both to users and non-users. The second was aimed at users and had questions regarding user behavior and interaction with the Capes Portal. The third was about user satisfaction with various aspects of the Portal. The last one was dedicated to the non-user of the Capes Portal of E-Journals. The non-user questionnaire had 9 questions. In the current work, questions 1, 2 and 9 were analyzed and presented. Question 1 asked the respondent about the motives for not using the Portal. Question 2 inquired about the use of other e-sources. Question 9 was about respondent's reaction about difficulties and barriers to use of the Portal. Other questions of the questionnaire dealt with non-user behavior or were qualitative questions which, for reasons of space, could not be included in this paper.

4 Presentation of Results

This session will present the results obtained from the analysis of the data. First the session presents an overview of the total number of respondents of the survey, followed by the analysis and discussion of the answers by the non-users to the questionnaire. The first questionnaire of the Web Survey had a total of 6,689 respondents. Of these, 1,075 (16.1 %) reported that they did not use the Capes Portal. Only 1,017 of these non-users continued in the Web Survey and answered the next questionnaire dedicated to the Portal non-users.

In Question 1 of the non-user questionnaire respondents had the option to mark twelve pre-established reasons for non-use of the Portal. Respondents could score as many options as judged necessary. Percentages reported below refer to total number of checked responses (1,536) not total number of respondents (1,017). The top five reasons for the non-use of the Portal, which account for 71.5 % of the options marked by Portal non-users were: not knowing about the Portal existence (reason indicated by 24.5 % of responses), the use of other resources (reason indicated by 22.3 % of responses), the preference for printed journals (reason indicated by 11.6 % of responses), the difficulty of access (reason indicated by 6.6 % of responses) and the impossibility of accessing the Website from home (reason checked by 6.4 % of the responses). The remainder of the reasons (It is time consuming/I don't have time, I don't have practice/I'm not able to use the computer, It is difficult to use, I don't need it, I'm not interested on it, I (don't like)/(am afraid of) computers and other) accounted for 28.5 % of the answers.

Question 2 of the questionnaire tried to identify other electronic information sources used by respondents. This question allowed the selection of multiple responses. Faculty had the option of selecting eight options already preset in the questionnaire. Percentages reported below refer to total number of checked options (2,657) not to total number of respondents (1,017). The four other sources most used by the non-users of the Portal, accounting for 73.3 % of the electronic sources used were: search engines on the Internet such as Google or Yahoo (source indicated by 28.1 % of responses); websites in general (source indicated by 18.7 % of responses); libraries on the Internet (source indicated by 14.8 % of responses); and online catalogs of libraries (source indicated by 11.7 % of responses). The remainder other sources (other e-journals in the internet; databases on CD-ROM or Internet; open archives and others) accounted for 26.7 % of the answers.

The ninth question of the questionnaire asked whether the respondent would use the Portal if problems or barriers were eliminated. Only 1.4 % of the Portal non-users said they would never use it in case the barriers they reported were remedied. It was also interesting to note that the response options representing a higher frequency of use of the Website if the barriers were eliminated (regularly every week and regularly every month), were the ones that had the highest number of responses (62.2 % of the responses). Both results show the importance placed by faculty on the Portal for university teaching and research.

5 Discussion and Conclusion

Results show that even the portion of the faculty that reported they do not use the Portal recognizes it as highly important and would use it if the problems they pointed out were eliminated. It is thus highly recommendable that attention be given to remedy these difficulties. The most important of these is not being informed about the Portal or even about its existence. Although there is already great effort on this area, this study recommends that new sources of electronic information, such as Portals or digital libraries, spend high efforts on visibility and be widely advertized to their potential users. Users should understand the contents and utility of the information resources present in the Portal to their teaching and research. Another important reason for non-use of the Portal is the preference for other e-resources such as search engines and other Web sites. In the case of the Portal, the user should be educated about the value and difference of the selective, expensive scientific databases and journals, freely available at the Portal, as compared to information on the Internet. Research about difficulties of the users in their interaction with the Portal is recommended to further understand non-use and underutilization of the resource.

The current study contributes to fill a gap in research about the non-user of digital libraries, considering the shortage of studies of non-users in the scientific literature. The study also may help in the elimination of barriers that hinder the use not only of the Capes Portal, but also of other portals and digital libraries, being, thus, of importance to understand the reasons why these electronic information systems are not used or underutilized.

References

1. Adegbore, A.M.: University faculty use of electronic resources: a review of the recent literature. PNLA Q **75**(4), 65–76 (2011)
2. Consonni, C.: Non-users'evaluation of digital libraries: a survey at the università degli studi di Milano. Int. Fed. Libr. Assoc. Inst. **36**(4), 325–331 (2010)
3. Cunha, A.Á.L.: Uso de bibliotecas digitais de periódicos: um estudo comparativo no Portal de Periódicos Capes entre áreas do conhecimento, thesis, Escola de Ciência da Informação, Universidade Federal de Minas Gerais (2009)
4. Le Coadic, Y.-F.: A ciência da informação. Briquet Lemos/Livros, Brasília (1996)
5. Maia, L.C.G.: Uso de periódicos eletrônicos: um estudo sobre o Portal de Periódicos Capes na Universidade Federal de Minas Gerais, thesis, Escola de Ciência da Informação, Universidade Federal de Minas Gerais (2005)
6. Martinez, M.L., Ferreira, S.M.S.P., Galindo, M.: Estudo de usabilidade do Portal de Periódicos da Capes: análise de perfil do usuário discente da UFPE. RBPG **8**(15), 61–107 (2011)
7. Nicholas, D., Rowlands, I., Williams, P.: E-journals, researhers – and the new librarians. Learn. Publ. **24**(1), 15–27 (2011)
8. Portal de periódicos Capes (2015). http://periodicos.capes.gov.br
9. Totterdell, B.: Libraries and their users. In: Harrison, K.E. (ed.) Prospects for British Librarianship, pp. 140–151. The Library Association, London (1976)

Czech Digital Library – Big Step to the Aggregation of Digital Content in the Czech Republic

Tomas Foltyn[1(✉)] and Martin Lhotak[2]

[1] National Library of the Czech Republic, Prague, Czech Republic
tomas.foltyn@nkp.cz
[2] Library of the Czech Academy of Sciences, Prague, Czech Republic
lhotak@knav.cz

Abstract. It is necessary to solve interoperability, compatibility of standards and even legal issues to secure aggregation of the digital content on the national level. Activities in the Czech Republic aim to create environment based on open source software to achieve such goals. New tools and systems are developed to secure complex digitization processes including digital data processing, work-flow monitoring, archiving and providing access to e-content in digital libraries. It is very cost effective to provide open source solutions for digital data production and dissemination on national level. Usage of the same solutions between culture heritage institutions results also in sharing the same data and metadata standards which is advantage in aggregation process.

Keywords: Digital libraries · Digitization · Aggregation · Digital data production · Communications protocols · Standards · Interoperability · Repository

1 Introduction

The interoperability of digital libraries and development of the open source solutions supporting digital document production and dissemination is the main topic of the development of the project "Czech Digital Library and Tools for the Management of Complex Digitization Processes"[1] which is supported by the Ministry of Culture of the Czech Republic. In this context interoperability makes possible to share and re-use the content (informational materials) from different digital collections and manage also the way of the production. The main requirement - to achieve interoperability - is reached by usage of the specific library communication protocols (e.g., OAI-PMH, Z39.50, SRU/SRW) and metadata standards (e.g., Dublin Core, METS, MODS, MIX, PREMIS, FOXML).

---

[1] Please see the official project webpage http://www.czechdigitallibrary.cz/, where all the general information about the project are available.

© Springer International Publishing Switzerland 2015
S. Kapidakis et al. (Eds.): TPDL 2015, LNCS 9316, pp. 351–354, 2015.
DOI: 10.1007/978-3-319-24592-8_33

2 Summary of the Project Goals

2.1 General Information

The goal of the project is to create the Czech Digital Library ("CDL") which will aggregate the content of digital libraries in the Czech Republic. It will serve both as a uniform interface for end-users and as a primary data provider for international projects, especially for Europeana, the UNESCO digital library etc. It will also be an important source of digital data generally and it is one of the main pillars needed to provide centralized digital services in the Czech Republic, as they are defined in the "Library Development Strategy of the Czech Republic for 2011 to 2015," approved by the Czech Ministry of Culture.[2] The whole system solution is composed from various independently working solutions used in the Czech Republic for many years (Fig. 1).

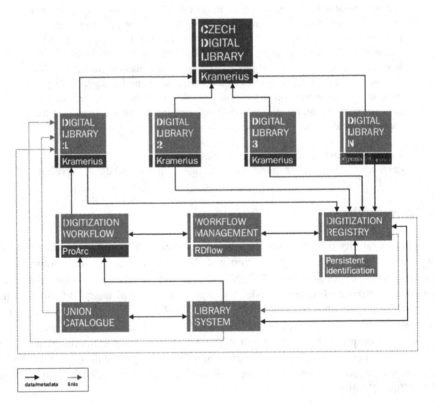

Fig. 1. Czech Digital Library - schema of the data flow

[2] All the information about the activities supported by the Ministry of Culture are available via website http://www.ukr.knihovna.cz/koncepce-rozvoje-knihoven-cr-na-leta-2011-2015-/.

2.2 KRAMERIUS Digital Library[3]

The Kramerius System is a software solution used as the main access point to digital documents coming from the modern documents digitization. It is primarily intended for digitized library collections, monographs and periodicals. It can be used to present other types of documents as well, e.g. maps, music sheets and old prints, or parts of documents, such as articles and chapters. The system is also suitable for the so-called born digital documents, the documents that originated in electronic form. Kramerius is continually adjusted to match the structure of the metadata standards proclaimed by the National Library of the Czech Republic.[4] The system provides an interface for end users, providing search in metadata and full text, generating multi-page PDF documents from selected pages, creating virtual collections and other operations on the stored collection of digital documents defined. It builds on the functionality of the previous versions of Kramerius ending marking 3.3.1. Open source Fedora Repository is used as the back-end solution. The third party technologies – e.g. Apache, Apache Solr, Postgres SQL were used more often during development. The system is based on Java technology and it can be run as web application in any of the J2EE container (e.g. Apache Tomcat). Almost libraries in the Czech Republic use Kramerius system to run their digital library. All installation hold currently more than 60 million pages together.

The open source Kramerius system will be the initial software solution for the Czech Digital Library. Ensuring the interoperability with various types of digital libraries and institutional repositories is necessary. Besides data harvesting from different instances of the Kramerius system, it is also necessary to arrange a connection with other systems (e.g., Dspace, Eprints, Digitool). It will also serve as OAI-PMH provider with the ESE profile support to share data with Europeana.

2.3 Digitization Registry[5]

The Digitization Registry[6] which was built formerly as a project of the Library of the Academy of Sciences and the National Library of the Czech Republic, is used as an interconnecting system and relevant source of information. It holds a large amount of information about digitized documents in the Czech Republic, including the identification of original printed documents, owner and location of the digital library where the digital document is available, persistent identification (e.g. Czech National Bibliography Number) and other relevant entries. The main aim of the national registry of the digitized documents is to avoid unwanted duplication to enable the sharing of digitization results throughout the Czech Republic and not waste time or money by scanning the same documents. These days almost 220 000 titles are involved in the application. Almost no

---

[3] More information about the Kramerius system can be seen in Foltyn, T: The Kramerius Systém; or Foltyn T.: Kramerius Information System.

[4] Valid standardization based on METS, MODS, ALTOxml and other worldwide used formats is available via http://www.ndk.cz/standardy-digitalizace.

[5] Please see Foltyn, T: Has It Been Already Digitized?; or Foltyn T.: Registrdigitalizace.cz.

[6] The Registry is publicly accessible online at http://www.registrdigitalizace.cz/.

duplicities has arisen from the everyday usage of the system. The Digitization Registry could also provide tools for digitization workflow management to simplify the process of monitoring the digitization. Very important is the fact that it cooperates with library catalogue systems as well as with digital document repositories.

In the context of interoperability and cooperation with library information systems, the Registry is designed to communicate and cooperate automatically with other library information systems as much as possible. It uploads bibliographic records of items chosen for digitization in batches exported from the Aleph catalogue in MARCXML. The Registry is able to harvest data from digital libraries via OAI-PMH to import data describing digitized documents. Finally, it sends information about completed digitization to library OPACs together with a link to digital documents. Information is subsequently sent from library OPACs to the Union catalogue of the Czech Republic.

2.4   ProArc – Digital Data Production System

ProArc is a digital document production and archiving system based on the Fedora Commons repository. Using the same production and archiving tool will enhance interoperability and data sharing between individual digitization projects. The rapid semi-automatic creation of the standard metadata is enabled by the production system. It is comprised of structural, descriptive and archival metadata, OCR and conversion to specific graphical formats. With regard to the archival part of the solution, standards for long term archiving, such as the OAIS model and archiving ISO standards, will be implemented.

3   Conclusion

Mutual interoperability between all developed systems and tools is accented in the frame of the project as well as interoperability with solutions already existing on the market. The aim is to share, use and reuse digital content as easily and effectively as possible and to provide access to all digitized documents in the Czech Republic under one roof.

References

1. Foltyn, T.: The Kramerius system – open source solution for digital libraries. In: Proceedings of the Third Workshop on Very Large Digital Libraries; Glasgow, Scotland (UK), 10 September 2010, Pisa 2010 (2010). ISBN 978-88628015-7
2. Foltyn, T.: Kramerius Information System as a Tool of Access to Digital Documents, PRESIDENTIAL LIBRARY COLLECTIONS, Digital Library series, Issue 3, Coordination and Standardization in the Sphere of Creation and Use of the National Information Resources (2012). ISBN 978-5-905273-25-4
3. Foltyn, T.: Has It Been Already Digitized? How to Find Information about Digitized Documents, Review of the National Center for Digitization, Faculty of Mathematics, Belgrade (2013). (Issue: 22, ISSN: 1820-0109)
4. Foltyn, T.: Registrdigitalizace.cz, IT LIB 2010/2, Bratislava (2010)

Mirpub v2: Towards Ranking and Refining miRNA Publication Search Results

Ilias Kanellos[1,2(✉)], Vasiliki Vlachokyriakou[1], Thanasis Vergoulis[1],
Georgios Georgakilas[3], Yannis Vassiliou[2], Artemis K. Hatzigeorgiou[3],
and Theodore Dalamagas[1]

[1] IMIS, Athena RC, Marousi, Greece
ilias.kanellos@imis.athena-innovation.gr
[2] National Technical University of Athens, Kesariani, Greece
[3] University of Thessaly, Volos, Greece

Abstract. In recent years, many articles studying microRNA (miRNA)
molecules and their connection to diseases have been published. However,
the wide range of literature in life sciences, raises a barrier in extracting
useful information from them. MirPub is a search engine that resolves
this issue, providing lists of articles related to particular miRNA terms
and useful filters to customise them. In this work, we extend mirPub
by utilising publication ranking methods to provide insights about the
importance of each publication. Moreover, we automatically identify the
species referred to in each publication to serve researchers studying particular species.

Keywords: microRNAs · Scientific databases · Publication ranking

1 Introduction

MicroRNAs (*miRNAs*) are small RNA molecules than inhibit the expression of
genes [1] and, thus, relate to many diseases, such as various types of cancer [9].
Because of their importance, many databases collecting microRNA-related data
have been developed in recent years [3,6,7].

MirPub [5] is a search engine for scientific publications related to miRNAs. It
uses an index on scientific publications collected from the U.S. National Library
of Health[1] to capture associations of miRNAs to scientific papers. MirPub aims
to provide a comprehensive set of publications related to any keyword that
describes a miRNA, by applying expansion rules on it. Currently mirPub indexes
20,690 papers associated to 31,984 distinct miRNA keywords.

Hitherto, mirPub provides the publications related to a set of miRNA terms
in descending chronological order. Since each term may relate to hundreds of publications, locating important information can be tedious. To address this issue,
mirPub v2 applies ordering based on publication ranking algorithms. Moreover,

---

[1] http://www.ncbi.nlm.nih.gov/pubmed.

© Springer International Publishing Switzerland 2015
S. Kapidakis et al. (Eds.): TPDL 2015, LNCS 9316, pp. 355–359, 2015.
DOI: 10.1007/978-3-319-24592-8_34

since researchers usually focus on studies related to particular species, mirPub v2 uses text mining to identify the species mentioned in each publication and provides a useful species filter that can be applied on the publication lists.

2 Ranking Publications

Ordering mirPub's lists of publications based on their contribution would facilitate focusing on important works. Traditionally, the importance of publications has been measured by counting citations. However, this measure cannot capture that citations made by important articles are more valuable than the others. To address this issue, PageRank [4], an algorithm originally introduced for ranking Web pages, has been recently applied on citation networks.

PageRank assigns a score $PR(p_i)$ to each publication p_i according to the following recursive formula:

$$PR(p_i) = \alpha \sum_{p_j \ cites \ p_i} \frac{PR(p_j)}{k_j} + (1 - \alpha)\frac{1}{n} \tag{1}$$

where n denotes the total number of publications in the citation network, p_j and k_j refer to a publication citing p_i and the total number of its references, respectively, and $\alpha \in [0,1]$ is a parameter of the algorithm. Thus, $PR(p_i)$ is the weighted sum of the scores of all p_j's that cite p_i plus a constant. The weighted sum captures the likelihood of someone reading p_i after following references to it from other articles, while the constant represents the probability of randomly selecting it. Based on the previous, we implemented PageRank for mirPub v2.

PageRank may misjudge important articles if they are not old enough to collect a representative number of citations. In [2], Hwang et al. proposed an adaptation that overcomes this by (a) favoring new articles, and (b) incorporating the impact factor (IF) of the article's journal in calculations. We refer to this algorithm as *YetRank*; based on it, we implemented two ranking schemes:

$$YR_1(p_i) = \alpha \sum_{p_j \ cites \ p_i} \frac{YR_1(p_j)}{k_j} + (1 - \alpha) \cdot IF(p_i) \cdot e^{\frac{t_{p_i} - t_{current}}{\tau}} \tag{2}$$

$$YR_2(p_i) = \alpha \sum_{p_j \ cites \ p_i} \frac{YR_2(p_j)}{k_j} + \beta \cdot IF(p_i) + \gamma \cdot e^{\frac{t_{p_i} - t_{current}}{\tau}} \tag{3}$$

where $IF(p_i)$ and t_{p_i} denote the impact factor of the journal and the year in which p_i was published, respectively, $t_{current}$ denotes the current year, and $\alpha, \beta, \gamma, \tau$ are tunable constants. Note that in Eq. (3) we set $\alpha + \beta + \gamma = 1$.

Both PageRank and YetRank1-2, were finally executed on the citation graph created for the mirPub papers[2]. The parameters of Eqs. (2) and (3) were tuned as follows: first PageRank was executed on mirPub's citation graph. Then, using different configurations, YetRank1-2 were executed on the subgraph of publications which are older than five years. The values for which the two rankings had the best spearman's rho correlation [8] were finally selected.

---

[2] MEDLINE's eLink API was used to collect citations.

3   Species Filter

Each miRNA official name has a prefix that encodes the organism in which it is found. For instance, in "hsa-let-7a-5p", "hsa" stands for "Homo sapiens", which means that the particular miRNA appears in humans. In the literature, the prefix is often omitted, since the organism referenced is obvious from the text.

During publication search for a miRNA name, mirPub also takes into account the name without the species prefix. Often, the majority of results relates to this "partial" name only. For example, among the 2,277 publications related to the name "cel-let-7", 2,200 refer to "let-7". A researcher looking for publications related to C. elegans (cel) will still have difficulty in locating this information. To address this issue, mirPub v2 provides an additional species filter. Its implementation was done following a two step approach:

- **Species Synonym Expansion.** In this step, a comprehensive set of synonyms was created for each species name. Initially, synonym sets contained only scientific names, given in binomial form. Such names are word pairs denoting the genus and species (e.g., "Homo sapiens"). Then, each set was expanded by adding an abbreviation of the scientific name (e.g. "Homo sapiens" yields "H. sapiens"). Next, common names (like "human") were identified and their singular and plural forms were included to each set.
- **Correlation Extraction.** Publication abstracts were probed for each item in the synonym sets. For each species a weighted count of the synonyms found was produced, with greater weights assigned to scientific names and their abbreviations. Then, *species weights* were calculated by dividing the weighted counts by the abstract's word length. Finally all results were normalised and the scores were interpreted as likelihoods for a species to be examined in a publication. Note that each abstract was correlated to all species whose synonyms were found in it, regardless of their likelihood scores.

4   User Interface and Demonstration

Figure 1 presents a snapshot of the user interface of mirPub v2. In particular, the publications related to the miRNA term "hsa-miR-594" are displayed. The user can order the list either chronologically or based on one of the ranking methods implemented, by selecting the corresponding option from the drop-down menu located on the right of the search box. Furthermore, she can keep only those publications that refer to a particular species of interest, simply by selecting the appropriate option from the menu on the right of the results list.

MirPub v2 is freely available online[3]. During the demonstration session, we are going to conduct queries using several miRNA names and exhibit (a) how different ranking schemes affect the order of results, and (b) how the species filter helps in narrowing down the search results.

---

[3] http://mirpub.imis.athena-innovation.gr/projects/diana/index.php?r=mirpub.

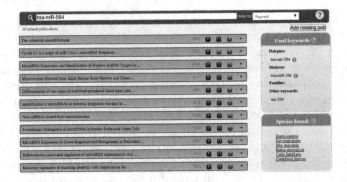

Fig. 1. The result page of mirPub v2 for "hsa-miR-594" keyword.

5   Conclusion

In conclusion, mirPub v2 improves mirPub's search engine in two aspects: (a) by providing an insight about the importance of the presented publications, and (b) by providing filtering based on the studied species of the publication.

Acknowledgements. This work was performed in the framework of MEDA project within GSRT's KRIPIS action, funded by Greece and the European Regional Development Fund of the European Union under the O.P. Competitiveness and Entrepreneurship, NSRF 2007–2013 and the Regional Operational Program of ATTIKI.

References

1. Carthew, R.W.: Gene regulation by microRNAs. Curr. Opin. Genet. Dev. **16**(2), 203–208 (2006)
2. Hwang, W., Chae, S., Kim, S., Woo, G.: Yet another paper ranking algorithm advocating recent publications. In: Proceedings of the 19th International Conference on World Wide Web, pp. 1117–1118. ACM (2010)
3. Kozomara, A., Griffiths-Jones, S.: miRBase: annotating high confidence microRNAs using deep sequencing data. Nucleic Acids Res. **42**, D68–D73 (2013). doi:10.1093/nar/gkt1181
4. Page, L., et al.: Pagerank: bringing order to the web, Stanford Digital Libraries Working Paper, vol. 72 (1997)
5. Vergoulis, T., et al.: mirPub: a database for searching microRNA publications. Bioinformatics **31**, 1502–1504 (2014). doi:10.1093/bioinformatics/btu819
6. Vlachos, I.S., et al.: DIANA-TarBase v7. 0: indexing more than half a million experimentally supported miRNA: mRNA interactions. Nucleic Acids Res. **43**(D1), D153–D159 (2015)
7. Xie, B., Ding, Q., Han, H., Wu, D.: miRCancer: a microRNA-cancer association database constructed by text mining on literature. Bioinformatics **29**(5), 638–644 (2013). doi:10.1093/bioinformatics/btt014

8. Zar, J.H.: Spearman rank correlation. In: Armitage, P., Colton, T. (eds.) Encyclopedia of Biostatistics. Wiley, Chichester (1998)
9. Zhang, L., et al.: Genomic and epigenetic alterations deregulate microrna expression in human epithelial ovarian cancer. Proc. Nat. Acad. Sci. USA **105**(19), 7004–7009 (2008)

A Proposal for Autonomous Scientific Publishing Agent

Adam Sofronijevic[✉], Aleksandar Jerkov,
and Dejana Kavaja Stanisic

University of Belgrade, University Library, Belgrade, Serbia
{sofronijevic,jerkov,kavaja}@unilib.bg.ac.rs

Abstract. A proposal for implementation of autonomous agent model in scientific publishing is presented along with general description of the business framework for its operations and ideas for underlying technological solutions. Drawing on general ideas that have been put forth in other industries the proposal deals with specifics of contemporary scientific publishing and pinpoints the aspects that may be improved by application of autonomous agent model. A proposal for the new approach to scientific publishing is based on available technologies providing possibilities for new ownership model. Software that owns itself and is an independent economic entity that runs operations of a scientific publisher and hires necessary human workforce is technologically viable. Some aspects of bitcoin and other concepts important for building the system described in the proposal are explained. The proposal for this novel approach to scientific publishing also deals with various economic and ethical issues that may arise in its use.

Keywords: Autonomous agent · Bitcoin · Open access · Scientific publishing · Tradenet

1 Introduction

Automation of intellectual jobs brought forth by the second machine age [1] changes many industries including scientific publishing. Software that creates structured texts not easily distinguished from those written by human authors [2] and new frontiers of automation of text analysis highlight the changing role of publishing industry. The loss of human jobs to machines in places where employees deal with structured data and structured business situation [3] raise many questions about the essence of work, leisure and human existence. On the other hand humans showed great resilience in communication with machines [4] and it seems like a plausible statement that the future of reading, writing and thus publishing will involve ever greater role for machines, either in relation to those textual forms intended for machines such as nano-publications [5] or mixed audiences such as fluid books [6], or in other roles related to more traditional textual forms where machines can be a part of different scenarios. This proposal will focus on one such new role for machines in publishing industry – a role of autonomous agent or autonomous software economic entity owning publishing business of its own.

© Springer International Publishing Switzerland 2015
S. Kapidakis et al. (Eds.): TPDL 2015, LNCS 9316, pp. 360–363, 2015.
DOI: 10.1007/978-3-319-24592-8_35

2 Autonomous Agent

The idea of the autonomous agent is fairly new and comes from the community developing Bitcoin concept. Many basic ideas presented in this text come from Mike Hearn or are articulated by him [7]. The basic idea behind the concept of the autonomous agent is that Bitcoin allows for economical autonomy of a machine – software code. This possibility did not exist in the past since economic transaction of two parties demanded an intermediary such as bank and these intermediaries did not want to work with entities other than humans. Bitcoin allows for direct peer transactions and does not recognize a difference between human or machine economic entity allowing for possibility of an autonomous agent in the first place.

Autonomous agent is therefore software that does not presuppose AI (Artificial Intelligence) features, but clever programming and prediction of various scenarios in which it can operate and achieve the programmed goals. Its main feature of economic independence is intrinsic, but also preprogrammed. The main goal of its operation is not to maximize profit for itself, but to achieve its operational goals in a sustainable way with profit margins adjusted to allow for sustainability.

Autonomous agent conduct it economic transactions through Bitcoin network since all other means of economic transactions available include intermediaries that exclude software as legitimate economic entity [8]. In operations of an autonomous agent Bitcoin can be approached as a currency in which direct payments can be made, but also as a conduit that allow for exchange of goods and services attached to Bitcoin software code in smart transaction involving smart products or smart services. Smart transactions and smart products present novel concept related to Bitcoin that use its features to allow for inclusion of agreement-like features to simple, peer financial transactions. Smart products and services intrinsically have Bitcoin processing capabilities allowing them to relate to an entity that owns them, further allowing a financial transaction to be conditioned in time or by functions provided by one side [9]. These guarantees a fair conduct for both sides involved in the transaction and is allowed for by Bitcoin features that can make possible exchanges of ownership that relates to exchange of currency occurring safely even in lightweight scenario of strangers' exchange without intermediaries [10].

Autonomous agent needs a specific additional infrastructure to function properly. Mike Hearn coined the name of this infrastructure as Tradenet. Tradenet is non existent as of March 2015, but technical preconditions exist for its creation. The functionality it should provide is a mix of a stock market and auction house. It provides framework in which bids for offers are placed on commoditized goods and services and in which providers of such goods and services can compete to sell their offerings. This allow for mass participation of both bidders and service and good providers. Also this provide for quick and smooth pairing of bidders and those making offers and conclusion of transactions. The described functionalities are needed in markets where there is a huge demand and supply and decision on commoditized goods and services exchange need to happen rather quick. Tradenet is envisioned as a global infrastructure similar to email or the web, serving multitude of markets globally with possibility of distinguishing spatial, temporal and other more specialized and good/service related aspects of bids and offers.

Autonomous agents conduct their operations through placing bids and/or making offers through Tradenet. Out of many proposed bids the best bid according to preprogrammed criteria of the entity making bidding wins and the transaction can be concluded. The transaction can include direct transfer of Bitcoins as currency of payment or be more complex and involve goods and/or services embedded within Bitcoin code. In both cases the transaction can be a smart one and involve smart goods/services meaning it is not just a simple currency exchange, but a contract involving conditions for both sides.

So far autonomous agents have been envisioned as autonomous economic entities capable of performing preprogrammed activities that does not involve the need for AI and are therefore inferior to humans in non structured situations. The operations they performed via Tradenet needs to be structured and involve structured – commoditized goods/services such as taxi drives, commoditized goods delivery or processing time.

3   Scientific Journals and Autonomous Agents

The idea of implementation of the concept of autonomous agent in scientific publishing does not aim for de-humanization of this industry or is it an attempt to totally quantify scientific outputs in every discipline. It is foremost an attempt to provide an alternative that deals in another way with issues of profit making, unfair access and work incentives policies in scientific publishing.

The preprogrammed aim of autonomous agent in this role would be to cater for sustainable publishing of the journal within predefined journal features. Some of them can be expressed quantifiably such as number of issues published per year or number of articles related to a certain topic. Other features are inherently qualitative such as scientific quality of the text or the level of its innovativeness. In order to make the concept of autonomous agent viable these features also need to be quantifiably expressed. One way is to involve human evaluators and combine them with methods already technologically available for automation of structure text production in scientific communication [3]. Algorithms for cross checking of human evaluation results, hiring of multiple human evaluators and most of all transparency of evaluation results allowing for human evaluators to be hired again by a scientific journal owned and run by autonomous agent will ensure human evaluators conformity to the task. Overall process of choosing the best offer on Tradenet, either an article, or peer reviewer or special issue editor, needs to retain enough human involvement in quantification process to be viable, but at the same time provide protection for autonomous agent against human misbehavior.

Commoditization of article slots in scientific journals and other elements involved in publishing process needs to be done in such a fashion to ensure enough space for all legitimate authors and a fair chance for publishing for all of them, involving thus several types of scientific journals according to several criteria, one of which might be how important the scientific contribution of the article is, the other may be the type of article proposed (literary review, research findings and others) and third may be the field of research. It is not possible to propose a comprehensive system for scientific journal commoditization so our intention is to raise the issue and perhaps inspire a project proposal that may deal with this issue deeper. The main point of our analysis is that for an independent agent model to be implemented in scientific publishing and more precisely in operation of a scientific journal a certain level of commoditizing of all elements that are involved in journal operation needs to be achieved prior to implementation phase.

4 Conclusions

The proposal for application of autonomous agent concept in scientific publishing industries is a general one and needs more details in order to provide for operationally viable model. The aim of this proposal is to highlight the new idea about the possibility of economically autonomous machine that builds on Bitcoin transaction framework and which can alleviate problems in those services and within those industries where high profit margins create problems for users and other stakeholders. The proposal represent a signal for discussion and project preparation for building basic infrastructure that may allow for implementation of this concept and its application for greater general public good in scientific publishing arena.

References

1. Brynjolfsson, E., McAfee, A.: The Second Machine Age: Work, Progress, and Prosperity in a Time of Brilliant Technologies. WW Norton & Company, New York (2014)
2. Sofronijevic, A.: Publishing against the machine: a new format of academic expression for the new scientist. In: Tokar, A., Beurskens, M.M., Keuneke, S.S., Mahrt, M.M., Peters, I.I., van Treeck, T.V.T., Weller, K.K. (eds.) Science and the Internet, pp. 251–262. Dusseldorf University Press, Düsseldorf (2012)
3. Sofronijević, A., Milićević, V., Ilić, B.: Creative management, intellectual capital and the race against the machine. In: Advances in Business-Related Scientific Research Conference - ABSRC 2012, Conference Proceedings (2012)
4. Branigan, H.P., Pickering, M.J., Pearson, J., McLean, J.F.: Linguistic alignment between people and computers. J. Pragmat. **42**(9), 2355–2368 (2010)
5. Mons, B., Velterop, J.: Nano-publication in the e-science era, In: Workshop on Semantic Web Applications in Scientific Discourse SWASD 2009 (2009)
6. Adema, J.: On open books and fluid humanities. Sch. Res. Commun. **3**(3), 16 (2012)
7. Hearn, M.: A talk on autonomous agents (2013). http://plan99.net/~mike/
8. Nakamoto, S.: Bitcoin: a peer-to-peer electronic cash system. Consulted **1**(2012), 28 (2008)
9. Hearn, M.: About smart contracts and smart property, newfination (2013). https://www.youtube.com/watch?v=zSWFqC-0hK8
10. Grinberg, R.: Bitcoin: an innovative alternative digital currency. Hastings Sci. Technol. Law J. **4**, 159 (2012)

Extracting a Topic Specific Dataset from a Twitter Archive

Clare Llewellyn[✉], Claire Grover, Beatrice Alex, Jon Oberlander,
and Richard Tobin

School of Informatics, University of Edinburgh, Edinburgh, UK
C.A.Llewellyn@sms.ed.ac.uk

Abstract. Datasets extracted from the microblogging service Twitter
are often generated using specific query terms or hashtags. We describe
how a dataset produced using the query term 'syria' can be increased in
size to include tweets on the topic of Syria that do not contain that query
term. We compare three methods for this task, using the top hashtags
from the set as search terms, using a hand selected set of hashtags as
search terms and using LDA topic modelling to cluster tweets and selecting appropriate clusters. We describe an evaluation method for accessing
the relevance and accuracy of the tweets returned.

Keywords: Social media · Topic modelling · Data selection

1 Introduction

This work compares three methods for extracting tweets to form a topic-specific
dataset from a Twitter archive. An evaluation method for assessing the relevance
of the set produced (to the topic specified) is described and results provided.

Many Twitter datasets are gathered for a particular need. The Twitter
streaming API allows the collection of Twitter data at the time it is produced.
This means that you need to know what you are looking for ahead of time. As
this is not always possible we describe methods for extracting data from a previously harvested Twitter dataset collected from the streaming API. We query
this data with specific search terms and then augment the extracted dataset
depending on the terms included in the original set. We describe and evaluate
three different methods for enriching this dataset, based upon commonly used
hashtags, hand selected hashtags and topic relevant tweets as identified by topic
modelling. Enriching a dataset brings about a requirement for testing the relevance of this data to the original search parameters. We describe an evaluation
technique that can be used to determine the relevance of the data.

2 Background

Twitter provides a streaming API giving access to up to one percent of the data
as it is produced. Users can either take a random sample or query using search

S. Kapidakis et al. (Eds.): TPDL 2015, LNCS 9316, pp. 364–367, 2015.
DOI: 10.1007/978-3-319-24592-8_36

words, phrases, hashtags, location bounding boxes or user IDs [4]. Previously it was possible to freely share sets of tweets between researchers, but changes to Twitter's terms of service mean that this is now not possible [5]. Instead it is possible to share the user id, tweet id and software for gathering those tweets directly from Twitter. Sharing Twitter data sets allows collaborative research, reproducibility of results and the use of Twitter as a research tool by non-technical researchers. One of the specific aims of the TREC microblog task is to encourage the re-use of Twitter data sets [4].

3   Methods

We did not know which search terms to query the Twitter API with ahead of time. We therefore investigated the best way to extract a topic-specific data set from a previously gathered Twitter archive provided by the ReDites research group [3]. This study investigates the best methods for extracting data relating to the conflict in Syria. Two events are studied and a week's worth of data has been selected associated with each of those events. In the first week, 1–8 March 2012 (2012 set), the UK embassy in Damascus was closed. In the second week, 29 August–4 September 2013 (2013 set), the UK Parliament voted not to authorise military action over chemical weapons use in Syria. Data was taken from the one percent stream limited to English tweets. We looked to select tweets that discussed Syria and Syria specific events. Initially we gathered all tweets that contained the term 'syria' in either upper or lower case, and we used this as a base set from which we could expand. Methods for increasing the size of the dataset are discussed below.

Method 1: Top Hashtags. From the tweets that contained the term 'syria' we extracted all of the hashtags. The top 40 hashtags from both time periods were selected and normalised to give 34. The hashtag terms (hashtags with the # removed) were used as search terms to gather more data. Not all tweets in the set were about the Syrian conflict, for example, the hashtag #UK collected tweets about various activities that were happening in the UK in the selected weeks.

Method 2: Hand Selected Hashtags. In order to make the dataset more focused on the Syrian conflict the hashtags about Syria were hand selected. The amount of content on the conflict varies between the datasets with the 2013 dataset being larger. Therefore, all hashtags that had a frequency over 10 from the 2012 set (this gave 60 in total) and all hashtags that occurred with a frequency above 20 in the 2013 set (giving 148 in total) were selected. Each hashtag was annotated by two human coders as either directly relating to the Syrian conflict or not. This included all locations, people and institutions from Syria or formed to deal with Syria or anything with any of those items incorporated into a compound term, for example 'norway4syria'. The human coders were in perfect agreement (Kappa 1.0) on which tags were related giving 32 which were used as search terms to gather more data.

Method 3: Topic Modelling. The clustering method used in this work is Latent Dirichlet Allocation (LDA) topic modelling [1]. It is used to identify patterns in text and thereby derive topics. A topic is formed from words that often co-occur, the words that co-occur more frequently across multiple documents are most likely to belong to the same topic. LDA provides a score for each document for each topic. We assign the document to the topic for which it has the highest score. This approach was implemented using the Mallet tool-kit [2]. The system provides a list of top words in each topic. The topics that were classed as relevant for this task were those which have 'syria' as one of these most frequent terms. Any tweets that were allocated one of these topics were classed as relevant. In this case the number of clusters generated was set to 15.

Table 1. Total number of tweets returned as relevant per set (size) and the percentages of tweets that were annotated as relevant per set per annotator (1 and 2) and inter-annotator agreement (Kappa)

Dataset	2012				2013			
	Size	1	2	Kappa	Size	1	2	Kappa
Full set	9988193				11272991			
Top hashtags	25753	9	8	0.936	231724	14	17	0.886
Hand selected hashtags	2555	95	91	0.695	23838	100	100	1.00
Topic modelling	2292	92	89	0.826	60613	61	57	0.876

4 Results and Discussion

The results are presented in terms of percentage of those that are relevant to the topic, and the F-score. The percentage that are relevant give an overview of the likely pollution of the dataset and the F-score gives an indication of accuracy for each method.

Relevance. The percentage of tweets that are relevant was calculated through a manual evaluation. There were 6 datasets: one for each of the 3 methods for each time period. For each set 100 tweets were randomly selected for manual examination. Each tweet was coded as relevant or irrelevant to the conflict in Syrian by two annotators. As can be seen in Table 1, inter-annotator agreement scores for this task show that in general there was high agreement. We can see in Table 2 the relevance of tweets extracted to the topic. The top hashtags approach gives very low relevance results. Therefore, while this approach gave a large data set it was not relevant to the topic. The hand selected hashtags method gives high relevance scores. Therefore, while the sets are fairly small in comparison with the other approaches they are relevant to the topic. The topic modelling approach provides a high level of relevance for the smaller 2012 set but lower scores and less relevant results for the larger 2013 set.

Table 2. Accuracy as shown by Precision, Recall and F-scores per set

2012	Precision	Recall	F-score
Hand selected hashtags	0.92	0.98	0.95
Topic modelling	0.76	0.80	0.78
2013			
Hand selected hashtags	0.95	0.66	0.78
Topic modelling	0.60	0.90	0.72

Accuracy. An F-score was calculated by comparing the automatically gener-ated results against a gold standard set. The tweets used to create the gold standard were extracted from the top hashtag set. This gave a set with a higher number of relevant tweets and, therefore, made the accuracy evaluation task diffi-cult and the results more robust. We randomly chose 1000 tweets from each time period which were then annotated as relevant or not. The highest F-score was for the hand selected approach for 2012 set as seen in Table 2. Both approaches showed a drop in F-score for the larger 2013 set. This drop in accuracy was smaller for the topic modelling than the hand selected hashtag approach. This was because while the precision score of the hand selected approach increased for the 2013 set the recall score decreased. The hand selected approach did select appropriate tweets but it also missed many, providing a relevant but small set. The opposite happens for the topic modelling approach. Overall, the F-scores for both datasets are lower but there was a lower drop in accuracy between the two sets. As a drop precision is balanced by a rise in the recall. Therefore while the topic modelling approach selected a larger set it was less relevant.

References

1. Blei, D.M., Ng, A.Y., Jordan, M.I.: Latent dirichlet allocation. J. Mach. Learn. Res. **3**, 993–1022 (2003)
2. McCallum, A.K.: MALLET: A machine learning for language toolkit (2002)
3. Osborne, M., Moran, S., McCreadie, R., Von Lunen, A., Sykora, M.D., Cano, E., Ireson, N., Macdonald, C., Ounis, I., He, Y., et al.: Real-time detection, tracking, and monitoring of automatically discovered events in social media (2014)
4. Soboroff, I., McCullough, D., Lin, J., Macdonald, C., Ounis, I., McCreadie, R.: Evaluating real-time search over tweets. In: Proceedings of ICWSM (2012)
5. Yang, J., Leskovec, J.: Patterns of temporal variation in online media. In: Proceed-ings of the Fourth ACM International Conference on Web Search and Data Mining, pp. 177–186. ACM (2011)

Author Index

Printed in the United States
By Bookmasters